Sport and Exercise Biomechanics

P. Grimshaw

School of Health Sciences, University of South Australia,
Adelaide, Australia

A. Lees

Research Institute for Sport and Exercise Sciences,
John Moores University, Liverpool, UK

N. Fowler

Department of Exercise and Sport Science, Manchester
Metropolitan University, Manchester, UK

A. Burden

Department of Exercise and Sport Science, Manchester
Metropolitan University, Manchester, UK

Taylor & Francis
Taylor & Francis Group

Published by:

Taylor & Francis Group

In US: 270 Madison Avenue,
New York, NY 10016

In UK: 2 Park Square, Milton Park
Abingdon, Oxon OX14 4RN

First published 2007
Reprinted in 2007

ISBN 1 8599 6284 X
978 1 85996 2848

Library of Congress Cataloging-in-Publication Data

Grimshaw, P. (Paul), 1961-
Sport and exercise biomechanics / P. Grimshaw ... [et al.].
p. cm.
Includes bibliographical references and index.
ISBN 1-85996-284-X (alk. paper)
1. Human mechanics. 2. Biomechanics. 3. Sports--Physiological aspects.
4. Exercise--Physiological aspects. I. Title.
QP303.G755 2006
612.7'6--dc22

2006019517

Editor: Elizabeth Owen
Editorial Assistant: Kirsty Lyons
Production Editor: Simon Hill
Typeset by: Phoenix Photosetting
Printed by: Cromwell Press Ltd

Printed on acid-free paper

10 9 8 7 6 5 4 3 2

Taylor & Francis Group
is the Academic Division of T&F Informa plc.

Visit our web site at http://www.garlandscience.com

CONTENTS

PREFACE

Over 2 years ago I was approached by Neil Messenger from Leeds University, UK who asked about my interest in being involved in a project to develop a book that provided a series of *Instant Notes for Sport and Exercise Biomechanics*. While the concept was not new to the traditional science areas it was certainly unique to the subject area of sport and exercise biomechanics. The thought of developing a text that could be used by both students and teachers alike was appealing and challenging. The text that is finally presented is essential for students in that it covers the fundamental areas of study in such a way that can be used in application or be expanded and developed at a higher research level. For the teacher it provides one single resource to plan and prepare more detailed lecture, laboratory and tutorial classes.

I wish to take this opportunity to thank Neil for his initial invitation, to say that I understand why he was not able to continue with it and finally to thank the other three authors (Adrian Lees, Neil Fowler and Adrian Burden) who helped to achieve its conclusion.

Paul Grimshaw

A1 ANATOMICAL DESCRIPTORS OF MOTION

Key Notes

Descriptions of motion	Superficial (close to surface), deep (away from surface), anterior (front), posterior (rear), medial (near mid-line), lateral (away from mid-line), superior (relative highest position), inferior (relative lowest position), proximal (near point of attachment to body), distal (furthest away from body attachment).
Joint movement patterns	Abduction (take away from mid-line), adduction (bring towards mid-line) internal–external rotation (lower leg inward and outward rotation about long axis), plantar- and dorsiflexion (pointing toes or bringing toes towards the shin), extension and flexion (straightening or bringing segments closer together), hyper-extension (excessive extension).
Ankle joint movement	Inversion and eversion (heel rolling outwards or inwards), pronation (complex tri-planar movement in foot involving eversion, abduction and dorsiflexion), supination (tri-planar movement in foot involving inversion, adduction and plantar-flexion).
Specific joint movement	Valgus (lower limb segment rotated about anterior–posterior axis through knee away from mid-line of body), varus (as for valgus but rotation towards mid-line), horizontal abduction and adduction (arm held out in front in transverse plane and then abducted or adducted), circumduction (rotation of a part or segment in a circular manner).

General terms	Parallel (equidistant and never intersecting), degrees of freedom (method used to describe movement or position), diagonal plane (a flat surface that is slanted), tension (to stretch or pull apart), compression (to squeeze together), elevate and depress (to raise up or push down).
	  Origin (starting or beginning point), insertion (anatomical origin), coordinate/s (a number or set of numbers corresponding to a system of reference), plane (a flat surface), perpendicular (at 90°).
	  Translate (change in position but without rotation), drawer (anatomical translation), anterior-drawer (drawer in an anatomical direction), rotate (move through an angle), vertical and horizontal (in a two-dimensional space usually upwards (in the y direction) and along (in the x direction)).

Coordinates	Abscissa (often the x axis), ordinate (often the y axis), intersect (cross each other).
Planes and axes of motion	Anatomical position (facing forwards, arms by side, feet forwards and parallel, palms forward and fingers extended), cardinal plane (plane passing through center of mass), sagittal plane (divides body or part into left and right portions), transverse axis (perpendicular to sagittal plane),

frontal plane (divides into front and rear portions), anterior–posterior axis (perpendicular to frontal plane), transverse plane (divides into upper and lower portions), longitudinal axis (perpendicular to transverse plane).

Coordinate systems

Global-laboratory coordinate system (fixed coordinate position in laboratory), local coordinate system (fixed coordinate system within body or segment), right-handed method of orientation (all coordinates in right-hand directions x, y, and z are positive).

Descriptions of motion

Anatomical descriptors of motion are essential for an understanding of biomechanics and it is important that many of the terms that are used in both the study of anatomy and biomechanics are explained in more detail.

Superficial describes the structures that are close to the surface of objects, whereas **deep** describes the structures that are not near the surface of the object. **Anterior** describes the front portion or part of a body, whereas **posterior** describes the rear or back portion of a body or structure. **Medial** movement describes movement in a direction that is towards the center line (mid-line) of the body or structure. **Lateral,** on the other hand, describes a movement that is away from the center line of the body or structure. The **medial part** of your knee would be the part that is nearest to the mid-line of the body (which is most likely to be the inside part of your knee) and the **lateral part** would be the part furthest away from the mid-line (which is likely to be the outside portion of your knee). **Superior** describes the higher position of a body or part that when standing would be the part furthest from the feet, whereas **inferior** would describe the lower part or portion that when standing would be the nearest part to the feet (hence we can see that it is a relative term). **Proximal** would be used to explain the closest point of attachment of a body or part to the rest of the body (e.g., the arm attachment to the trunk) and **distal** would explain the furthest point away from the attachment of the body or part to the rest of the body. In the case of the forearm the elbow would be at the **proximal end** and the wrist would be the **distal end** from the shoulder joint point of attachment to the body. *Fig. A1.1* helps to identify some of these terms in more detail.

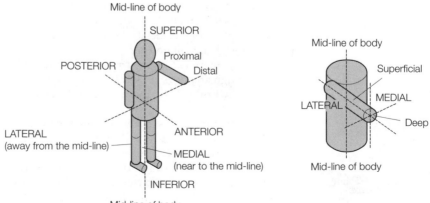

Fig. A1.1. Anatomical descriptors of motion

Joint movement patterns

Abduction involves the movement (usually as a rotation) of a body part or segment away from the body, while **adduction** involves the bringing of this segment or body part closer to the body. **Internal rotation** or movement involves the rotating (either clockwise or anti-clockwise) of a limb or segment towards the mid-line of the body, whereas **external rotation** involves the rotation of this part away from the mid-line of the body. These terms, however, can be confusing; for example, during the internal and external rotation of the lower leg about the long axis it is possible to see that the anterior part of the leg will rotate towards the mid-line of the body, whereas the posterior part of the lower leg will rotate away from the mid-line. **Plantar-flexion**, usually most commonly expressed in relation to the ankle joint (because of the reference to the plantar surface of the foot), involves the movement causing a pointing of the toes downwards. **Dorsiflexion** involves the opposite (again in relation to the ankle) where the toes are brought towards the shin and in an upward rotational movement or direction. Similar movements occur at the wrist but these are more conventionally referred to as flexion and extension. **Extension** is defined as an extending (that is, straightening out) of the limb or segment whereas **flexion** involves the bringing of the segments that are being flexed closer together. Extension at the knee joint would be straightening your leg, whereas flexion at the knee joint would be bringing your lower leg segment and upper leg segment (thigh) closer together. Often the term **hyper-extension** is also used in the context of these movements. In this case hyper-extension would be an excessive amount of extension (i.e., above that normally seen in the joint or structure). However, these movements can be confusing and they are dependent upon the structure that is either extending or flexing. For example, flexion at the hip joint would be where the upper leg segment was moved in the direction towards the trunk (*Fig. A1.2* helps to illustrate hip flexion and extension in more detail).

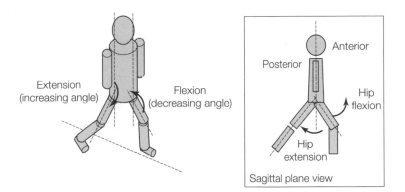

Fig. A1.2.   Hip flexion and extension

Ankle joint movement

Inversion at a joint or structure refers to the rolling outwards (laterally) while **eversion** involves a rolling inwards (medially) of the structure or segment. These two terms are also often confusing and are best described with reference to a structure or segment. For example, inversion of the calcaneus (or heel bone in the foot) would be when you rolled this part of the foot over on the outside of your ankle. Conversely, eversion of the calcaneus would be rolling the ankle (or heel bone) inwards. *Figs A1.3* and *A1.4* help to illustrate this in more detail. **Pronation** and **supination** are complex movements that involve motion in three planes and about three axes of rotation simultaneously. These movements are often described at the sub-talar joint which is in the foot. However, pronation and supination can also occur at the wrist joint. Pronation and supination at the sub-talar joint in the foot involve: **pronation**: calcaneal **eversion**; ankle **dorsiflexion** and **forefoot abduction**; **supination**: calcaneal **inversion**; forefoot **adduction** and ankle **plantar-flexion**.

These complex movements will be described in more detail later in this text. *Figs A1.3, A1.4* and *A1.5* help to illustrate some of these movements in more detail.

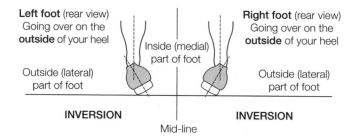

Fig. A1.3.   *Inversion of the calcaneus (heel bone in the foot)*

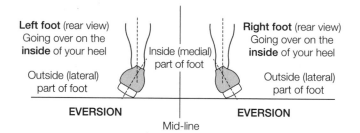

Fig. A1.4.   *Eversion of the calcaneus (heel bone in the foot)*

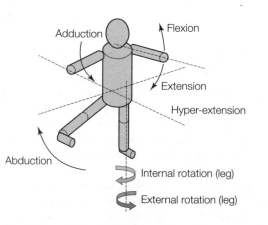

Fig. A1.5.   *Anatomical descriptors of motion*

Specific joint movement

In addition to these movements there are also a number of specific definitions that are unique to certain body segments or joints. For example, **valgus rotation** is usually made with reference to the knee joint where the lower leg is moved in a lateral but rotational manner (i.e., away from the mid-line of the body). **Varus rotation** is the opposite of valgus rotation and is where the lower limb segment is rotated about the knee in a medial direction towards the mid-line of the body. Both these actions would occur about an anterior–posterior axis that is presented through the knee joint. Similarly, **horizontal abduction** and **horizontal adduction** tend to be stated with reference to the shoulder joint where the limb is first moved to a horizontal position (i.e., straight out in front of the body) and then the limb is either abducted or adducted (moved away from or towards the mid-line of the body). **Circumduction**, again often made with reference to the shoulder joint, is where the limb is held out in front (horizontally) and where it is rotated in a circular pattern (circumducting). This rotation involves a combined movement of flexion/extension and adduction/abduction but with no shaft rotation. This movement (circumduction) can also be made with reference to many other joints and structures (e.g., the fingers can easily circumduct). *Fig. A1.6* shows some of these movements in more detail.

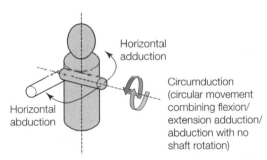

Fig. A1.6.   Anatomical descriptors of motion

General terms

In biomechanics there are many different descriptive terms that are used to help describe movement patterns and parameters within the body. However, in addition there are also a number of common terms that are used together with the many anatomical descriptors. Some of those that are used in such association are as follows: **parallel** can be described as being equidistant and not intersecting (e.g., parallel lines never converge); **degrees of freedom** is a term that is used widely to describe the number of coordinates it takes uniquely to specify the position of a system (the movement of the knee could be described with six degrees of freedom); a **diagonal plane** is a flat or level surface that is slanted in an oblique (neither parallel, perpendicular nor right angular) direction; **tension** is classified as the act or process of stretching something, whereas **compression** is the act or process of squeezing something; **elevation** or **to elevate** is the event of raising something upwards and **depression** or **to depress** is the event of pushing something downwards.

In many examples within biomechanics it is important that we understand the movement of the body through both two-dimensional (width and height or x and y directions) and three-dimensional space (width, height and depth or more commonly termed x, y, and z directions). In order to assist in our understanding of this movement it is therefore necessary to be able to define the space in which we are moving. Again, a number of descriptors are helpful in our understanding of movement in this space and these can be summarized as follows: the term **origin** refers to the point that is the start/beginning and it can also be classified as a reference point to which other movements are relative. In the case of a two-dimensional (2D) plane or movement the origin is often the intersection point of the x and y axes (i.e., where the horizontal (x) and vertical (y) axes cross). This would be the reference point that has the coordinates of 0, 0 (x and y coordinates). In anatomical terms this point of origin is often referred to as the point of **insertion** for muscle, tendon or ligament. The word **coordinates** refers to the set of numbers (two numbers in a two-dimensional space and three numbers in a three-dimensional space) that describe a point of location. The term **plane** refers to a flat two-dimensional surface and an **axis** refers to a straight line that often passes through a body, part or segment and is usually used to describe rotation. In this context it is important to clarify that the axis does not necessarily have to pass through the body or segment and it can be located elsewhere. **Translate** or **translation** is the word that can be used to mean the change in position of body parts or segments without rotation (as in the case of translatory motion along a straight line in a single plane). In anatomical terms this is often referred to as a **drawer** of a structure or joint. For example, an **anterior drawer** of your tibia would be to translate the bone (lower leg) in an anterior direction in a straight line. **Rotate** or **rotation** means to move the limb or segment through an angle about a joint or axis or rotation (movement of your lower leg about your knee joint would be an act of rotation). **Vertical** defines the upward direction or in the case of a flat two-dimensional surface it would be the y direction (upward), and **horizontal** defines the direction that is along the x axis (again with reference to a two-dimensional surface).

Coordinates

In a two-dimensional example (such as the page you are reading) we have two dimensions of space. Vertical and horizontal (or height and width) are terms that are often used to express two-dimensional space. The pages of this book will have a vertical distance (height) and a horizontal distance (width). In this context we often use x and y axes to represent the two-dimensional space we are considering. The x axis would be drawn in the horizontal direction and the y axis would be drawn in the vertical direction (although they can be used to describe whichever direction is required and it is not always the case that the x axis represents the horizontal direction). The x axis is often termed the **abscissa** and the y axis the **ordinate**. The point at which the two axes **intersect** (cross) is called the **origin** and it is important to point out that these two axes would always be expressed perpendicular (at 90°) to each other. *Fig. A1.7* identifies this configuration in more detail.

Planes and axes of motion

In three-dimensional (3D) space a third axis is needed to describe the movement and this is usually described as the z axis. This axis also acts through the origin but is perpendicular to both the x and y axes described previously. *Fig. A1.8* shows the third axis and the **planes** (flat 2D surfaces) that are created from the configuration of these three axes in more detail.

The three planes of motion that have been created from the three axes of motion can also be translated to the human body. In this case the origin of these planes and axes is usually expressed at the center of mass of the body. Using the example of the body that is shown in *Figs A1.9, A1.10* and *A1.11*, this is at

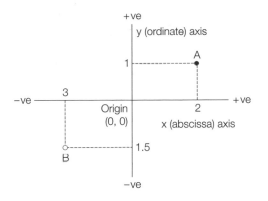

Point A has the (x, y) coordinates (+2, +1)
Point B has the (x, y) coordinates (–3, –1.5)

Fig. A1.7. Two-dimensional plane and axes of motion (not to scale)

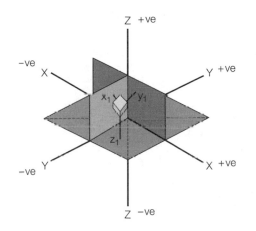

Fig. A1.8. Three-dimensional planes and axes of motion

approximately the mid-point between the two hips. It is important to note that in these figures the body is presented in what is called the **anatomical position** (facing forward, arms by the side, feet forwards and parallel, and palms forwards with fingers extended). However, in many cases, within human movement the body will be in different positions and it will be important to be able to describe the movement in these positions relative to the three **cardinal planes** (a plane that passes through the center of mass of the body), and axes of motion described.

The **sagittal plane** runs from a front to back and superior to inferior (top to bottom) orientation. The plane in this example divides the body into equal left and right portions. The **transverse axis** of rotation is **perpendicular** (at 90°) to the sagittal plane (*Fig. A1.9*). A typical movement in this plane and about this axis of rotation would be a **somersault**.

The **frontal plane** runs from a side to side and superior to inferior orientation. The plane in this example divides the body into equal front and rear portions. The **anterior–posterior axis** of rotation is perpendicular to this plane (*Fig. A1.10*). A typical movement in this plane and about this axis of rotation would be a **cartwheel**.

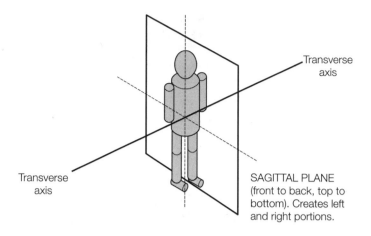

Example movement in this plane and about this axis = SOMERSAULT

Fig. A1.9.   Three-dimensional planes and axes of motion

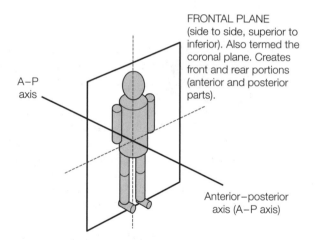

Example movement in this plane and about this axis = CARTWHEEL

Fig. A1.10.   Three-dimensional planes and axes of motion

The **transverse plane** runs from a side to side and anterior to posterior orientation. This plane in this example divides the body into equal upper and lower portions (superior and inferior parts). The **longitudinal axis** of rotation is perpendicular to the transverse plane (*Fig. A1.11*). A typical movement in this plane and about this axis of rotation would be a **pirouette** (as in ice skating).

Many different sagittal, frontal, and transverse planes can pass through different individual body parts. Each individual set of planes (i.e., sagittal or frontal or transverse) are parallel to each other. All sagittal planes are perpendicular (at 90°) to all frontal planes that are perpendicular to all transverse planes. **Anatomical axes** are described as lines that are perpendicular to defined planes of motion. Again, it is important to point out that all anterior–posterior axes are perpendicular to all transverse axes that are perpendicular to all longitudinal axes. Motion by limbs often takes place in several planes and about several axes

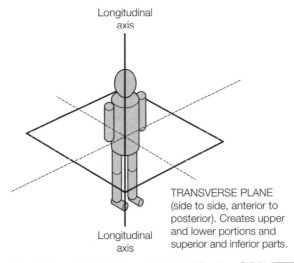

Longitudinal
axis

Longitudinal
axis

TRANSVERSE PLANE
(side to side, anterior to
posterior). Creates upper
and lower portions and
superior and inferior parts.

Example movement in this plane and about this axis = PIROUETTE

Fig. A1.11.   Three-dimensional planes and axes of motion

of rotation; for example, as we have said pronation and supination are tri-planar movements that are often described at the sub-talar joint in the foot. However, other anatomical movements can sometimes be described with reference to one plane and one axis of rotation: for example, knee/elbow or shoulder flexion or extension would occur in the **sagittal plane** and about the **transverse axis** of rotation; abduction and adduction movement at most joints and valgus and varus rotation at the knee would occur in the **frontal plane** and about the **anterior–posterior axis** of rotation; internal and external rotation of the knee joint (or more precisely internal and external rotation of the tibia/fibula or femur) and horizontal abduction and adduction would occur in the **transverse plane** and about the **longitudinal axis** of rotation.

Coordinate systems

Within any 3D data collection in biomechanics it is important to be able to specify a certain coordinate system that is used to explain and clarify movement patterns. Two common coordinate systems that are used are either a **global** or **laboratory coordinate system** and a **local coordinate system.** In both examples the **right-handed** method of configuration is used. The right-handed system of configuration can be defined with reference to both *Figs A1.8* and *A1.12.* In *Fig. A1.8* it is possible to see that the right-hand corner of the 3D configuration has certain x, y, and z specifications (i.e., all the coordinates located in this 3D space have a positive value). In the right-handed system of 3D configuration the coordinates are arranged in this manner (all positive and all in a right-handed direction). This is shown more specifically in *Fig. A1.12* where the x, y, and z right-handed configuration is presented in isolation.

The **global coordinate system** (GCS) (also known as the inertial reference system) is presented when the object space is defined during 3D data capture. The system is right handed and is used to define the fixed coordinate position within the laboratory. This position is then used to define all other positions within the data capture process that follows. The **local coordinate system** (LCS) is used to describe the position within a body or segment. This coordinate system would

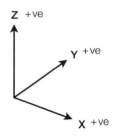

Fig. A1.12.   Right-handed coordinate system

stay within the body or segment during movement. Again this is a right-handed method of orientation with the center of the LCS usually placed at the center of mass of the body. *Fig. A1.13* helps to identify this relationship between the GCS and the LCS in more detail.

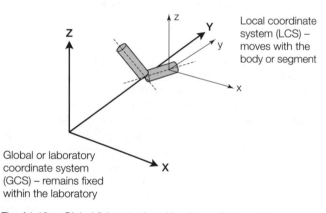

Local coordinate system (LCS) – moves with the body or segment

Global or laboratory coordinate system (GCS) – remains fixed within the laboratory

Fig. A1.13.   Global (laboratory) and local coordinate system

A2 MECHANICAL DESCRIPTORS OF LINEAR MOTION

Key Notes

Biomechanics	Is the study of forces and the effects of these forces on living things.
Kinematics and kinetics	These are subdivisions of mechanics that are concerned with displacement, velocity and acceleration (kinematics) and forces that cause or result from motion (kinetics).
Linear and angular motion	Linear motion (or translatory motion) is concerned with movement along a line that is either straight or curved and where there is no rotation and all body parts move in the same direction at the same speed. Angular motion involves movement around an axis of rotation.
Scalar quantity	A quantity that is represented by magnitude (size) only.
Vector quantity	A quantity that is represented by both magnitude and direction.
Distance and displacement	The term distance is classified as a scalar quantity and is expressed with reference to magnitude only (i.e., 14 miles). Displacement is the vector quantity and is expressed with both magnitude and direction (i.e., 14 miles north-east).
Speed and velocity	Speed is the scalar quantity that is used to describe the motion of an object. It is calculated as distance divided by time taken. Velocity is the vector quantity and it is used to also describe the motion of an object. It is calculated as displacement divided by time taken.
Acceleration	Is defined as the change in velocity per unit of time. It is calculated as velocity divided by time taken.
Average and instantaneous	Average is the usual term for the arithmetic mean. The sample mean is derived by summing all the known observed values and dividing by their number (i.e., how many of them there are). For example over a 26 mile race the average speed of the athlete was 14 miles per hour (mph). Instantaneous refers to smaller increments of time in which the velocity or acceleration calculations are made. The smaller the increments of time between successive data points the more the value tends towards an instantaneous value.

Biomechanics   **Biomechanics** is broadly defined as the study of forces and their affects on living things. In mechanics there is use of a further subdivision into what is known as **kinematic** and **kinetic** quantities. Biomechanics and mechanics are used to study human motion. This section is concerned with linear (i.e., translational – where all the points move in the same direction in the same time and without rotation) kinematics. *Fig. A2.1* helps to illustrate the definition of biomechanics and kinematics in more detail.

Human movement or motion can be classified as either **linear or angular motion**. Most movements within biomechanics are a combination of translation and rotation. This leads to a description that is termed **general motion**. Linear motion (or translation) is movement along a line which may be either straight or curved and where all the body parts are moving in the same direction at the same speed. This can be classified as either rectilinear motion (motion in a straight line) or curvilinear motion (motion in a curved line). Angular motion (which will be discussed in the next section) involves movement around an axis (either imaginary or real) with all the body parts (or individual body parts) moving through the same angle at the same time. *Fig. A2.2* identifies these types of motion in more detail.

Kinematics and kinetics

Linear kinematics is concerned with the quantities that describe the motion of bodies such as **distance**, **displacement**, **speed**, **velocity**, and **acceleration**. These

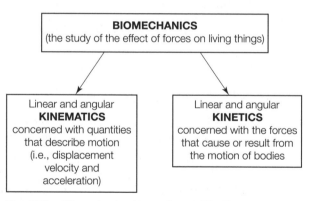

Fig. A2.1.   Biomechanics, kinematics and kinetics

Linear motion

Angular motion

Somersault

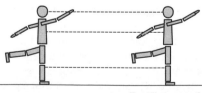

Rectilinear motion

High bar swing

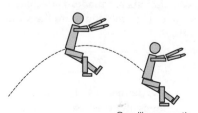

Curvilinear motion

Fig. A2.2.   Different types of motion

quantities can be classified as either scalar or vector quantities. **Scalar quantities** are represented by magnitude (size) only, whereas **vector quantities** are represented by both magnitude and direction. Hence, vector quantities can be presented mathematically or graphically on paper by scaled straight lines or arrows. For example, **speed** is defined as the **distance traveled** per unit of **time** and as such it is a scalar quantity (i.e., no direction is specified).

$$Speed = \frac{Distance\ traveled}{Time\ taken}$$

Ex 1. **If an athlete ran 14 miles in 1 hour and 15 minutes what was the athlete's average speed?**

$$Speed = \frac{Distance}{Time}$$

$$= \frac{14\ miles}{1\ hour\ 15\ minutes}$$

Convert the time component to one common quantity (i.e., hours)

$$= \frac{14\ miles}{1.25\ hours}$$

$$= 11.2\ miles\ per\ hour\ (mph)$$

This would represent the average speed of this athlete over the whole 14 mile running activity. Hence the measure of speed in this case is a scalar quantity and is expressed in magnitude only (i.e., 11.2 mph). In this example we could have expressed speed in many different units, for example meters/second (m/s) or kilometers per hour (kph). **See if you can convert an average speed value of 11.2 mph into units of metres/second (m/s)?** *Figs A2.3* **and** *A2.4* **show the solution to this problem which present both the direct conversion of 11.2 mph to m/s and the revised calculation in m/s for the athlete described in this example.**

Scalar and vector quantities

In example 1 we can see that the athlete covered a distance of 14 miles but we do not know whether this was in a straight line, in a series of curves, or indeed in a circle starting and finishing at the same point. In this context the term speed is

1 mile = 1609.344 meters
1 hour = 60 minutes = 60 × 60 seconds = 3600 seconds

11.2 miles = 11.2 × 1609.344 m = 18024.652 m

$$Speed\ in\ m/s = \frac{18024.652\ m}{3600\ s}$$

Speed = 5.0068 m/s

Average speed of 11.2 mph = 5.0 m/s (to 1 decimal place)

Fig. A2.3.   Converting an average speed of 11.2 mph into the units of m/s

1 mile = 1609.344 meters
1 hour = 60 minutes 1 minute = 60 seconds

14 miles = 14 × 1609.344 m = 22530.76 m

1.25 hours = 1.25 × 60 min × 60 s = 4500 s

$$\text{Average speed in m/s} = \frac{22530.76 \text{ m}}{4500 \text{ s}}$$

Average speed = 5.0068 m/s

Average speed of athlete = 5.0 m/s (to 1 decimal place)

Fig. A2.4. Calculation in m/s for athlete described

used because there is no directional component specified. However, if we now reword this example it is possible to express the solution as a **vector quantity** such as **velocity**. Vector quantities are expressed with reference to both magnitude and direction and in the case of the runner in example 1 this can be restated as follows.

Ex 2 If an athlete covered a displacement of 14 miles in a straight line in a north-east direction in a time of 1 hour 15 minutes, what would be the athlete's average velocity over this time period?

Distance and displacement

Note: in this example the term **distance** has been replaced with the term **displacement**, which is used to express a directional component (i.e., straight line north-east direction). Although the result would be of the same magnitude (because the athlete covered the same distance/displacement in the same time) the quantity would be a vector quantity because there would now be a directional component to the solution. This vector quantity could now be expressed graphically to scale by an arrow on a piece of paper or by mathematical representation. *Fig. A2.5* illustrates this in more detail.

$$\text{Velocity} = \frac{\text{displacement}}{\text{time}}$$

$$= \frac{14 \text{ miles}}{1.25 \text{ hrs}}$$

$$= 11.2 \text{ mph}$$

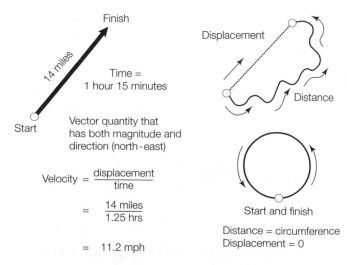

Fig. A2.5. Defining the terms distance and displacement

Speed and velocity

Often within biomechanics it is useful to be able to express both **speed** and **velocity** components. Sometimes it is only the average speed that is of interest (such as, for example, when an athlete runs a marathon race (26.2 miles or 26 miles 385 yards) and the coach is interested in getting a quick and simple measure of how the race was performed overall). As this **average speed** would be presented over a 26 mile running distance it does not really describe the specific details of the race but it may be useful for training. Similarly, during the long jump take-off phase it is interesting to be able to know exactly what the vertical and horizontal velocities are at the point of take-off. Such information would allow the coach or scientist to be able to work out the angle of take-off and observe whether the athlete jumped with a flat, long trajectory or a high, shorter one. Both these aspects (speed and velocity) are equally important for the under-standing of sport, exercise, and general human movement.

Both **speed** and **velocity** can be uniform or non-uniform quantities. **Uniform** describes motion that is constant over a period of time (i.e., constant velocity or speed (no acceleration or deceleration)) and **non-uniform** describes varying or changing velocity or speed over time (i.e., with some acceleration or decelera-tion). In human motion it is usually the knowledge of non-uniform motion that is more beneficial to the athlete, coach, scientist, and student of biomechanics. For example, in the case of our runner in example 1, who covered 14 miles in 1 hour 15 minutes, it would be more beneficial to know what changes in the runner's speed or velocity occurred throughout the activity. Such information would have important training and performance implications and would be as valuable in a sprint race lasting no more than 10 seconds (i.e., 100 m sprint) as it would be in a marathon event lasting several hours.

Linear velocity and **acceleration** are important quantities within biomechanics that are used to describe and analyse the motion of human bodies. *Fig. A2.6* illustrates a series of 100 m sprint data from a university level athlete.

From consideration of *Fig. A2.6* it is possible to see that the athlete covered the 100 m displacement (horizontal displacement in a straight line along a track) and that this 100 m displacement is divided into 10 m sections or intervals. For example, the first 10 m was covered in 1.66 seconds and the second 10 m in 1.18 seconds (or 20 m in 2.84 seconds (cumulative time)). It is possible to see from this

Disp. (m)	Cumulative time (s)	Time (s)	Average velocity (m/s) 10 m intervals
10	1.66	1.66	6.03
20	2.84	1.18	8.47
30	3.88	1.04	9.62
40	5.00	1.12	8.92
50	5.95	0.95	10.50
60	6.97	1.02	9.80
70	7.93	0.96	10.40
00	8.97	1.04	9.62
90	10.07	1.10	9.09
100	11.09	1.02	9.30

Average horizontal velocity over 100 m = 100/11.09 = 9.01 m/s

Fig. A2.6.   Sprint data for university level 100 m athlete

data that the athlete covered the whole 100 m displacement in 11.09 seconds. We can now use this data to determine **average velocity** over smaller increments (such as every 10 m interval). Such information would provide us with a bio-mechanical description of the whole 100 m event. The presentation and analysis of this velocity can be seen from the consideration of the calculations and data identified in *Figs A2.7, A2.8* and *A2.9*. Note: it is important to point out that this is expressed as velocity (a vector quantity) because we have a directional compo-nent (i.e., horizontal displacement along a straight 100 m track) and even though we are considering the velocity (average) over much smaller increments (i.e., 10 m intervals) it is still an average velocity over that horizontal displacement interval or section. In this context taking even smaller time intervals will eventually lead to an "**instantaneous**" value for the calculation of speed or velocity. Such analysis provides a more detailed biomechanical breakdown of the event of the 100 m sprint race.

This data (average velocity of the whole 100 m, specific velocity for each 10 m section of the race or "instantaneous" values for even smaller time or displace-ment intervals) could be compared with values for Olympic and World athletic performances or indeed to other athletes within the club or university. Obviously

Average velocity over first 10 m

$$0\text{–}10\text{ m} = \frac{10\text{ m}}{1.66\text{ s}} \quad = 6.03\text{ m/s}$$

Average velocity between 10–20 m

$$10\text{–}20\text{ m} = \frac{10\text{ m}}{1.18\text{ s}} \quad = 8.47\text{ m/s}$$

Average velocity between 20–30 m

$$20\text{–}30\text{ m} = \frac{10\text{ m}}{1.04\text{ s}} \quad = 9.62\text{ m/s}$$

Fig. A2.7.   Velocity calculations (example 10 m intervals) for 100 m sprint data of university level athlete

1. During first second of motion (5.0 m) the velocity increased rapidly

2. During the next 4.75 seconds the velocity increased to maximum value of about 10 m/s which was achieved at 60 m

3. Maximum velocity (around 10 m/s) maintained for about 1 second to 70 m

4. Velocity decreased steadily from 10 m/s to 9.2 m/s over the last 30 m

'He/she who slows down the least wins the sprint race'

Fig. A2.8.   Analysis of velocity data

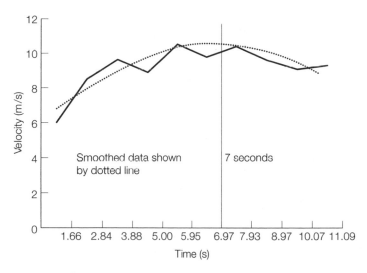

Fig. A2.9.   Graphical presentation of velocity data

such knowledge of individual and comparative performances would have important training and performance implications for both the athlete and the coach.

Acceleration

Acceleration is defined as the **change in velocity per unit of time** and it is usually measured in meters per second squared (m/s²). This means that the velocity of an object will increase/decrease by an amount for every second of its motion. For example, a constant (uniform) acceleration of 2.5 m/s² indicates that the body will increase its velocity by 2.5 m/s for every second of its motion (2.5 m/s for 1 second, 5.0 m/s for 2 seconds, 7.5 m/s for 3 seconds and so on). *Figs A2.10, A2.11,* and *A2.12* show the calculation and presentation of some acceleration data for the university 100 m sprint performance used in the previous example.

Acceleration is defined as the change in velocity per unit of time
(rate of change of velocity)

$$\text{Acceleration} = \frac{V - U}{t_2 - t_1}$$

U = velocity of the object at time t_1
V = velocity of the object at time t_2
U = initial velocity
V = final velocity

Positive acceleration	Negative acceleration
When the velocity increases over a time period (speeding up)	When the velocity decreases over a time period (slowing down)

Fig. A2.10.   Acceleration defined

Analysis of 100 m university sprinter (acceleration)

Acceleration between 0 and 7 seconds

$$a = \frac{10.51 - 0 \text{ m/s}}{7.0 - 0 \text{ s}} \qquad = 1.50 \text{ m/s}^2$$

Acceleration between 0 and 11 seconds

$$a = \frac{9.21 - 0 \text{ m/s}}{11.0 - 0 \text{ s}} \qquad = 0.83 \text{ m/s}^2$$

Acceleration between 7 and 11 seconds

$$a = \frac{9.21 - 10.51 \text{ m/s}}{11.0 - 7.0 \text{ s}} \qquad = -0.33 \text{ m/s}^2$$

Fig. A2.11.   Acceleration calculations for selected time intervals (reading values from the graph of velocity vs. time)

From consideration of these figures it is possible to see that the athlete is both accelerating and decelerating throughout the activity. If we now look at the velocity versus time graph (shown in *Fig. A2.9*) we can see that it is possible to read values directly from this graph for specific time points (i.e., 7 seconds into the race). Between 0 and 7 seconds we can see that there is an average positive acceleration of +1.50 m/s² (i.e., indicating the athlete is on average speeding up over this period of time). Between 0 and 11 seconds (almost the whole race) the athlete has an average horizontal acceleration of +0.83 m/s². However, more detailed analysis (over smaller time intervals) shows that the athlete is actually decelerating (slowing down) between 7 and 11 seconds in the activity (–0.33 m/s²). This data provides valuable biomechanical information for the athlete and coach that can be used to improve performance. As an alternative to reading specific time points from the graph we can use the velocity calculations that we have already (i.e., the velocity values for each 10 m displacement). In this context, the following example determines the acceleration between the velocity points of 10.50 and 8.92 m/s (approximately between the 40 and 50 m points).

Acceleration of the athlete between velocity points of 10.50 m/s and 8.52 m/s
Using the formula for acceleration

$$\text{Acceleration (a)} = \frac{v - u}{t_2 - t_1}$$

$$= \frac{10.50 - 8.92 \text{ m/s}}{5.95 - 5.00 \text{ = s}}$$

$$= +1.66 \text{ m/s}^2 \text{ (average acceleration over this time)}$$

Note: in the context of the graph it can be seen that the values that are plotted are between the points of displacement or time (i.e., indicating an average between two points that is expressed at the mid-point). In addition, considering that velocity is a vector quantity, the positive and negative sign would represent the directional component. A positive velocity value would indicate movement along the 100 m track towards the finish line, whereas a negative value for horizontal velocity would indicate movement back towards the start (which in a 100 m sprint race would not usually happen). However, in terms of acceleration, a positive value would indicate speeding up (accelerating) and a negative value slowing down (decelerating). In this example the velocity and acceleration signs (positive and negative) are independent. However, it is also possible to have a negative acceleration value when the object is speeding up (increasing velocity or accelerating). For example, in the case of acceleration due to the gravity of the earth the acceleration is often expressed as –9.81 m/s². This indicates a *downward* (towards the earth) acceleration of 9.81 m/s² (i.e., an object will speed up (increase its velocity) as it falls towards the center of the earth (see section on gravity within this text)). However, in the case of acceleration in the horizontal direction (as in the example of our 100 m sprinter) a negative acceleration value would indicate a deceleration (slowing down) of the athlete.

Finally, in terms of biomechanics it is useful to be able to present all of this data in a series of graphs. In order to analyse performance, the coach and the athlete can use the graphs for displacement/time, velocity/time and acceleration/time. *Fig. A2.12* (1–3) presents graphs for the data calculated for the 100 m university level sprinter used in our example. Note that the acceleration data is presented for 10 m intervals between velocity values, as is the data for velocity (i.e., between displacement values). The data is presented both as raw values and smoothed (using a curve of best fit) between data points.

From consideration of these graphs, it is possible to see that the velocity data indicates the athlete increases velocity from the start and reaches a peak at around the 60 m point in the race (or at about 7 seconds). At this point the athlete manages to hold this peak velocity for about 1 second to 70 m before it then begins to fall towards 100 m. This is confirmed by the acceleration/time graph, which shows positive (increasing velocity) values up to 60 m. Although it appears that the acceleration/time graph is decreasing during this section, the values are still all positive and are hence indicating acceleration or speeding up. The acceleration/time graph then passes through zero (which at this point would indicate no acceleration), as the athlete would have constant horizontal velocity for this brief 1-second period. Next, the acceleration/time graph becomes

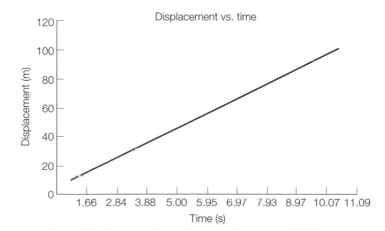

Fig. A2.12(1).   *Displacement/time, velocity/time, and acceleration/time graphs of data for 100 m university level athlete over 10 m intervals (best fit straight line shown)*

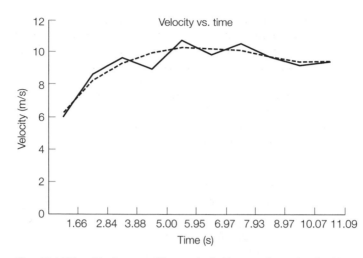

Fig. A2.12(2).   Displacement/time, velocity/time, and acceleration/time graphs of data for 100 m university level athlete over 10 m intervals (smoothed data indicated by dotted line)

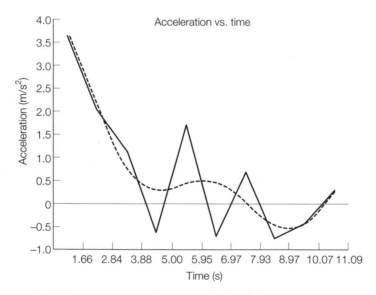

Fig. A2.12(3).   Displacement/time, velocity/time, and acceleration/time graphs of data for 100 m university level athlete over 10 m intervals (smoothed data indicated by dotted line)

negative, indicating a deceleration or slowing down (i.e., from about 70 to 100 m). Hence, the statement made by many athletics coaches and biomechanists of *"he/she who slows down the least wins the sprint race"* appears to be true of our 100 m university level sprinter. This characteristic speeding up (increasing horizontal velocity) to a peak at around 60 m, holding this speed for about 1 second and then slowing down as they approach 100 m is typical of many 100 m performances at many different levels (from amateur to Olympic athlete). Hence, it is obvious that such biomechanical analysis may have important implications for both training and performance.

Application

From the following (*Fig. A2.13*) set of data taken from two different world record 1500 m freestyle swimming performances (Kieran Perkins 1994 and Grant Hackett 2001) calculate the average horizontal velocity and acceleration over each 100 m displacement (distance) interval. Also see if you can provide a brief analysis of each swimmer's race. Note: in this context in may be important to qualify that the displacement in a swimming event such as this is technically zero (i.e., the athlete starts, swims 50 m (down the pool length), turns, and then returns to the start again). Hence the term distance and speed are probably more appropriate in this application.

Disp (m)	1994 Perkins	2001 Hackett
100	54.81	54.19
200	1:52.91	1:52.45
300	2:51.48	2:51.29
400	3:50.37	3:50.18
500	4:49.04	4:48.82
600	5:48.51	5:47.45
700	6:47.72	6:45.96
800	7:46.00	7:44.47
900	8:45.28	8:43.05
1000	9:44.94	9:41.78
1100	10:44.63	10:40.56
1200	11:44.50	11:39.51
1300	12:44.70	12:38.51
1400	13:44.44	13:37.89
1500	14:41.66 WR	14:34.56 WR

Fig. A2.13.   Two sets of world record 1500 m freestyle swimming data shown over 100 m intervals

A3 MECHANICAL DESCRIPTORS OF ANGULAR MOTION

Key Notes

Angular motion	Is where all the parts of a body (i.e., all parts on a rigid object or all parts on a segment of the human body) move through the same angle.
Angular kinematics	Describes quantities, such as angular displacement, angular velocity, and angular acceleration.
Angular displacement and distance	Angular displacement is the difference between the initial and the final angular position of a rotating body (it is expressed with both magnitude and direction). For example, 36 degrees anti-clockwise. Angular distance is expressed with magnitude only (i.e., 2.4 radians).
Degrees and radians	Units that are used to measure angular displacement (where a circle = 360 degrees or 2π radians). 1 radian is approximately 57.3 degrees.
Angular velocity and angular acceleration	Angular velocity is the angular displacement divided by the time taken. Angular acceleration is defined as the rate of change of angular velocity and is calculated by angular velocity (final – initial) divided by the time taken.
Clockwise and anti-clockwise rotation	Clockwise rotation is movement in the same direction as the hands of a clock (i.e., clockwise) when you look at it from the front. Clockwise rotation is given a negative symbol (–ve) for representation. Anti-clockwise rotation is the opposite movement to clockwise rotation and it is given a positive symbol (+ve) for representation.
Absolute and relative angles	An absolute angle is the angle measured from the right horizontal (a fixed line) to the distal aspect of the segment or body of interest. A relative joint angle is the included angle between two lines that often represent segments of the body (i.e., the relative knee joint angle between the upper leg (thigh) and the lower leg (shank)). In a relative angle both elements (lines) that make up the angle can be moving.
Included angle and vertex	An included angle is the angle that is contained between two lines that meet or cross (intersect) at a point. Often these lines are used to represent segments of the human body. The vertex is the intersection point of two lines. In human movement the vertex is used to represent the joint of interest in the human body (i.e., the knee joint)

Angular motion   **Angular motion** is rotatory movement about an **imaginary** or **real axis** of rotation and where all parts on a body (and the term body need not necessarily be a human body) or segment move through the same angle. **Angular kinematics**

describes quantities of angular motion using such terms as angular displacement, angular velocity and angular acceleration. *Fig. A3.1* identifies two examples of angular motion in more detail.

Angular distance or **displacement** (scalar or vector quantity) is usually expressed in the units of **degrees** (where a complete circle is 360 degrees). Similarly, **angular velocity** and **angular acceleration** are often expressed as degrees per second (°/s) and degrees per second squared (°/s² or degrees/second²) respectively. However, it is more convenient within human motion to use the term **radian**. The value for **1 radian** represents an angle of approximately **57.3°**. *Fig. A3.2* defines the term radian and its relationship with angular degrees of movement.

As with the terms used to describe linear motion, within angular motion there exists both scalar and vector quantities. However, it is often possible and more easily understandable to describe angular movement using such definitions as clockwise or anti-clockwise rotation. Again, positive and negative signs can be used to denote the different directions (e.g., **clockwise rotation** may be assigned a **negative sign** and **anti-clockwise rotation** a **positive sign** which is the common

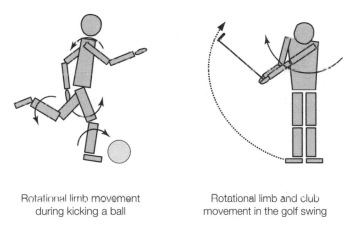

| Rotational limb movement during kicking a ball | Rotational limb and club movement in the golf swing |

Fig. A3.1. Angular/rotational movement within human motion

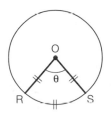

Circle with center O

OR = OS = the radius of the circle

When the distance RS (the arc length of the circle) is equal to the radius of the circle the angle θ is approximately 57.3 degrees (or 1 radian)

Within a circle (360°) there are *exactly* 2π radians

π = 3.142 (to three decimal places)
Hence 2 × 3.142 = 6.284 radians
360/6.284 = 57.3 (to one decimal place)

1 radian = 57.3 degrees

Fig. A3.2. The definition of 1 radian

convention used within biomechanics). *Figs A3.3* and *A3.4* help to indicate scalar and vector quantities and the directions of angular motion.

Considering *Fig. A3.4* it is possible to see the actions employed by the leg in kicking a soccer ball. The upper leg segment (thigh) moves with an anti-clockwise rotation between position 1 and position 2. The lower leg segment also moves in an anti-clockwise rotation between these positions. Note that these two actions happen simultaneously and in association with the linear (forward translation) movement of the whole body. From such a description and knowing the time taken for this movement it would be possible to calculate the angular velocity of each of these segments in kicking this soccer ball.

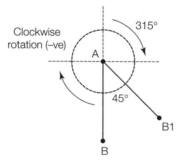

The "arm" AB moves in a clockwise rotation through 315 degrees (i.e., 5.5 radians). The distance (scalar quantity) covered by the arm is 315 degrees whereas the displacement (vector quantity) is only 45 degrees (anti-clockwise). However, in this example, to calculate the average angular velocity of the arm it would be necessary to use the distance (angle) that the arm has moved through (i.e., 315 degrees clockwise in this case).
Note: by giving the distance value a direction (clockwise) it is a vector quantity

Distance   = 315° (5.5 radians)
Displacement =   45° (0.76 radians) anti-clockwise

Fig. A3.3.   Angular movement

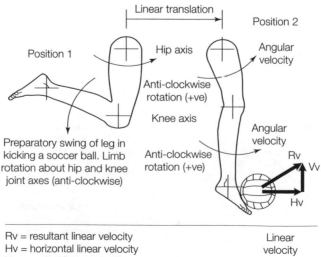

Rv = resultant linear velocity
Hv = horizontal linear velocity
Vv = vertical linear velocity

Fig. A3.4.   Limb rotation in kicking a ball

Angular velocity Considering the individual action of the upper leg segment in this example (*Fig. A3.5*) as taken from *Fig. A3.4*, we can see that the upper leg segment (represented by a single line in *Fig. A3.5*) moves anti-clockwise through 30 degrees (10° before the vertical line and 20° after the vertical line). If we know that the upper leg moved through this angle in 0.5 seconds, it is possible to calculate the average angular velocity of this limb segment. *Fig. A3.6* shows the calculation of average angular velocity depicted by the **symbol** ω (the Greek letter omega) in more detail.

It is important to point out that for every part (or point on the limb) that is along the limb segment shown in *Fig. A3.5*, the average angular velocity will be the same. All the parts along this limb travel through the same angle of 30 degrees in 0.5 seconds (10° before the vertical line and 20° after the vertical line) and as such their average angular velocities (ω) will be determined using the formula:

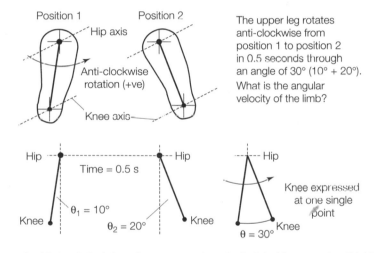

Fig. A3.5.   *Calculation of average angular velocity (ω) of the upper leg (thigh)*

Angular velocity (ω) is given by the following equation:

$$\omega = \frac{\text{angular displacement (degrees or radians)}}{\text{time taken (seconds)}}$$

Where angular displacement = angular movement between the initial and final angular position (which is 30° anti-clockwise (+ve) in this case)

$$\omega = \frac{30°}{0.5\text{ s}}$$

$$= 60°/s$$

average angular velocity =

1.05 radians/s

Note: it is important to point out that this is the average angular velocity of every point along the limb or segment

This is the average angular velocity (anti-clockwise) of the upper leg (thigh) during the kicking of a soccer ball

Fig. A3.6.   *Calculation of average angular velocity (ω) of the upper leg*

ω = $\dfrac{\text{angular displacement}}{\text{time taken}}$ (measured in degrees or radians)
                                                    (measured in seconds)

= angular velocity
   (measured in degrees/second or radians/second (°/s or rads/s))

In this example, **angular displacement** is defined as the **difference** between the **initial and final angular position** of the object or segment (in either a clockwise or anti-clockwise rotation).

Angular acceleration

Angular acceleration, as depicted by the **symbol** α (the Greek letter alpha), is calculated by dividing the **angular velocity (ω)** by the **time taken**. It is defined as the rate of change of angular velocity and is expressed between two points of interest (i.e., an interval of time (t_1 and t_2) or position 1 and 2 in our example). In the case of the example shown in *Fig. A3.5*, it is possible to see that the average angular acceleration can be calculated by using the angular velocity and the time taken for the movement between the two positions. This is shown in more detail in *Fig. A3.7*.

Again, it is important to point out that, as for angular velocity, all the parts along this limb segment will have the same angular acceleration as they all have the same angular velocity. The angular displacement (rotation) is the same for a part that is far away from the axis of rotation (which is the hip joint axis of rotation in this case) as it is for a part that is close to the axis of rotation.

Considering *Fig. A3.4*, when kicking a ball it is possible to see that the upper segment of the leg (the thigh) rotates about the hip joint (or hip axis of rotation) in an anti-clockwise direction (hence making it a displacement (vector quantity) because we now have a directional component). Similarly, and at the same time, the lower segment of the leg (the shank) rotates about the knee joint (knee axis of rotation) also in an anti-clockwise direction. Both these actions occur simultaneously and contribute to the average angular velocity and angular acceleration generated by the limb in kicking this ball. *Fig. A3.8* shows the angular displacement (and linear translation forward of the body) of the two segments (upper leg (thigh) and lower leg (shank)) working together in this example (ignoring the movement of the foot).

Following this it is now possible to consider the movement of the lower leg and determine its average angular velocity and acceleration. From *Fig. A3.9* we can

Average angular acceleration (α) is given by the following equation:

$$\alpha = \frac{\text{angular velocity (final – initial)}}{\text{time taken} \quad (t_2 - t_1)} \quad \begin{array}{l} \text{(degs/s or rads/s)} \\ \text{(seconds)} \end{array}$$

In this example (*Fig. A3.5*) the initial angular velocity of the limb was zero (position 1) and the final angular velocity was 60°/s (position 2). Hence average angular acceleration can be calculated as follows:

$$\alpha = \frac{60°/s - 0°/s}{0.25\ s - 0\ s}$$

$$= 240°/s^2 \text{ or } 4.19 \text{ rads/s}^2$$

Note: for every point on the limb segment (points near to the axis of rotation (which is the hip joint) or points far away from the axis of rotation) the angular acceleration will be the same

Fig. A3.7.   Calculation of angular acceleration (α)

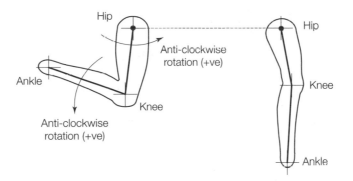

Fig. A3.8. Angular movements of limb segments in kicking a ball (showing only upper and lower leg segments)

see that the lower leg (during this simultaneous action with the upper leg) traveled through an angle of 105° in an anti-clockwise direction (100° before the vertical line and 5° after the vertical line). Similarly, because it is attached to the upper leg at the knee joint (which is the axis of rotation for this segment) all this happens in the same time of 0.5 seconds. *Fig. A3.10* shows the average angular velocity and angular acceleration calculations for the lower leg segment. Again, it is important to point out that all parts of the body along this lower leg segment (represented by a line in this example) will have the same average angular velocity and the same average angular acceleration. All the parts along this lower leg segment (depicted as a line in *Fig. A3.9*) travel through the same angle (105°) in the same time (0.5 s).

Summation of speed principle

Considering *Figs A3.8* and *A3.9* it is possible to see that the upper leg segment (hip to knee (thigh)) and the lower leg segment (knee to ankle (shank)) are linked together and they and move in one simultaneous action from position 1 to position 2. Although the angular displacements of the two segments are different (upper leg segment moves through 30° rotation anti-clockwise and lower leg segment moves through 105° rotation anti-clockwise), both limbs are

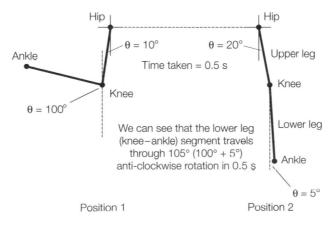

Fig. A3.9. Angular movements of limb segments in kicking a ball (only upper and lower leg shown)

Lower leg moves through 105° (100° + 5°) in a time of 0.5 s

$$\text{Average angular velocity } (\omega) = \frac{\text{displacement (angular)}}{\text{time taken}}$$

$$= \frac{105}{0.5}$$

$$= \textbf{210°/s} \text{ (or 3.66 rads/s)}$$

$$\text{Average angular acceleration } (\alpha) = \frac{\text{angular velocity } (\omega)}{\text{time taken}}$$

$$= \frac{210 - 0}{0.5 - 0}$$

$$= \textbf{420°/s}^2 \text{ (or 7.33 rads/s}^2)$$

Fig. A3.10.   Average angular velocity and average angular acceleration of lower leg segment in kicking a ball (from Fig. A3.9)

attached to each other. At the same time (in this example) the whole body (or whole leg in this case) moves forwards with a linear translation (as you would expect when you attempt to kick a ball). The **summation of speed principle** which has been widely discussed within the biomechanical literature suggests that such a movement is initiated from the larger segments and is then transferred to the smaller segments. For example, in throwing a ball, movement is first initiated from the legs, transferred through the hips to the shoulders and then on to the elbow, wrist, hand, and fingers. As each part of the body approaches extension (and often peak linear and angular velocity) the next part begins its movement. Although this certainly appears to be true for the actions of kicking and throwing (as anyone who has kicked or thrown a ball will know) the biomechanical research on this topic is not conclusive as to the exact mechanism for the generation of final velocity at the point of contact or ball release (because the actions are multi-planar and three-dimensional).

Absolute and relative joint angles

Within biomechanics a joint angle can be expressed as two lines that intersect at a point. The intersection point is termed the **vertex** and the joint angle can be the angle that is contained between the two lines (**the included angle**). The two straight lines usually represent the segments of the body (e.g., the upper leg segment and the lower leg segment in our example) and the vertex the joint center (which would be the knee joint in this case). **Absolute joint angles** are determined from the right horizontal to the distal end of the segment of interest. **Relative joint angles** are the angles that are contained between segments and these are the **included angle** between the longitundinal axes of the two segments. In the description of relative angles within the human body it is possible that both lines (or segments) will be moving. However, when discussing absolute angles, one of the lines (i.e., the right horizontal) is fixed and does not move. *Fig. A3.11* illustrates the difference between absolute angles and relative joint angles in more detail.

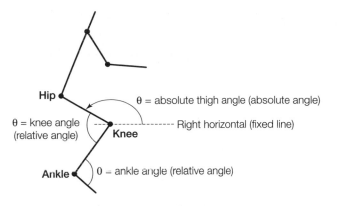

Fig. A3.11.   Absolute and relative joint angles

A4 THE RELATIONSHIP BETWEEN LINEAR AND ANGULAR MOTION

Key Notes

Linear and angular movement	The linear and angular components of movement are linked by a mathematical relationship. Specific formulae exist that are used to show the linear translation caused by an object that is rotating.

The angular movements of arms, segments and implements causes the linear motion of an end point that is applied to objects such as soccer balls, golf balls, and tennis balls.

Objects and points on a rotating arm (segment)	In order to examine the angular displacement of an object that is located on a rotating arm (rigid body) it is necessary to join two points on the object with a line. The new line on the object will have the same angular displacement as the original line segment (rigid body). A point does not have a rotational orientation and hence can not be expressed as having angular displacement. However, a collection of points (i.e., a line) or arm/segment can be used to determine this angular displacement.

Linear distance and displacement of a point on a rotating arm	The linear distance covered by a point on a rotating arm is the arc length (curve of motion). The linear displacement of a point on a rotating body is expressed by determining the chord length. The linear distance (arc length) is calculated by the formula $s = r\,\theta$ where θ is expressed in radians. A point that is further away from the axis of rotation will cover a greater linear distance (arc length) than a point that is nearer to the axis of rotation. Hence this point (which is further away) will also have greater linear velocity and acceleration.

Relationship between linear and angular velocity and between linear and angular acceleration of objects/points on a rotating arm	The formula $\mathbf{v} = \boldsymbol{\omega}\,\mathbf{r}$ is used to relate the average linear velocity of a point or object on a rotating arm with the average angular velocity of the arm.

The formula $\mathbf{a} = \boldsymbol{\alpha}\,\mathbf{r}$ is used to relate average linear acceleration of a point or object on a rotating arm with the average angular acceleration of the arm.

Instantaneous linear velocity and acceleration	Instantaneous velocity or acceleration is the velocity or acceleration expressed at an instant in time. Average velocity or acceleration is determined between successive time points. Instantaneous is defined as occurring at a given instant or limit as the time interval approaches zero. As the time interval between points gets smaller (i.e., approaches zero) the value determined for velocity and acceleration tends towards an instantaneous value.

Tangential	The linear velocity and acceleration of a point on a rotating body/arm/segment acts at a tangent to the curve (arc of motion). A tangent is defined as a line that touches a curve at a point. This line has the same gradient as that of the curve at that point. It acts perpendicular (at 90°) to the rotating arm or segment.

Linear and angular movement

The **linear** and **angular components** of movement are linked by a **mathematical relationship**. Specific formulae exist that show how the linear translation of points on a rotating object can be determined. Often within biomechanics it is necessary to understand and apply this relationship. For example, in the case of the soccer kick it is the angular movement of the leg that creates the resultant linear velocity (and horizontal and vertical components) that is applied to the ball in order to give it trajectory and movement. Similarly, in golf it is the angular movement of the arms and the club that imparts resultant linear velocity to the golf ball to give it an angle of take-off and a parabolic flight path. *Fig. A4.1* shows an element of this angular–linear relationship.

In *Fig. A4.1* the arm (AB) moved from position 1 to position 2 in 0.45 seconds. The **angular displacement** in this example was 35° (i.e., **change in angular position**). The **average angular velocity** of the arm AB can be determined by the **change in angular position** (angular displacement) divided by the **time taken**.

ω = angular displacement   (between position 1 and position 2)
    ——————————
    time taken

= 35°
  ——
  0.45 s

Average angular velocity = **77.8 °/s (or 1.36 rads/s)**

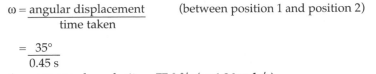

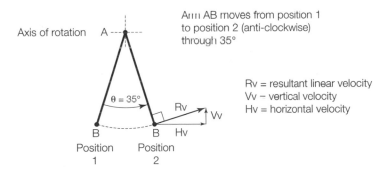

Fig. A4.1.   Linear and angular components of motion

This value represents the average angular velocity of the arm AB between position 1 and position 2. The angular displacement is angle that is formed between the final and the initial position of the line/segment/arm (i.e., 35° in this example). Any point on this line AB (or segment) will move through the same angle in the same time. Within biomechanics, however, it is often the case that we refer to points or objects on a rotating line or segment. In order to express the angular displacement of an object that is on a rotating line or segment it would be necessary to choose any two points that are on the object and join them with a line. The new line on the object will rotate (if the original line/segment/arm is a rigid body) in exactly the same manner as the original segment or line that the object is located on. *Figs A4.2* and *A4.3* help to illustrate this in more detail.

In *Fig. A4.3* it is possible to see that all the points (parts) on a rotating body (A, B and, C collectively in this case (i.e., the additional line)) will travel through the same angle (60° anti-clockwise). Since the angular displacement of the original line/segment is also 60°, and it moves through this angle in the same time, the

Objects on a rotating arm (segment)

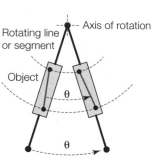

An object (rectangle) is positioned on a rotating line. Two points on the object are joined by a line (new line). The rotation of the object is the same as the rotation of the line/segment (providing the line or segment rotating is a rigid body). Thus θ experienced by the object (and the new line) = θ experienced by the rotating line/segment (original line)

Fig. A4.2.   Points on a line or segment that is rotating.

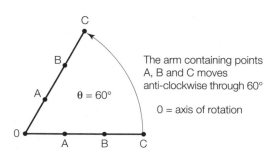

The arm containing points A, B and C moves anti-clockwise through 60°

0 = axis of rotation

The angular displacement covered by *all parts* on a rotating body is equal (i.e., 60° anti-clockwise in this case). Providing it is a rigid body. *Note*: In describing angular movement it is not technically correct to refer to these elements as points because it is not possible for a point to have a rotational orientation. However it is possible for a collection of points to have rotational orientation (i.e., the arm/line containing points A, B and C)

Fig. A4.3.   Relationship between linear and angular components of motion.

angular velocities of all the parts on the arm will be the same. In this context it is important to indicate that in terms of angular displacement we are technically referring to a collection of points or parts (i.e., the arm/line A, B, and C) as it is not possible for a single point to have an angular orientation (see *Fig. A4.3*). However, if we were to consider the linear distance/displacement and linear velocities (technically linear speed without the directional component) of each point A, B, and C (which we can because a point can have a linear orientation) we can determine that the linear distances covered by each of these points and hence their linear velocities will be different (because each will have a different radial (radius) distance from the point of rotation). The further away the point is from the center of rotation (as in the case of point C in *Fig. A4.3*) the greater will be the linear distance/displacement covered by that point.

Linear distance and displacement of rotating points

Linear distance/displacement of a point on a rotating arm
The point B on the rotating arm in *Fig. A4.4* will move through a **linear distance** that can be expressed by determining the **length of the arc**. However, the **linear displacement** of the point B can be determined by calculating the **length of the chord** that is contained from the position of point B at the beginning of the movement to the position of point B at the end of the movement. A **chord** is defined as a straight line connecting two points on a curve or surface that lies between the

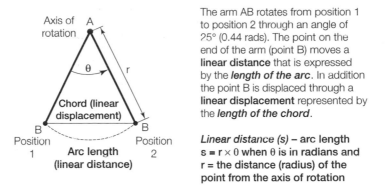

The arm AB rotates from position 1 to position 2 through an angle of 25° (0.44 rads). The point on the end of the arm (point B) moves a **linear distance** that is expressed by the *length of the arc*. In addition the point B is displaced through a **linear displacement** represented by the *length of the chord*.

Linear distance (s) – arc length
s = r × θ when θ is in radians and r = the distance (radius) of the point from the axis of rotation

Fig. A4.4.   *Linear distance/displacement of a rotating object/point*

two points. To calculate the **linear distance** covered by the rotating point (i.e., the arc length) we can use the relationship that is presented in *Fig. A4.4* that links the angle (only when it is expressed in radians) with the distance of the point from the axis of rotation (the radius). In *Fig. 4.4*, if point B was located 0.34 m from the axis of rotation (i.e., the radius) what would be the linear distance covered by the point if the arm rotated through an angle of 25° (0.44 radians)? The following formulae (also shown in *Fig. 4.4*) is used to calculate the linear distance (i.e., the length of the arc) moved by the point in this example.

Linear distance moved by a point on a rotating arm/line/segment (Fig. A4.4)

$$s = r\,\theta$$

where
s – linear distance (arc length)
θ = angle or angular displacement (expressed in radians only)
r = distance of point from axis of rotation (radius)

For point B located 0.34 m from axis of rotation and moving through 25° (0.44 radians)

s = r θ
s = 0.34 × 0.44
s = 0.15 m (to two decimal places) – arc length or linear distance moved by point B

In order to calculate the length of the **chord** (or the **linear displacement**) it is necessary to also use the distance of the point from the axis of rotation (the radius) and the angular displacement (i.e., the angle θ that the arm has been displaced through). However this relationship is not as straightforward as the calculation for linear distance (the arc length) because although the **chord length** is directly proportional to the radius it is not directly proportional to the angle or angular displacement. *Fig. A4.5* shows the calculation for chord lengths for angles up to 90°. For angles greater than 90° there are more complex tables of chords that are used within the area of mathematics (i.e., Ptolemy's table of chords).

From *Fig. A4.6* it is possible to see that **a point that is closer** to the **axis of rotation will travel with less linear distance** (arc length) **than a point that is further away from the axis of rotation** (when both points are located on the same line/segment/rotating arm). This is the same application as with muscles and their points of attachments to bones. The muscle (muscle tendon and muscle

O = axis of rotation
OC rotates anti-clockwise to OA

AC = chord

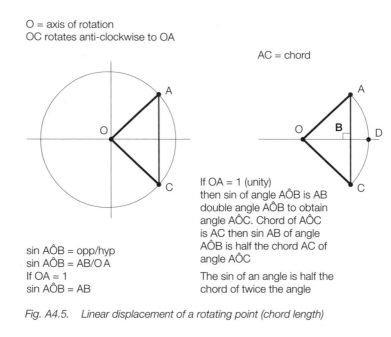

If OA = 1 (unity)
then sin of angle AÔB is AB
double angle AÔB to obtain
angle AÔC. Chord of AÔC
is AC then sin AB of angle
AÔB is half the chord AC of
angle AÔC

sin AÔB = opp/hyp
sin AÔB = AB/OA
If OA = 1
sin AÔB = AB

The sin of an angle is half the
chord of twice the angle

Fig. A4.5.   Linear displacement of a rotating point (chord length)

O = axis of rotation
OA = 0.46 m
OB = 0.67 m
anti-clockwise rotation 22°

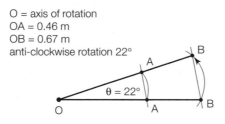

What is the linear distance covered by each point (A and B)?

Fig. A4.6.   Linear distance moved by points acting at different distances from the axis of
rotation

contraction which causes linear movement) will move with a small amount of
linear distance (i.e., a small contractile element) but it will cause a large amount of
movement at the end of the segment or arm (i.e., flexing your arm in a bicep curl
causes a large linear movement at the end of the arm (i.e., at the hand) and a small
linear movement at the muscle point of attachment (which is ideal because the
muscle is only able to move a small linear distance in contraction)). In *Fig. A4.6*
point A is located 0.46 m from the axis of rotation (O) and point B is located 0.67
m from the axis of rotation. If the angular displacement of the limb (rotating arm)
in this case is 22° (0.38 radians) anti-clockwise what is the linear distance (arc
length) covered by point A and point B?

Linear distance covered by point A

s = r θ (where θ is expressed in radians)
s = 0.46 × 0.38
s = 0.175 m

Linear distance covered by point B

$s - r\,\theta$

$s = 0.67 \times 0.38$

__$s = 0.255$ m__

All points (parts) on this rotating arm in *Fig. A4.6* will move through the same angle (angular displacement) in the same time. Hence all these points (or collection of points) will have the same average angular velocity. However, point A has moved through a linear distance of 0.175 m and point B has moved through a linear distance of 0.255 m. Both these movements occurred in the same time and therefore both points will have different average linear velocities (because B has moved through 0.255 m in the same time that A has moved through 0.175 m).

Relationship between linear and angular movement

Considering Fig. A4.7 it is possible to see that we can take the formula used to determine the linear distance ($s = r\theta$) moved by a point on a rotating body (the arc length) and by algebraic manipulation we can develop the formula that is used to link average linear velocity (v) with average angular velocity (ω). Fig. A4.7 shows that by algebraic manipulation we can derive that the relationship between average linear velocity and average angular velocity is as follows:

$$v = \omega\, r$$

where
v = average linear velocity
ω = average angular velocity
r = radius or distance from point of rotation to point of interest

By algebraic manipulation we can take the formula used to determine linear distance and develop the equations to be used to calculate the average linear and average angular velocity components of rotational motion

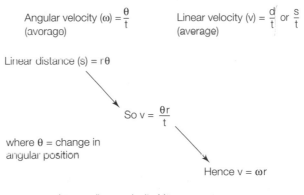

Angular velocity (ω) = $\dfrac{\theta}{t}$
(average)

Linear velocity (v) = $\dfrac{d}{t}$ or $\dfrac{s}{t}$
(average)

Linear distance (s) = $r\theta$

So v = $\dfrac{\theta r}{t}$

where θ = change in angular position

Hence v = ωr

Average linear velocity (v) = average angular velocity (ω) × radius (r)

Fig. A4.7.   Calculation of average angular velocity

Average and instantaneous values

As we have seen the **average linear velocity** (or speed without a directional component) of a point on a rotating arm is derived from the equation of **average angular velocity** multiplied by **the radius** or distance of the point of interest from the axis of rotation (v = ω r). This is expressed as an average value because it is determined from the movement between two points (i.e., an angular displacement). However, the linear velocity of point B at any **instant in time** throughout the movement will act at a **tangent** to the curve (arc of movement). This will be expressed as velocity at an instant and is know as **instantaneous velocity**. In this context it is expressed as a **tangential linear velocity** and is given the expression v_T. A **tangent** is defined as a line that touches a curve at a point. The line will have the same gradient (slope of a line measured as its ratio of vertical to horizontal change) as that of the curve at that point. The tangent will act at 90° (perpendicularly) to the rotating arm that the point is contained on (*Fig. A4.8*). The **direction** of the velocity of this point will be **perpendicular** to the rotating arm (radius of the point) and at a **tangent** to the curve (circular path of the object).

Considering *Fig. A4.8* it is possible to see that at any instant in the rotational movement of the arm AB the point B will have a tangential linear velocity (a velocity that acts at a tangent to the curve (arc) of rotation). This linear tangential velocity will be derived from how much linear distance (the arc length) the point has moved through in a given time or from the angular velocity in a given time. The smaller the time intervals the more the value will tend towards an instantaneous value. **Instantaneous** is defined as **occurring at a given instant** or limit as the time interval approaches zero. In biomechanics it is important to understand and determine this linear velocity so we can assess how effective our rotational movements (such as in the golf swing) are in relation to producing linear motion (such as the resultant linear velocity of the golf ball).

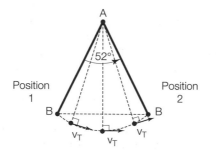

The arm AB rotates anti-clockwise from position 1 (about axis A) to position 2 through 52° in 0.25 s. At each instant in time in this rotational movement point B will have tangential linear velocity. The average angular velocity is measured by the angular displacement covered by the arm (52°) over the time taken (0.25 s)

Instantaneous linear (tangential) velocity v_T of point B at various instants in time throughout the rotational movement of the segment/arm

Fig. A4.8.   Relationship between linear and angular components of motion

Angular and linear acceleration

As with velocity, a mathematical relationship exists that will link **average linear acceleration** with average angular acceleration and this is portrayed in *Fig. A4.9*. From *Fig. A4.9* we can see that **average angular acceleration** is expressed as **change in angular velocity over time** (or the rate of change of angular velocity). This is portrayed as $\alpha = \omega/t$. In order to link average linear acceleration with average angular acceleration we use the same method of algebraic manipulation that we used for determining the relationship between angular and linear velocity. Thus the formula that links average linear acceleration with average angular acceleration is as follows:

$$a = \alpha r$$

$$\text{Average angular acceleration } (\alpha) = \frac{\omega \text{ (change in angular velocity)}}{t \text{ (time)}}$$

$$\text{Average linear acceleration } (a) = \frac{v \text{ (linear velocity)}}{t \text{ (time)}}$$

From *Fig. A4.7*
Average linear velocity (v) = average angular velocity (ω) × radius (r)

Hence
Average linear acceleration (a) = $\dfrac{\omega r}{t}$ $\longrightarrow$ $a = \alpha r$

Average linear acceleration (a) = average angular acceleration (α) × radius (r)

Fig. A4.9. Calculation of average angular acceleration

where
a = average linear acceleration
α = average angular acceleration
r = radius or distance of point of interest from axis of rotation

In this context it is important to also clarify that if this **linear acceleration** was to be determined at an **instant in time** then it would also be classified as an **instantaneous value** that is acting at a tangent to the curve (or arc of motion). This would be represented by the expression a_T or **tangential linear acceleration**. Similarly as with linear tangential velocity this acceleration would act in the direction of the tangent to the curve at that instant in time.

As we have seen from section A3, linear and angular movement have an important relationship with each other. In human movement we use rotational motion of an arm to generate linear translation of a point. For example, hitting a tennis ball requires rotational movement of the arm and racket; throwing a basketball requires rotational movement of the upper body, arms, and hands, and obviously in golf we use rotational displacements of the club and arms to generate high linear velocity of the club head and hence the golf ball (in excess of 45 m/s in professional golfers). Therefore, a good understanding of this relationship is valuable towards providing an effective knowledge of human movement.

A5 GRAPHICAL PRESENTATION OF KINEMATIC DATA – NUMERICAL DIFFERENTIATION

Key Notes

Numerical differentiation	Numerical differentiation is the name given to a method for calculating the rate of change of one variable with respect to another, usually time. It does this using data collected during an experiment. In sport and exercise biomechanics the variables most widely used are displacement and velocity. The rate of change of displacement with respect to time is called velocity while the rate of change of velocity with respect to time is called acceleration.
Gradient of a curve	The gradient of a curve representing data gives the rate of change and is calculated from the slope between two data points. The process is best illustrated graphically.
Average and instantaneous velocity	The gradient of a displacement-time curve gives the average velocity. If the time interval between the two data points reduces to a very small value the average velocity becomes the instantaneous velocity.
Positive and negative gradients	A positive gradient indicates a positive rate of change. For a displacement-time curve this represents a positive velocity, in other words an increase in velocity. A negative gradient indicates a negative velocity and means the object is traveling in the negative direction with respect to the measuring axes.
Points of minima, maxima, and inflection	The displacement-time curve will have points of inflection and localized minima and maxima. These indicate something special is happening to the motion of the object. Points of minima and maxima indicate the object has zero velocity. Points of inflection indicate a minimum or maximum velocity has been reached.
Finite difference method for numerical differentiation	The finite difference method is an algorithm for performing numerical differentiation. In practice this is a simple method and is based on the equation for average velocity (when calculating velocity from displacement data) or average acceleration (when calculating acceleration from velocity data).

Numerical differentiation

The biomechanical study of human motion requires an understanding of the precise relationship between the changes in position (**displacement**), how fast the body is moving (**velocity**), and indeed how the velocity itself is changing (**acceleration**). In section A2 it is shown that the **average velocity** of any moving object is given by the **change in displacement divided by the time** over which the change takes place. If displacement is represented by the letter s and time by the letter t, the average velocity between instant 1 and instant 2 may be determined from the equation:

$$\textbf{average velocity} = v_{av} = \frac{s_2 - s_1}{t_2 - t_1} \qquad \textbf{(A5.1)}$$

The average velocity is also called the rate of change of displacement. Remember, velocity is a vector quantity and therefore this represents the average velocity in a specific direction; if the direction is not specified or unimportant to the situation then the above equation is preferably termed the **average speed**.

Fig. A5.1a graphically represents the displacement of a moving object plotted against time. From this it can be seen that the equation for the average velocity between s_1 and s_2 is in fact the same equation that gives the **slope or gradient** of the line between the points marked A and B, which correspond to the times t_1 and t_2 respectively. Similarly, the **gradient** of the line between points C and D must be the **average velocity** of the object over the smaller time interval δt. (Note: δ is the Greek lower case letter delta and is often used in mechanics to indicate a small change in some quantity, in this case a small change in time.)

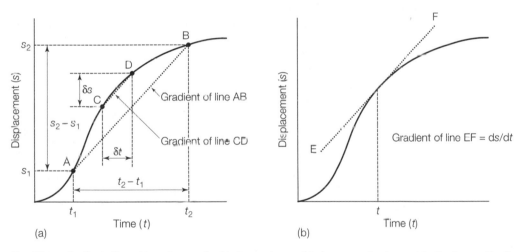

Fig. A5.1. An illustration of how the gradient between two points becomes the tangent to the line as the time interval between the two points reduces to zero

Instantaneous values

Knowing the average velocity of a moving object is only of limited value; usually of greater interest is the velocity at a particular point in time and to know this it is necessary to know the object's instantaneous velocity. The **instantaneous velocity** is the velocity that exists at any point in time and is given by the point on the curve in Fig. A5.1a at that time. As the change in time becomes smaller and smaller, the average velocity becomes the instantaneous velocity as the gradient of line C–D becomes the **tangent** to the curve at that instant in time (Fig. A5.1b, line E–F). Mathematically this is represented as the instantaneous velocity (v)

$$v = \frac{ds}{dt}$$

and is said to be the **differential of displacement**, s, with respect to time, t.

Following similar reasoning, and given that the average acceleration is

$$\text{average acceleration} = a_{av} = \frac{v_2 - v_1}{t_2 - t_1}$$ (A5.2)

the **instantaneous acceleration** is given by the gradient of the tangent to the velocity curve at that instant in time and therefore the instantaneous acceleration, a, is said to be the differential of velocity, v, with respect to time, t. This is written mathematically as:

$$a = \frac{dv}{dt}$$

As this term contains velocity, which is itself a rate of change of displacement with respect to time, acceleration is said to be the **second differential of displacement** with respect to time.

The sign of the gradient

In *Fig. A5.1a*, the average velocity between time t_1 and t_2 will be positive because s_2 is greater than s_1 and therefore subtracting s_1 from s_2 will produce a positive result. The **gradient** of the line between A and B is a **positive gradient**. Similarly, the gradient of the tangent to the curve in *Fig. A5.1b* is positive.

Consider now *Fig. A5.2*. Here s_2 is less than s_1 therefore subtracting s_1 from s_2 will produce a **negative gradient** and the velocities will also be negative. Because velocity is a vector quantity, its sign tells us about the direction of travel. In *Fig. A5.1*, the object is moving away from the reference point (i.e., its displacement is increasing from zero). In *Fig. A5.2*, however, the object's displacement is decreasing: it is getting closer to the origin. The negative sign of the velocity tells us that the object is now moving in the opposite direction.

Acceleration may also be either positive or negative but whilst the sign of the velocity is only dependent upon the direction of motion, the sign of an object's acceleration is dependent upon whether the object is accelerating or decelerating. For example, a ball thrown vertically into the air will be moving in a positive direction but as it is slowing down its acceleration will be negative (i.e., decelerating). When the ball reaches the apex of its flight and falls back to earth the magnitude of its velocity will now be increasing but in a negative direction (i.e., its velocity is negative) but its acceleration will still be negative (*Fig. A5.3*).

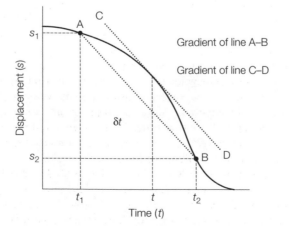

Fig. A5.2. An example of a negative gradient and negative tangent

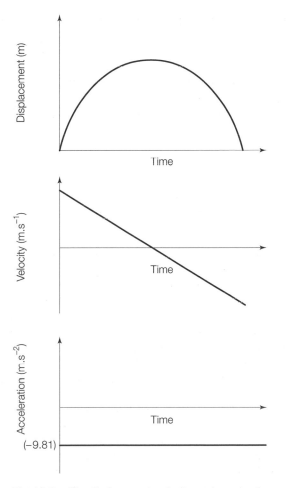

Fig. A5.3.   The displacement, velocity, and acceleration profiles for a projectile

Points of maxima, minima and inflection

Sometimes, when plotting the motion of an object on a displacement time graph, we see localized **points of maximum** (point A, *Fig. A5.4a*) or **minimum** (point B, *Fig. A5.4a*) displacement (localized maxima and minima). At these points the gradient of the curve is neither positive nor negative because the tangent is horizontal. Here the velocity must be zero.

Points of inflection may also occur. Points of inflection occur when the curve moves from a concave to convex (point C, *Fig. A5.4b*) or from convex to concave (point D, *Fig. A5.4b*). These represent localized maximum and minimum gradients respectively and hence points of maximum or minimum velocity. Following the same reasoning, points of inflection on a velocity time graph must indicate local maximum or minimum acceleration.

A special case is the projectile flight of *Fig. A5.3*. At the point of inflection of the displacement, the tangent is horizontal and indicates a change in direction of the projectile from an upward motion to a downward motion. *Fig. A5.5* represents the flexion angle, angular velocity and angular acceleration of the knee joint during a normal walking stride from heel strike to heel strike. Note that the

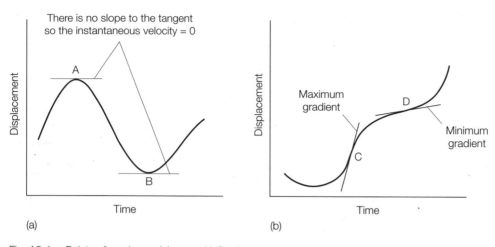

(a)   (b)

Fig. A5.4.   Points of maxima, minima, and inflection

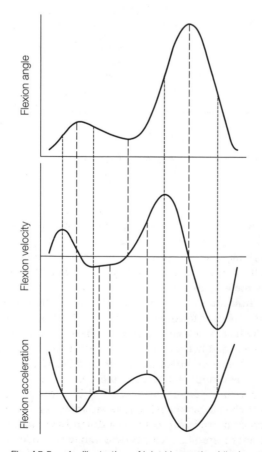

Fig. A5.5.   An illustration of joint kinematics (displacement, velocity, and acceleration)

angular displacement, velocity, and acceleration curves follow the same rules as linear displacement, velocity, and acceleration curves. Note also that the points of inflection on the displacement curve indicate local maximum or minimum velocity and therefore points of zero acceleration.

Numerical differentiation – the finite difference method

Differentiation is a mathematical process that quantifies the change in one variable with respect to another. In this case, displacement and velocity with respect to time. Therefore, differentiation of a displacement–time curve allows the determination of the rate of change of displacement (i.e., the instantaneous velocity). Similarly the differentiation of a velocity–time curve allows the determination of the rate of change of change of velocity (i.e., the instantaneous acceleration).

In sport and exercise biomechanics, experimental data consist of a series of discrete data values, so to obtain the instantaneous velocity (or acceleration) it is necessary to use a method of **numerical approximation**. There are a number of **numerical differentiation methods** we can use but the simplest (and most frequently used) technique is the **finite difference method**.

Table A5.1 represents the displacement-time data for an athlete during a 100 m sprint race. The displacement from the start at each one second interval is recorded in column two. The true instantaneous velocity of the athlete can never be known but if the time interval was sufficiently small it may be possible to estimate the velocity of the athlete during the race using the average velocity between the known data points. That is, we use the equation A5.1:

$$v = \frac{s_{i+1} - s_i}{t_{i+1} - t_i}$$

where v is the approximated instantaneous velocity, and s_i and t_i are displacement, and times at instant i, and s_{i+1} and t_{i+1} are the displacement and times at the next data value. For example, between time 0.0 s and 1.0 s the velocity is given by:

$$v = \frac{s_1 - s_0}{t_1 - t_0} = \frac{4.20 - 0.00}{1 - 0} = 4.20 \text{ m.s}^{-1}$$

It is important to note that the velocity value is ascribed to the mid-point between the two displacement values (i.e., for displacement at times 0 s and 1 s the velocity is attributed to 0.5 s). This is because the velocity is assumed to be constant between the two displacement points so it makes sense to pick a point mid-way between them. Acceleration may be calculated in a similar way and the acceleration value is ascribed to the mid-point between the two velocity values. Thus, end point data are lost when velocity is calculated and that even more data points are lost when acceleration is determined. The calculated velocity and acceleration for the whole race is given in Table A5.1.

The estimate of the instantaneous velocity becomes better and better the smaller the value of t becomes. It is common practice in sport and exercise biomechanics to use video to collect images of performance. This gives the possibility of reducing the time interval to 0.04 s (for a video image rate of 25 Hz). With special cameras or analysis equipment to access every video field (equal to an image rate of 50 Hz) a higher image rate (and so a smaller time interval) can be achieved. Under these circumstances the "approximated" instantaneous velocity will be close to the actual instantaneous velocity.

Table A5.1.   A worked example of numerical differentiation for a 100 m race

t (s)	s (m)	v (m.s^{-1})	a (m.s^{-2})
0.0	0.00		
0.5		4.20	
1.0	4.20		3.03
1.5		7.23	
2.0	11.43		1.68
2.5		8.91	
3.0	20.34		1.51
3.5		10.42	
4.0	30.76		1.10
4.5		11.52	
5.0	42.28		0.61
5.5		12.13	
6.0	54.41		0.16
6.5		12.29	
7.0	66.70		−0.29
7.5		12.00	
8.0	78.70		−0.75
8.5		11.25	
9.0	89.95		−1.20
9.5		10.05	
10.0	100.00		

A6 GRAPHICAL PRESENTATION OF KINEMATIC DATA – NUMERICAL INTEGRATION

Key Notes

Numerical integration

Numerical integration is the name given to a method for calculating the total change of one variable with respect to another, usually time. It does this using data collected during an experiment. In sport and exercise biomechanics the variables most widely used are acceleration and velocity. The integration of acceleration with respect to time gives the total velocity change. The integration of velocity with respect to time gives the total displacement change. There are several algorithms for performing this calculation but a common one is the Trapezium Rule. This process is best illustrated graphically.

Area under a curve

The area under a data curve between two points in time the gives the total change in that variable from the first point in time to the second point in time. For a velocity–time graph, the area under the curve between two points in time gives the distance traveled during this time period.

Accurate and estimated displacements – the Trapezium Rule

The area under a velocity-time curve can be broken down into small slices representing the distance traveled during each small time interval. Each of these slices can be represented by a trapezium (a rectangular shape with one side longer than the other). The area of a trapezium is easily calculated and so the total area under the curve is given by the sum of the areas of all of the trapezia. This area is an estimate of the total area so the distance computed represents an estimate of the total distance traveled. As the time interval for each small slice reduces, the estimate becomes better and ultimately becomes an accurate value.

Integration

In section A5 it has been shown that the process of differentiation allows the velocity to be determined from displacement, and acceleration to be determined from velocity. **Integration** is the reverse of differentiation and allows the determination of velocity from acceleration and displacement from velocity. In other words:

Differentiation allows	*Displacement*	⇨	*Velocity*	⇨	*Acceleration*
Integration allows	*Acceleration*	⇨	*Velocity*	⇨	*Displacement*

Whilst differentiation measures the gradient of the appropriate curve at a given instant, **integration measures the area under a curve**. This is demonstrated in the velocity–time graphs presented in *Fig. A6.1.*

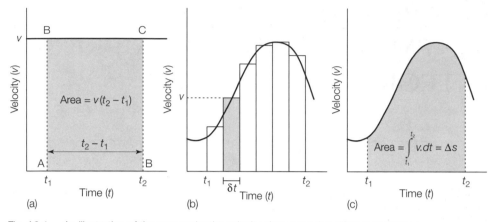

Fig. A6.1. An illustration of the area under the velocity–time curve for different conditions

Fig. A6.1a represents an object moving with constant velocity. Between two points in time, t_1 and t_2, we know from section A5 that:

$$v = \frac{(s_2 - s_1)}{(t_2 - t_1)} \qquad \text{(A5.1)}$$

Where v is the object's velocity, $s_2 - s_1$ is the change in displacement (Δs) between instants 1 and 2, and $t_2 - t_1$ is the change in time. Therefore:

$$s_2 - s_1 = v(t_2 - t_1) = \text{area of the rectangle ABCD}$$

The velocity data in *Fig. A6.1b* is more complex. This may be approximated by a number of smaller rectangles of width δt. The area of a single rectangle (given by $v \times \delta t$) must be approximately equal to the change in displacement over the time δt. The area under the curve, and hence the change in displacement over the period from t_1 to t_2 can then be approximated by adding together the areas of all such rectangles between t_1 and t_2. This is only an approximation as the velocity is assumed to be constant during each small time interval δt, but the approximation gets closer to reality the smaller δt is, and is exact if $\delta t = 0$.

The ideal case, when the time intervals become so small that their sum reflects the exact area under the curve, is illustrated in *Fig. A6.1c*, and this equals the change in displacement that occurs between t_1 and t_2. This is written as:

$$\Delta s = \int_{t_1}^{t_2} v.dt$$

Where Δs is the change in displacement (note: Δ is the Greek upper case letter delta and is often used in mechanics to indicate a change in some quantity), v is velocity, and t is time. The symbol $\int$ is the symbol for integration. The letter d indicates the variable over which the change is being measured (in this case time, t) and t_1 and t_2 are termed the limits of integration.

Using similar reasoning, it can be shown that the area under an acceleration time curve between two points in time must be equal to the change in velocity that occurs and hence:

$$\Delta v = \int_{t_1}^{t_2} a.dt$$

where a is acceleration.

Numerical integration

In sport and exercise biomechanics the data that are most commonly available for which the process of integration is appropriate are data from force platforms and accelerometers. The force data from force platforms can be used to compute acceleration following Newton's second law ($F = ma$). The process of integration enables the velocity to be obtained, and the process can be repeated on this velocity data in order to obtain displacement. Integration of this type of data is best done using numerical integration. The term **numerical integration** describes a process of finding the **area under the data curve**, represented by the thin columns in *Fig. A6.1b*, which are summed together to provide an approximation for the true area under the curve. There are a number of numerical integration techniques that can be used but the most commonly used is the **trapezium rule**.

Fig. A6.2 illustrates a velocity–time graph in which the velocity data have been sampled at equal intervals Δt. The curve may be represented by a series of trapeziums. A trapezium is a rectangular shape with one side longer than the other. The area under the curve may then be considered to be equal to the sum of the areas of the trapezia. The area of a trapezium is equal to half the sum of its two sides multiplied by its base. If, for a set of discrete data, the base of each trapezium is equal to the time between samples (Δt), and the sides are defined by the magnitude of adjacent samples (v_i and v_{i+1}), the area under a single trapezium $= \Delta t.(v_i + v_{i+1})/2$, and for n samples, the total area under the velocity curve is given by:

$$\text{Area} = \Sigma\ (\Delta t.(v_i + v_{i+1})/2) \qquad \text{for i=1 to n–1} \qquad \textbf{(A6.1)}$$
$$= \Delta s$$

and is equal to the **change in displacement** (Δs) from t_1 to t_n. The term Σ means sum all terms between the stated limits, here from 1 to n–1. Similarly, if we were working with acceleration–time data,

$$\text{Area} = \Sigma\ (\Delta t.(a_i + a_{i+1})/2) \qquad \text{for i=1 to n–1} \qquad \textbf{(A6.2)}$$
$$= (\Delta v)$$

and is equal to the **change in velocity** (Δv) from t_1 to t_n. An example is given in *Fig. A6.3* which shows how the increase in distance can be calculated based on the above equations.

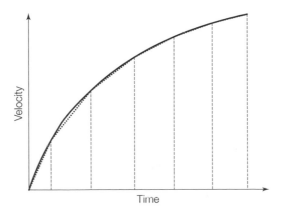

Fig. A6.2.   Instantaneous (solid line) and trapezoidal approximation (dotted line) showing that as the interval reduces the two curves become similar

Problem

Determine the distance traveled (i.e., the change in displacement Δs) of the sprinter over the first 3 s given the velocity data in the table

Time (s)	Velocity (m/s)
0.0	0
0.5	2.2
1.0	3.3
1.5	4.0
2.0	4.6
2.5	5.1
3.0	5.5

Solution

From the Trapezium Rule: $\Delta s = \Sigma(\Delta t.(\mu_i + \mu_{i+1})/2)$ for $i = 1$ to 6

$\Delta s =$ ((0.0 + 2.2)/2) × 0.5
      + ((2.2 + 3.3)/2) × 0.5
      + ((3.3 + 4.0)/2) × 0.5
      + ((4.0 + 4.6)/2) × 0.5
      + ((4.6 + 5.1)/2) × 0.5
      + ((5.1 + 5.5)/2) × 0.5
      = 0.55 + 1.375 + 1.825 + 2.15 + 2.425 + 2.65

therefore

$\Delta s = 10.825$ m

Fig. A6.3.   Example of numerical integration

The Trapezium Rule is only an approximation of the true area under the curve because it assumes that the curve between two adjacent samples is a straight line. If the curve is not a straight line this will result in an error. These errors will be reduced if the time interval between samples is reduced. (The narrower the trapezium, the more closely the curve between the sides of the trapezium will approximate to a straight line). Although some error will always persist, for most practical purposes in biomechanics this method of integration is considered suitable.

A7 UNIFORMLY ACCELERATED AND PROJECTILE MOTION

Key Notes

Introduction	Newton's Second Law of Motion dictates that bodies which experience a constant force also accelerate at a constant rate. The most common example of this occurring on Earth is when a body is airbourne; where the attractive force between the body and the Earth provides an acceleration equal to -9.81 m·sec^{-2}.
Effects of constant acceleration	A body subjected to a constant acceleration will experience a linear change in velocity and a curvilinear change in position when viewed over time. At any point in time the motion of a body (e.g., its position or velocity) that is accelerating constantly can be calculated using one of Galileo's equations of uniformly accelerated motion. These equations can be used to find, for example, the height raised by an athlete's center of gravity (CG) during a jump.
Projectile motion	A projectile is a body that is unsupported (i.e., a ball in flight) and only affected by the forces associated with gravity and air resistance. Projectiles generally have both horizontal and vertical velocity components during flight. If air resistance can be ignored, the horizontal velocity remains constant and the vertical velocity is affected by the constant acceleration due to gravity; which results in the projectile having a parabolic flight path.
Maximizing the range of a projectile	For a body that lands at the same height that it was projected from, its range is dependent upon both its velocity and angle of projection at take-off. More specifically, range is proportional to the square of the take-off velocity; so higher velocities will result in proportionally greater gains in range. At any velocity, a take-off angle of 45° will result in the greatest range.
Projectiles with different take-off and landing heights	For projectiles that are released from and land at different heights the optimal angle of projection is dependent upon both the take-off height and velocity. In the more common situation in sport, where the height of take-off is greater than landing (e.g., shot putt), the optimal angle is always less than 45°. The smaller the distance between take-off and landing heights, the closer the optimal angle gets to 45°. Similarly, for greater velocities, the optimal angle approaches 45°.

Introduction   In the examples of movements that occur with the body in contact with the ground (e.g., the take-off phase of the standing vertical jump (SVJ)) the acceleration of the body is rarely constant, or uniform, because of the changing forces that act on it. Section B describes such forces and explains the effect that they have on the motion of the body.

However, in situations where the forces acting on the body are constant, it experiences a constant acceleration. An obvious example of this is when a body is in flight (e.g., the time when a shot putt or a long jumper is airborne) and the only force acting on it is attractive or gravitational force that exists between it and the Earth (this force is further explained in section B). This is assuming that the effect of air resistance (see section D), is negligible; which it can be for bodies of large mass traveling at low speeds. The acceleration that a body experiences as a consequence of the gravitational force varies slightly depending on its position on Earth (it is slightly greater at the poles than the equator) but is generally agreed to be equal to 9.81 m·sec^{-2}. It should also be referred to as negative (i.e., –9.81 m·sec^{-2}) because the acceleration acts in a downwards direction, towards the surface of the Earth. However, other constant acceleration situations can occur when a body is not airborne. For example, a cyclist who stopped pedaling on a flat road would experience a fairly constant horizontal deceleration. Similarly, providing it was traveling up or down a smooth incline, a bobsleigh would also experience an approximately constant deceleration or acceleration.

Effects of constant acceleration

When a body is moving in one direction in a straight line under constant acceleration (e.g., a car experiencing approximately constant acceleration at the start of a race) its velocity increases in a linear fashion with respect to time and thus its position changes in a curvilinear (exponential) manner, as shown in *Fig. A7.1*. The situation is more complicated when a body moves in two directions, again in a straight line. An example of this is when someone jumps directly up and then lands back in the same place (e.g., a SVJ), and experiences the constant acceleration due to gravity during both the ascent and descent. In this situation the velocity of the body decreases linearly to zero at the apex of the jump and then increases in the same manner until landing. Their position changes in a curvilinear fashion, as shown in *Fig. A7.2*.

Equations of uniformly accelerated motion

The changes in position and velocity of a constantly accelerating body were first noted by an Italian mathematician called Galileo in the early 17th century. Galileo also derived the following equations that can be used to generate the curves shown in *Figs A7.1* and *A7.2*, and therefore to describe the motion of bodies experiencing constant acceleration.

$$v_2 = v_1 + at \qquad\qquad\qquad\qquad\text{(A7.1)}$$
$$d = v_1 t + \tfrac{1}{2}at^2 \qquad\qquad\qquad\text{(A7.2)}$$
$$v_2{}^2 = v_1{}^2 + 2ad \qquad\qquad\qquad\text{(A7.3)}$$
$$d = \tfrac{1}{2}(v_1 + v_2)t \qquad\qquad\qquad\text{(A7.4)}$$

The equations include linear kinematic variables that are defined as follows:

$$v_1 = \text{initial velocity}$$
$$v_2 = \text{final velocity}$$
$$d\ = \text{change in position or displacement}$$
$$t\ = \text{change in time}$$

Applications of equations of uniformly accelerated motion

Sport and exercise biomechanists often wish to analyze the motion of a body whilst it experiences constant acceleration. It may be important to know, for example, how high somebody jumped, what velocity they would experience after

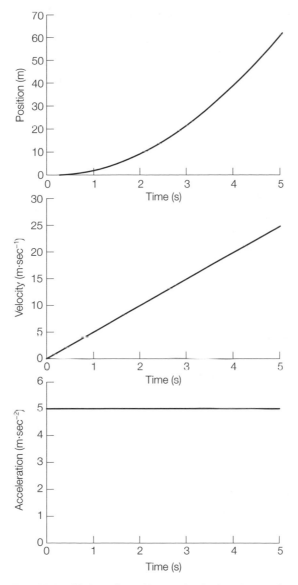

Fig. A7.1. Horizontal position and velocity of a body when experiencing a constant acceleration

a certain time, or how long it would take them to reach that velocity. Any of the equations above (A7.1–A7.4) can be used to answer such questions. For example, consider someone performing a SVJ with a take-off velocity of their center of mass (c of m) of 2.4 m·sec⁻¹. What would be the displacement of their c of m between the instant of take-off and the highest point of their c of m (i.e., how high do they jump)? The best way to answer this question is to break it down into a series of steps:

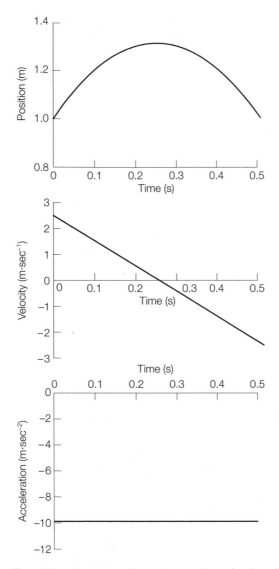

Fig. A7.2.   *Vertical position, velocity and acceleration of the CG during the flight phase of a standing vertical jump*

Step 1. Decide which of the equations (A7.1–A7.4) is appropriate.

In this case we have the take-off velocity (v_1) and we also know that the c of m will have a velocity of zero when it reaches the apex of its flight before the person falls back to the ground; so $v_2 = 0$. We also know that the acceleration of the c of m during flight is –9.81 m·sec⁻². Thus the only equation that we can use to determine the displacement (d) of the c of m is number A7.3. All of the other equations either do not include what we wish to find (i.e., equation A7.1 does not include d) or include variables that aren't available to us (i.e., equations A7.2 and A7.4 include t).

Step 2. Rearrange the equation, if necessary.

Sometimes the variable that you want to find is already isolated on the left-hand side of the equation and the equation does not need to be rearranged. However, in our situation, we do need to rearrange the equation to isolate d. The stages involved in this are shown in the box below:

$$v_2^2 = v_1^2 + 2ad$$

Subtract v_1^2 from both sides of the equation:

$$v_2^2 - v_1^2 = 2ad$$

Divide both sides of the equation by $2a$:

$$\frac{v_2^2 - v_1^2}{2a} = d$$

Step 3. Insert the known variables and calculate the unknown variable.

$$d = \frac{v_2^2 - v_1^2}{2a}$$

$$d = \frac{0^2 - (2.4)^2}{2a}$$

$$d = \frac{-5.76}{}$$

$$d = 0.29 \text{ m}$$

Thus, the c of m rose by 0.29 m between the instant of take-off and the highest point of the jump.

Suppose that the equipment required to obtain take-off velocity (i.e., a force platform or video camera) was not available but that the time that the person spent in the air was able to be recorded using (e.g., a jump mat); and was 0.489 sec. Assuming that the time the c of m spent rising the instant after take-off was the same as the time that it spent lowering the instant before touchdown, equation A7.2 can be used to calculate the displacement of the c of m instead. Thus v_1 becomes the velocity of the c of m at the apex of the flight phase (i.e., zero), t is half of the flight time (i.e., $0.488/2 = 0.244$ sec), and therefore d is the displacement of the c of m between this point and landing. The equation does not need rearranging so d can be calculated by inserting the known variables into equation A7.2:

$$d = v_1 t + \frac{1}{2} at^2$$
$$d = (0 \times 0.244) + (\frac{1}{2} \times -9.81 \times (0.244^2))$$
$$d = -0.29 \text{ m}$$

Unlike the (upward) displacement that was calculated before using the take-off velocity, this displacement is negative as it is the downward displacement of the c of m between the apex and landing.

Projectile motion   In the examples described above, the body in question was moving in the same direction (e.g., the horizontal motion of a car during the start of a race) or along the same line (e.g., the vertical motion of the c of m during a SVJ). However, in many situations (e.g., the flight of a soccer ball or javelin) the body has both a horizontal and vertical component of velocity at the point of release or take-off, and thus moves horizontally and vertically in flight. A body or object that is unsupported (i.e., in flight) and only affected by the forces associated with gravity and air resistance is known as a projectile. If air resistance is ignored, as it often is for bodies of relatively large mass traveling at low speeds, the flight path or trajectory of a projectile follows that of a parabola, which is symmetrical about its highest point. The greater the vertical in relation to horizontal component of velocity that the projectile has at release or take-off, then the more peaked its trajectory will be, as shown in *Fig. A7.3.*

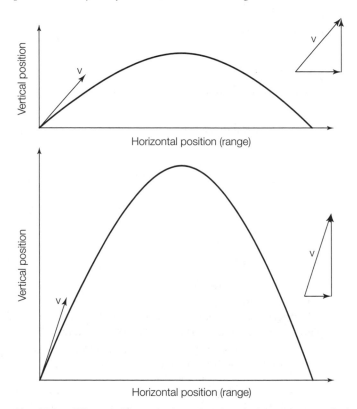

Fig. A7.3.   Effect of different horizontal and vertical components of take-off velocity on the trajectory of the projectile

Horizontal and vertical acceleration of a projectile

If the effect of air resistance is neglected there are no other horizontal forces acting on a projectile and its horizontal acceleration is zero. Thus, a shot putt or someone doing a standing broad jump will have the same horizontal velocity when landing as they do at release or take-off. Conversely, the vertical motion of a projectile is affected by the gravitational force which, as stated above, on Earth provides an acceleration of -9.81 m·sec^{-2}. Thus, if a projectile has a positive velocity at take-off or release, the effect of this downward acceleration will be to decrease this velocity to zero at the apex of the trajectory. The projectile then gains negative (i.e., downward) velocity until landing. If the projectile lands at the same height that it was released from, then it will have the same magnitude of velocity at the start and end of its trajectory. *Fig. A7.4* shows how the pattern of both horizontal and vertical velocity of a projectile alter throughout its trajectory.

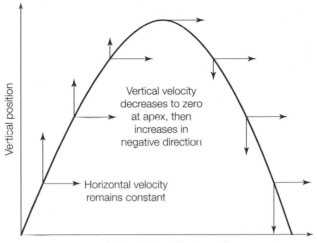

Fig. A7.4.   Change in horizontal and vertical velocity of a projectile during flight

Maximizing the range of a projectile

Often in sport, the aim is to maximize the range, or horizontal displacement, of a projectile (e.g., soccer goal kick). The range of a projectile that lands at the same height that it is released from is given by the following equation, which can be derived from the equations of uniform acceleration:

$$R = \frac{v^2 \sin 2\theta}{g} \qquad (A7.5)$$

where:

$R =$ range or horizontal displacement
$v =$ resultant take-off or release velocity
$\theta =$ take-off or release angle, defined as the angle between the horizontal component of velocity and the resultant velocity vectors at take-off or release
$g =$ acceleration due to gravity (i.e., –9.81 m·sec⁻²)

From equation A7.5 it is evident that, for a given take-off angle, the range of a projectile will increase in proportion to the square of the take-off velocity, as shown in *Fig. A7.5*.

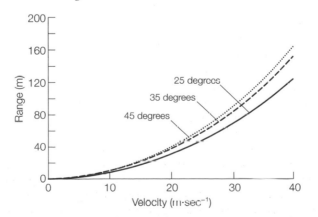

Fig. A7.5.   Effect of take-off velocity on the range of a projectile, at three different take-off angles

The same equation also reveals the optimum take-off angle for a projectile (i.e., the angle that will result in the greatest range). The sine of 90° is equal to 1, and any angle either smaller or larger than this will result in a sine of the angle that is less than 1. Thus as equation A7.5 contains the expression "sin 2θ", a take-off angle of 45° will result in a value for this expression of 1 and, therefore, produce the optimum range; as shown in *Fig. A7.6*. For any given take-off velocity a take-off angle that is a particular number of degrees less than 45° will result in a range that is identical to that produced by an angle that is the same number of degrees greater than 45°. For example, as shown in *Fig. A7.6*, if the take-off velocity of a projectile is 20 m·sec⁻¹ its range will be 38.3 m if the take-off angle is either 35° or 55°.

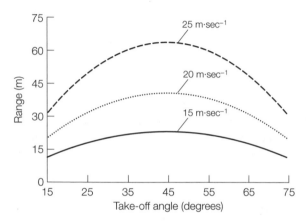

Fig. A7.6.   Effect of take-off angle on the range of a projectile, at three different take-off velocities

Calculating the height and flight time of a projectile
Often in sports biomechanics it is the height (H) that a projectile reaches (e.g., volleyball blocker) or the time (T) that it is in the air (e.g., a soccer pass) that is of greater importance. These variables can be calculated using equations A7.6 and A7.7, which also can be derived from the equations of uniform acceleration:

$$H = (v \sin \theta)^2 \qquad \text{(A7.6)}$$

$$T = \frac{2v \sin \theta}{g} \qquad \text{(A7.7)}$$

Thus, for a volleyball blocker who has a take-off angle of 80° and velocity of 3.2 m·sec⁻¹, the height raised by the c of m during flight can be calculated by inserting the known variables into equation A7.6:

$$H = \frac{3.2^2 \times \sin^2 80}{}$$

$$H = \frac{10.24 \times 0.97}{19.62}$$

$$H = 0.51 \text{ m}$$

Using equation A7.7, the flight time of a soccer ball kicked with a velocity of 16 m·sec^{-1} at an angle of 56° can be calculated:

$$T = \frac{2 \times 16 \times \sin 56}{9.81}$$

$$T = \frac{32 \times 0.829}{9.81}$$

$$T = 2.70 \text{ sec}$$

Projectiles with different take-off and landing heights

Most of the projectiles used as examples in the previous sections landed at the same height as they were released from (e.g., a soccer goal kick or pass that is not intercepted by another player). However, in many sports the projectile is more commonly released from a greater height than it lands at (e.g., shot putt, long jump) or less frequently lands at a greater height than is was projected from (e.g., basketball free throw). Equations A7.5–A7.7 can only be used in situations where the release height is the same as the landing height, and two new equations (A7.8 and A7.9) are needed to calculate the range and flight time of a projectile that has different release and landing heights:

$$R = \frac{v^2 \sin \theta \cos \theta + v \cos \theta \sqrt{(v \sin \theta)^2 + 2gh}}{g} \qquad \textbf{(A7.8)}$$

$$T = \frac{v \sin \theta + \sqrt{(v \sin \theta)^2 + 2gh}}{g} \qquad \textbf{(A7.9)}$$

In such situations the optimal angle of take-off or release is no longer 45°, as it is for projectiles that have the same take-off and landing heights. For projectiles that have a higher take-off than landing height (e.g., shot putt) the angle that will result in the greatest distance is always less than 45°. Conversely, bodies that land higher than they are released from (e.g., basketball free throw) have an optimal release angle of more than 45°. The actual optimal angle of projection in either situation is dependent on both the difference in height between take-off and landing, and the take-off velocity; as shown in equation A7.10:

$$\cos 2\theta = \frac{gh}{v^2 + gh} \qquad \textbf{(A7.10)}$$

In the more common situation experienced by long jumpers and shot putters and so on, the optimal angle of take-off or release decreases with the difference between take-off and release height. For example, for a given release velocity, shot putters with a high release height will have a lower optimal release angle than those athletes who release the shot from a lower height. The optimal angle also depends on the take-off or release velocity of the projectile. The higher the velocity the closer the angle gets to, but never reaches, 45°. *Fig. A7.7* shows the effect that both the height and velocity of release have on the optimal release angle. This figure shows that at low velocities (less than 5 m·sec^{-1}) small changes in velocity have a large affect on the optimal angle of projection. At higher velocities, that are more realistic for shot putters (10–15 m·sec^{-1}), *Fig. A7.7* also shows that the same changes in velocity or release height have a much smaller influence on the optimal angle of release.

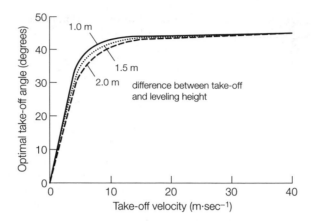

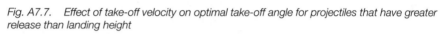

Fig. A7.7. *Effect of take-off velocity on optimal take-off angle for projectiles that have greater release than landing height*

B1 FORCES

Key Notes

Attractive and contact forces

Forces can be divided into either attractive forces or contact forces. An example of an attractive force is the gravitational force of attraction that exists between any two objects that have mass. An example of a contact force would be the contact of the foot with a soccer ball or the contact of the foot with the ground during running. All forces will produce or alter motion although this alteration in motion (i.e., acceleration or deceleration) will not always be apparent. The pushing of a book across a table will not occur unless you apply enough force to overcome the frictional force between the book and the table.

External and internal forces

Forces can also be categorized as both external and internal forces. An external force would be the force on the tennis ball that is exerted by the racket, whereas an internal force would be the forces occurring in the elbow joint when you hit the ball with the racket.

Force and inertia

An application of force is required to cause a change in movement of an object (i.e., an acceleration or deceleration). Inertia is the reluctance of an object to change its state of motion. Inertia is directly related to mass. The more mass an object has the more inertia it has. If you push a massive book across a table it will take more force to accelerate the book at a rapid rate than it would to accelerate a less massive book at the same rate.

Mass and weight

Mass is the measure of the number atoms or molecules in an object. Mass is relatively constant over time. If you have a mass of 75 kg on Earth you will have a mass of 75 kg wherever you are. Weight is the measure of force acting on an object. It is dependent on the position of the object on the planet and also on which planet the object is located on. An object on Earth will weigh more than the same object on the Moon.

Vector quantities

Vector quantities have both magnitude and direction. A force can be expressed as a vector quantity. This can be either mathematically expressed or graphically represented on paper by a line drawn to scale. The length of the line would be the magnitude of the force and the position/angle/orientation of the line would be its direction.

Resolution of forces

Several forces acting simultaneously on an object can be resolved into one force. This resolution of force can be carried out by a mathematical method or by using a graphical technique called "tip to tail." Forces can be resolved in both two and three dimensions. The resolution of forces is critical for an understanding of both performance and injury prevention.

Attractive and contact forces

The consideration of forces within human movement can be broadly defined into two categories: attractive and contact forces. **Attractive forces**, such as in the case of gravity, are the result of any two masses acting upon each other. In the case of the Earth's gravitational force, the earth is acting on the human body. Similarly, although, with much less affect, the human body is acting on the Earth. **Contact forces** can be used to describe most other forms of force encountered within human movement, such as the frictional force between the foot and the ground when walking. Contact forces can involve both "pushes" and "pulls" and they cause some form of change in direction or movement (speeding up or slowing down). However, it is important to point out that although **all forces will produce or alter motion**, this motion is not always apparent. Some examples of contact forces within human movement include: the reaction force between the feet and the ground during landing from a jump, the impact force between two players colliding in soccer, the force applied to a tennis ball by the racket of the tennis player or the force exerted across the knee joint when the quadriceps muscle contracts during movement.

Within the human body these forces can be further defined into being either external or internal forces. **External forces** include all the forces that exist outside of the body: such as in the case of kicking a soccer ball and **internal forces** include all those that act inside the body: such as the forces across the anterior cruciate ligament in the knee when a player is tackled in rugby. Often many of the external forces are responsible for the internal forces that are experienced by the muscles, bones, joints, ligaments, and tendons. Both external and internal forces act at various places all over the body during movement. Without the existence of these forces we would not be able to move with any acceleration. Similarly, to achieve a specific performance such as in the case of sprinting 100 m in less than 10 seconds requires the careful execution and control of force. However, such forces can also cause injury and a more thorough understanding of them will help in both the improvement in performance and future prevention of injury. Consider *Fig. B1.1* and see if you can identify some of the different external and internal forces that are occurring in these activities.

Fig. B1.1.   Forces acting within sport

Force and inertia Any change in a body's motion (considered without rotation at this stage) that is brought about by the application of a force will incorporate a change in speed and/or a change in direction (i.e., changing its velocity which was identified in section A2). In order to start an object moving we need to apply a force to the object. **Inertia** is defined as the reluctance of an object to undergo any change in velocity (that is to either change its current state of velocity or begin any state of movement). The amount of inertia possessed by a body is directly proportional to the amount of mass possessed by the body. **Mass** is defined as the quantity of matter (atoms and molecules) present in a body (the term body is used to describe both the human body and any objects associated with it). Inertia is directly proportional to mass, which is measured in kilogrammes (kg). The mass of a body remains relatively constant over time (relative of course to how much you eat and drink in the case of the human body) and it is the same for an object that is on the Moon as it is for the same object when it is on the Earth. For example, if you have a mass of 55 kg on the Earth you will also have a mass of 55 kg on any other planet or indeed a mass of 55 kg when you are not on any planet at all.

In the context of understanding the term inertia, imagine trying to push a book across a table surface. Initially the book will be stationary but as you apply a force the book will begin to move (accelerate). The resistance you feel to your "pushing" of the book is a measure of the frictional force that exists between the book and the table. The frictional force is derived from the weight of the book. The more mass the book has, the more weight the book will have. Since the inertia possessed by the book is related to how much mass the book has, the more force you will need to apply to accelerate the book at a greater rate. At the beginning of this action you may notice that you applied a force and yet the book did not move. This would be because you did not apply enough force to overcome the frictional force between the book and the table. Next, place another book on the same table but this time use a book that is much heavier. Now try the experiment again. This time you will see that you need to exert a much greater force to accelerate this new book across the table at the same rate as the original book. The new book has more mass and hence more inertia, and thus a greater reluctance to change its current state of motion (i.e., accelerate across the table).

This same understanding of inertia and force applies to the movement of the human body. If you try to push over (or accelerate) another individual you will feel a resistance to your efforts to this "pushing". This resistance will be dependent upon the mass of the person you are trying to push over and the frictional force between them and the ground. Trying to push over someone who is 110 kg is much more difficult than trying to push over someone who is 52 kg (although the relative heights and positions of their respective centre of gravity will also have an effect on this exercise).

Mass and weight As we have already observed mass is the term used to describe the quantity of matter in an object (a measure of the number of atoms and molecules in the object) and it is relatively constant. **Weight** is the effect of the Earth's gravitational force acting on a body (again the term body can be used for the human body or any object). Mass and weight are different quantities and the units of measurement for each quantity are also different. Mass is measured in kilogrammes and weight (by virtue of the fact that it is a measure of the force acting on a body) is measured in Newtons (N). The unit of 1 Newton (named after the

English mathematician Isaac Newton 1642–1727) is derived from the force required to accelerate a mass of 1 kg at a rate of 1 m/s^2.

$$1 \text{ Newton (N)} = 1 \text{ kilogramme (kg)} \times 1 \text{ m/s}^2$$

The terms mass and weight are different, and should not be used to describe the same quantity. How often have you heard people say: How much do you weigh? The correct answer to this would be to work out the force acting on your body (your weight) by virtue of the fact that you are being pulled to the center of the Earth by the gravitational force of the planet. Since, the acceleration due to gravity at sea level is given as 9.81 m/s^2 we should, if we know our mass, be able to accurately work out our weight (*Fig. B1.2*).

However, if you were asked this question while you were standing on the Moon then the answer would be very different. The Moon has a much smaller mass than that of the earth and, therefore, it will have a much smaller gravitational affect on your body. Although, you will still have the same mass on the Moon as you did on the Earth you will actually weigh much less.

The product of mass multiplied by the acceleration (acceleration due to gravity in the calculation to determine weight) of the object is a measure of the force.

$$F = ma$$

Where

F = force (measured in Newtons (N))
m = mass (measured in kilogrammes (kg))
a = acceleration (measured in meters per second squared (m/s^2))

And, as we have seen, this equation is often re-written to express the calculation of the weight of a body that is being acted upon by gravitational acceleration.

$$W = mg$$

Where

W = weight (measured in Newtons (N))
m = mass (measured in kilogrammes (kg))
g = acceleration due to gravity (measured in meters per second squared (m/s^2))

Now, think back to the experiment of pushing the book across the table; once the book is moving it will have accelerated or be accelerating (i.e., it was stationary before you pushed it and now it is moving across the table – hence its velocity has changed – and it therefore must have accelerated). The product of the mass of the book multiplied by the acceleration possessed by the book as it changed its velocity

Determine the weight of a 75 kg person

Weight = force acting on a person by virtue of the gravitational pull of the planet Earth and at sea level this is expressed as an acceleration due to gravity which is 9.81 m/s^2

Weight = mass × acceleration due to gravity

For the 75 kg person
Weight = 75 kg × 9.81 m/s^2
Weight = 735.75 Newtons (N)

Fig. B1.2. Calculation of weight

will be a measure of the amount of force that you exerted (or are currently exerting) on the book to move it. Note: to keep the book accelerating you will need to continue to apply force ($F = ma$). This same understanding is applicable to many aspects of human movement: If you accelerate your leg/foot down onto the ground quickly you will feel a larger force than if you move it towards ground contact slowly, or if you hit a ball with a heavier racket or bat you will produce more force acting on the ball and again you will feel a force in a reaction on your arms. Such, understanding of force and its relationship with mass and acceleration is a very important concept in human movement and will be expanded in more detail in the sections concerned with impulse and Newton's laws of motion.

Vector quantities   Since a force has both magnitude (the amount of force you exert or is exerted upon you) and direction (the specific direction in which the force occurs or is applied) it can be expressed either mathematically or graphically on paper using straight lines. A vector quantity, such as force, can be identified as an arrow that has both magnitude (length) and direction (angle – position). *Fig. B1.3* illustrates this vector representation of a force in more detail.

A vector quantity has both magnitude and direction and, as we have seen, force (because it is a vector quantity) can be represented by a single line with an arrow indicating a position and direction. The length of the line (usually drawn to a scale) is the magnitude (size) of the force. When two forces are acting vertically upwards the composition (or resultant) of these two forces can be defined as the summation of their magnitudes. Similarly, when two forces are acting horizontally (in the same direction) the resultant can be expressed as the sum of their two parts. The same principle applies for any number of forces that are acting parallel to each other and in the same direction (see *Fig. B1.4*). Similarly, if two forces are

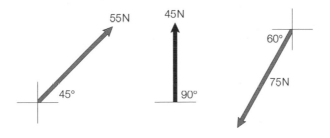

The length of the line (when drawn to a scale) would represent the magnitude of the force and the angle – position would represent the direction of the force

Fig. B1.3.   *Force expressed as a vector quantity*

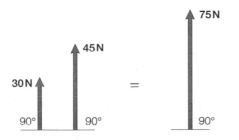

Fig. B1.4.   *Composition of force expressed as vector quantities (not drawn to scale)*

acting in direct opposition to each other the sum of their respective parts (i.e., upward minus downward) will indicate the resultant of the two and the direction in which it occurs (i.e., positive or negative value).

However, imagine trying to move a box that is placed on a table: if you push the box with ONE force it will generally (provided you apply enough force) move off in the direction of this single force. But if you push the box with TWO forces it will now move off in the resultant direction of the two forces. Similarly, it will move off with an acceleration (changing from rest to movement) that is proportional to the magnitude of the resultant of the two forces that you are applying to the box (see *Fig. B1.5*).

In order to determine the resultant of these two forces (or any number of forces) it is necessary to either solve this problem mathematically or graphically. Graphically the solution is achieved by drawing each force to a specific scale on a piece of paper and using the "tip to tail" method of resolution. This method (which works for any number of co-planar forces) is achieved by carefully drawing each force with the next force drawn on the tip of the previous force. In addition, it does not matter in what order you draw the force vectors, as the solution (resultant) will always be the same. In this context, it is important to point out that the resolution of forces will also work for forces that are acting in several planes simultaneously (i.e., three-dimensional). *Fig. B1.6* illustrates the graphical "tip to tail" method in more detail.

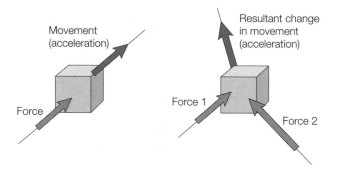

Fig. B1.5.   Resultant of force application

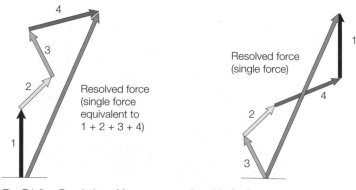

Fig. B1.6.   Resolution of forces vectors (graphical solution)

These force vectors can also be solved mathematically using trigonometry and this involves first resolving the forces into a single vertical and single horizontal component. Next, the vertical and horizontal components are resolved into one resultant force. *Figs B1.7a–h* illustrate this method in more detail.

Using the same forces as presented in *Fig. B1.6*

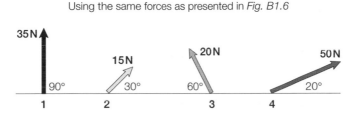

Express all the forces at a single point

Fig. B1.7a. Mathematical resolution of force application (composition of force vectors solution)

Note: that forces opposing each other should be subtracted and forces acting in the same direction summated. In this case ALL forces have an upward component (with force **1** being perfectly vertical). However, force **3** has a left component while forces **2** and **4** have a right component. Hence they should be subtracted. Note force **1** has no left or right component

Fig. B1.7b. Mathematical resolution of force application (composition of force vectors solution)

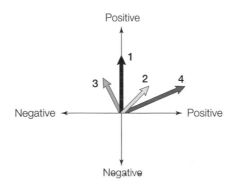

Note: in this configuration it is possible to divide the diagram into positive and negative components as identified above. This process will help determine the direction of the magnitude of the resultant force at the end of the calculation

Fig. B1.7c. Mathematical resolution of force application (composition of force vectors solution)

Note: all forces are
acting upwards
hence all positive

1 **35N** at 90°
2 **15N** at 30°
3 **20N** at 60°
4 **50N** at 20°

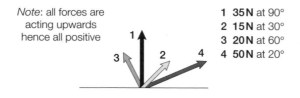

Sum of all forces (vertically) = F sin θ

F sin θ

= 35 sin 90° + 15 sin 30° + 20 sin 60° + 50 sin 20°

= (35 × 1.0) + (15 × 0.5) + (20 × 0.866) + (50 × 0.342)

= 35 + 7.5 + 17.32 + 17.1

= **76.92N** (vertical component of all the above forces)

Fig. B1.7d.   Mathematical resolution of force application (composition of force vectors solution)

Note: force **3** is acting
left and forces **2** and **4**
are acting right hence
consider **3** as negative

1 **35N** at 90°
2 **15N** at 30°
3 **20N** at 60°
4 **50N** at 20°

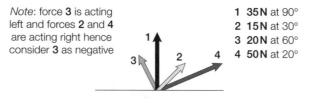

Sum of all forces (horizontally) = F cos θ

F cos θ

= 35 cos 90° + 15 cos 30° + (−20 cos 60°) + 50 cos 20°

= (35 × 0) + (15 × 0.866) + (−20 × 0.5) + (50 × 0.939)

= 0 + 12.99 + (−10) + 46.95

= **49.94N** (horizontal component of all the above forces)

Fig. B1.7e.   Mathematical resolution of force application (composition of force vectors solution)

Note: the following two equations for vertical (**a**) and horizontal (**c**)
resolution of forces were derived from the following trigonometric
functions in a right-angled triangle (**b** would equal the force vector)

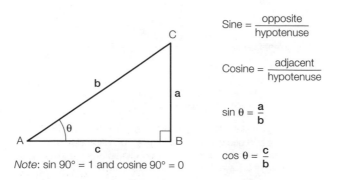

$$Sine = \frac{opposite}{hypotenuse}$$

$$Cosine = \frac{adjacent}{hypotenuse}$$

$$\sin θ = \frac{a}{b}$$

$$\cos θ = \frac{c}{b}$$

Note: sin 90° = 1 and cosine 90° = 0

Fig. B1.7f.   Mathematical resolution of force application (composition of force vectors solution)

Determination of the resultant force (using Pythagoras' theorem)

Resultant force = $\sqrt{FV^2 + FH^2}$

Where

FV = vertical force

FH = horizontal force

$=$

$R = \sqrt{FV^2 + FH^2}$

$= \sqrt{76.92^2 + 49.94^2}$

$= \sqrt{5916 + 2494}$

$= \mathbf{91.71\,N}$

Angle of application of
resultant force

$\tan \theta = \dfrac{FV}{FH}$

$=$

$\tan \theta = \dfrac{FV}{FH}$

$\tan \theta = \dfrac{76.92}{49.94}$

$\tan \theta = 1.54$

$\theta = INV\ TAN\ (1.54)$

$\theta = \mathbf{57.0°}$

Fig. B1.7g. Mathematical resolution of force application (composition of force vectors solution)

It is now possible to see that the graphical representation of the resultant force
is the same as the mathematical representation of the resultant force

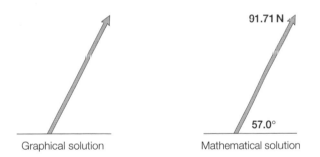

Graphical solution

91.71 N

57.0°

Mathematical solution

Fig. B1.7h. Mathematical resolution of force application (composition of force vectors solution)

Application

Consider the forces acting underneath the foot as a person walks or runs along the ground. *Fig. B1.8* identifies the forces that are acting in the sagittal plane and frontal plane (three-dimensional) during walking when a person's foot hits the ground. From consideration of *Fig. B1.8* it is possible to see that there are two forces acting in the sagittal plane and two forces acting in the frontal plane (with the vertical force being common to both planes). The forces in the sagittal plane are classified as the **vertical force** (acting straight upwards) and the **anterior–posterior force** (acting posteriorly (as a braking force) when the foot hits the ground at heel strike when it is moving forwards). In the frontal plane it is possible to also see another force, which is classified as the **medial–lateral force** (depending on whether it is going medially or laterally with respect to the foot) is also acting. In the same diagram the resolution (resultant) of all these three forces is also shown. This force is known as the **ground reaction force** and it is the force that is acting at a specific direction and with a specific magnitude. This is the force that can be important for injury considerations. The ground reaction force is the resultant force, which is derived from the composition of the three planar forces described previously. Speed of running, running shoes, type

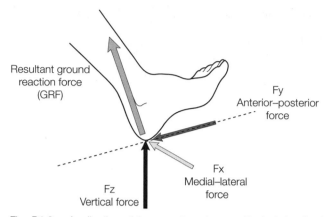

Fig. B1.8. Application of force vectors (composition) during heel strike with the foot and ground during walking

of running style, type of surface of contact, type of foot contact, and previous injury can all affect these forces. From both an improvement in performance and injury prevention perspective it is important that the development and attenuation of these forces is fully understood by the student of biomechanics.

B2 NEWTON'S LAWS OF MOTION – LINEAR MOTION

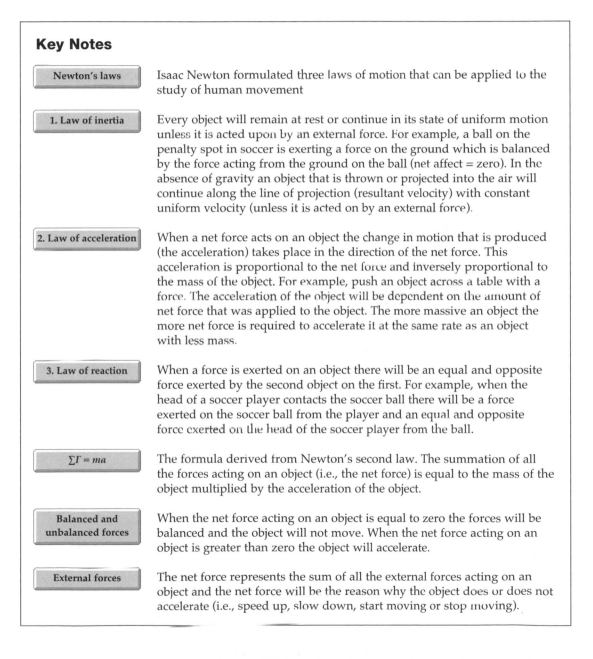

Key Notes

Newton's laws	Isaac Newton formulated three laws of motion that can be applied to the study of human movement
1. Law of inertia	Every object will remain at rest or continue in its state of uniform motion unless it is acted upon by an external force. For example, a ball on the penalty spot in soccer is exerting a force on the ground which is balanced by the force acting from the ground on the ball (net affect = zero). In the absence of gravity an object that is thrown or projected into the air will continue along the line of projection (resultant velocity) with constant uniform velocity (unless it is acted on by an external force).
2. Law of acceleration	When a net force acts on an object the change in motion that is produced (the acceleration) takes place in the direction of the net force. This acceleration is proportional to the net force and inversely proportional to the mass of the object. For example, push an object across a table with a force. The acceleration of the object will be dependent on the amount of net force that was applied to the object. The more massive an object the more net force is required to accelerate it at the same rate as an object with less mass.
3. Law of reaction	When a force is exerted on an object there will be an equal and opposite force exerted by the second object on the first. For example, when the head of a soccer player contacts the soccer ball there will be a force exerted on the soccer ball from the player and an equal and opposite force exerted on the head of the soccer player from the ball.
$\sum F = ma$	The formula derived from Newton's second law. The summation of all the forces acting on an object (i.e., the net force) is equal to the mass of the object multiplied by the acceleration of the object.
Balanced and unbalanced forces	When the net force acting on an object is equal to zero the forces will be balanced and the object will not move. When the net force acting on an object is greater than zero the object will accelerate.
External forces	The net force represents the sum of all the external forces acting on an object and the net force will be the reason why the object does or does not accelerate (i.e., speed up, slow down, start moving or stop moving).

Newton's Laws   Isaac Newton (1642–1727) formulated three laws of motion that created the basis of Newtonian mechanics and which can be directly applied to human movement and the study of biomechanics. These are summarized as follows:

Law 1: The law of inertia
Every object will remain at rest or continue with uniform motion unless it is acted upon by an unbalanced force.

Law 2: The law of acceleration
When a force acts on an object the change of motion (momentum) experienced by the object takes place in the direction of the force, is proportional to the size of the force and inversely proportional to the mass of the object.

This law indicates that if a net external force acts on an object it will accelerate (i.e., speed up, slow down, start moving, or stop moving) in the direction of the net external force. This acceleration is proportional to the net external force and is inversely proportional to the mass of the object.

Law 3: The law of reaction
Whenever an object exerts a force on another there will be an equal and opposite force exerted by the second object on the first.

In order to understand these laws in more detail, and in particular their relevance to human movement, it is necessary to consider an application of each of the three laws separately.

1. **The law of inertia**: application to human movement.
Newton's first law states that a body at rest will remain at rest and a body in motion will continue in motion unless it is acted upon by an unbalanced external force. In relation to human movement it is important to divide the understanding of this law into three components.

Unbalanced forces
Bodies that are <u>not</u> moving (i.e., stationary)
Bodies that are moving (i.e., in motion)

Note: the term body can be applied to the human body or any external body associated with the human body such as a soccer ball, a basketball, a tennis racket or a javelin.

Unbalanced forces

Consider the experiment outlined in section B1, where you were asked to place a book on a table and then apply a force to overcome friction and cause the book to move across the table. In this experiment the force you applied to the book must have been greater than the force offered in resistance. If these two forces (your effort and the frictional resistance) were equal (i.e., balanced) the book would not have moved. When considering **balanced and unbalanced** forces it is important to understand the term net force. The term net is defined as the final number (subject to no more deductions or calculations) and in this case it refers to the summation (positive and negative) and result of all the forces acting on an object. Remember, from section B1, that forces are vector quantities and have both magnitude and direction, and it is possible that two forces act in an opposite direction to each other. These forces would be summated (i.e., they have positive and negative signs) to produce the net effect. If the net effect is zero then the force system is **balanced** and there is no movement (or no acceleration). If the net effect is not zero then the forces are **unbalanced** and movement (acceleration) will take place (except of course when the force you exert on the book does not overcome the frictional effects and the book remains stationary). In terms of you pushing a book across a table you will have to overcome the frictional force before you can move it and then once it is moving you will have to overcome the inertial effects to keep the object accelerating. In addition, there is the force that acts to oppose this movement that is externally provided from air resistance (although in the case of a book being moved across a table this will be negligible (so small it is not considered significant)). In both cases (moving and not moving) it will still be necessary to overcome the frictional force that exists between the book and the table. *Fig. B2.1* helps to illustrate this understanding in more detail.

The same principle applies when trying to move any object within human movement. For example, if you wanted to lift a barbell containing weights (in this example the term weight is used to describe the weights attached to the bar) in an upward vertical direction, you will need to exert a force on the barbell that is large enough to overcome the gravitational effects acting on the barbell and weights (because you are trying to move it vertically upwards and gravity will continually oppose this movement – by pulling it downwards).

It is important to identify that gravity is only an external force when you are trying to move an object vertically (i.e., upwards or downwards). If there is no vertical movement to your action then gravity will not act as an external force (although technically all objects on this planet are subjected to the vertical force of gravity even when they are not moving).

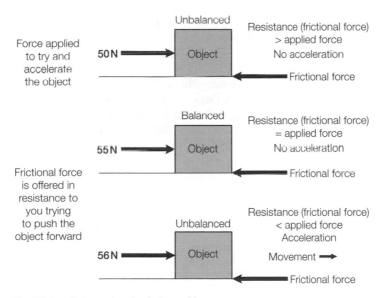

Fig. B2.1. Balanced and unbalanced forces

Bodies that are not moving (i.e., stationary)

In this context Newton's first law states that an object that is *not* moving will remain in a non-moving state (at rest) providing it is not acted upon by an unbalanced external force. In human movement **it is difficult to see how this law can directly apply to any situation.** For example, all objects that are on this planet will be subjected to the vertical external force of gravity. If we did not have the ground on which to stand, we would accelerate towards the center of the Earth at a rate of approximately 9.81 m/s² (which is stated as 'approximately' because this acceleration varies slightly depending on where you are on the surface of the planet (i.e., in relation to the center of the Earth)). When we try to jump off the ground, gravity will immediately pull us back down. Although, this effect is actually happening all the time it is more obvious as soon as we are in the air (unless of course you are able to exert enough force to overcome the force of gravity and get away from the Earth's gravitational pull – such as in the case of a rocket and space shuttle traveling into space). Hence, in this static example of Newton's first law, it is difficult to see how it can apply. Consider the following examples: a soccer ball placed on the penalty spot, the book placed on the table, the hurdle on the track in a 400 m hurdle race or the human body sat in a chair, and it is obvious that all these are continuously subjected to the force of gravity (i.e., they all have weight).

Bodies that are in motion (i.e., moving)

1. **The law of inertia**: application to human movement.

In order to see how Newton's first law of motion applies to human movement in objects that are in motion it is useful to use the example of the long jumper. During a long jump an athlete will leave the ground with both vertical and horizontal velocity. This combination of velocities determines the angle of take-off, the resultant velocity and primarily the distance jumped by the athlete. This horizontal and vertical velocity produces a **projectile motion** (subjected to only the external force of gravity) of the athlete during the flight phase, which is illustrated in more detail in *Fig. B2.2*.

As the athlete leaves the ground he/she will have both vertical and horizontal velocity. Once in the air the athlete will be a projectile and the flight path of the athlete will already be pre-determined. The parabolic flight path (see *Fig. B2.2*) will be a result of the combination of vertical and horizontal velocities present at the take-off point. Since the vertical motion of a body is affected by gravity, it is necessary to consider the velocities (motions) separately in order to see an application of Newton's first law of motion during this movement.

In respect of the vertical motion of the athlete (or vertical velocity), the athlete will travel both upwards and downwards while at the same time traveling forwards (horizontal velocity). Gravity will affect the vertical component thus allowing the athlete to reach a peak in the parabolic flight path while constantly being pulled back towards the ground (at a rate of 9.81 m/s²). However, because the

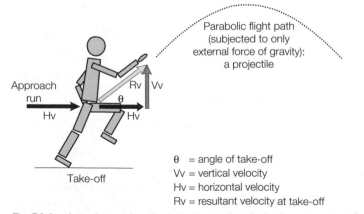

Fig. B2.2.   Long jump take-off – horizontal and vertical velocities and parabolic flight path

athlete also has horizontal motion (or velocity) we can see that the athlete travels forward and has the characteristic parabolic curved flight path (see *Fig. B2.3*).

The body in flight during the long jump is considered to be a projectile (like a ball in soccer, or a javelin in athletics). In this example it could be argued that other forces (as well as gravity) act on these objects during flight. For example, air resistance will also affect the parabolic flight path of the long jumper. However, for this example we can consider this air resistance to be negligible. In the context of the horizontal motion of the long jumper it is possible to see that the athlete will travel in a straight line (this can be seen more clearly when viewed from above (plan view) in *Fig. B2.4*). In addition, during this straight line motion the athlete

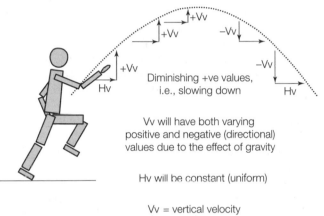

Diminishing +ve values, i.e., slowing down

Vv will have both varying positive and negative (directional) values due to the effect of gravity

Hv will be constant (uniform)

Vv = vertical velocity
Hv = horizontal velocity

Fig. B2.3.   Long jump take-off – constant horizontal velocity

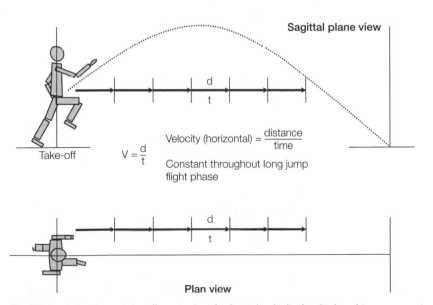

Sagittal plane view

Take-off

$V = \dfrac{d}{t}$

Velocity (horizontal) = $\dfrac{\text{distance}}{\text{time}}$

Constant throughout long jump flight phase

Plan view

Fig. B2.4.   Long jump take-off – constant horizontal velocity (sagittal and transverse plane (plan) views)

will also travel with a constant horizontal velocity. Hence, according to Newton's first law of motion, in the absence of an external force (as gravity will only affect the vertical component) the body will continue in a straight line with uniform (constant) velocity.

Although, it does not seem logical that Newton's first law can apply to the long jumper in flight because the athlete will actually come to rest (a stop) in the sand, it is in fact correct. The horizontal velocity of the long jumper is constant and the path of the athlete will be in a straight line (when viewed from above). The jumper will not speed up nor slow down (horizontally) and no matter what the athlete does during the flight phase the path will be pre-determined from the point of take-off. The reason the athlete comes to rest (in the sand) is because the vertical component of the jump is affected by gravity and this will pull the athlete back towards the ground as soon as they leave it. Eventually, the athlete will hit the sand pit and stop, and it is the force from the sand pit on the athlete that would stop that motion. Therefore, in the absence of gravity the long jumper would continue to travel both upward and horizontally in a straight line with a constant velocity (note: upward and horizontally because the athlete would follow the resultant take-off velocity vector; i.e., which is upward and horizontal).

All projectiles that are thrown with horizontal and vertical velocity and that are only subjected to the external force of gravity will have a parabolic flight path that is pre-determined and they will all obey this law. A soccer ball when kicked, a basketball when thrown at the hoop, a tennis ball hit across court and even as simple as a pen that is thrown a short distance, will all obey and demonstrate Newton's first law of motion. For a practical example, consider when you are traveling on a bus and the bus suddenly comes to a stop. In this case you will continue forward with the same velocity towards the front of the bus even though the bus has stopped. Hence, in order to stop yourself from traveling forwards you will need to hold on to something like a hand rail (i.e., thus applying an external force). According to Newton's first law you continued forward with uniform velocity until you were acted upon by an external force (i.e., the gripping of the hand rail to stop yourself moving forward). The exact same situation applies when you are holding a cup of coffee and someone walks into you and the coffee is spilt. In this example, the coffee continues in its state of rest and is spilt because both you and the cup move in another direction. In this example the body (you) and the cup are attached to each other and essentially move together (i.e., your body, your arm, your hand, and the cup). However, the coffee although it is in the cup, acts independently and continues its state of motion (i.e., at rest). Hence, the coffee is split and Newton's first law of motion has provided a scientific reason why this has happened.

2. **The law of acceleration**: application to human movement.
Newton's second law of motion states that when a force is applied to an object (and the result is a net force of greater than zero (i.e., unbalanced)) the change of motion in the object (i.e., change in velocity (acceleration)) is proportional (as one quantity increases in value so does the value of the other quantity) to the force applied to the object. This movement takes place in a straight line and in the direction in which the net force was applied. In addition, the law also states that this change in motion (acceleration or rate of change in velocity) is inversely proportional (as one value increases the other will decrease) to the mass of the object. *Fig. B2.5* helps to illustrate this law with a diagram.

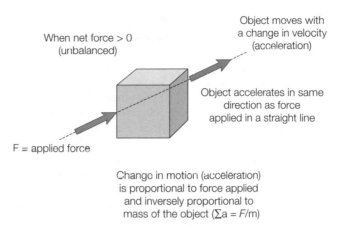

Fig. B2.5.   Newton's second law of motion

This law is the most powerful of Newton's three laws of motion because it allows us to use this in the calculation of dynamics (movement). **For example, how do the velocities of objects change when forces are applied to them?** Using Newton's second law gives us the opportunity to calculate these changes. In the understanding of this law it is, however, important to identify that the force applied to an object only causes a change in velocity (an acceleration or deceleration) and it does **not** maintain this velocity.

In order to put Newton's second law of motion into a formula that we can use in understanding human movement it is necessary to identify the following equation:

$$\sum F = ma$$

Where
$\sum F$ = net external force (N)
m = mass of the object (kg)
a = acceleration of the object (m/s²)

Using the previous example of the long jumper (*Fig. B2.4*), which was used to illustrate Newton's first law, we observed that the horizontal velocity of the athlete during flight was constant. In addition, we also learned that once the athlete had left the ground (at take-off) the flight path was pre-determined (fixed). **How can this example be used to illustrate Newton's second law?** (Consider *Fig. B2.6*.)

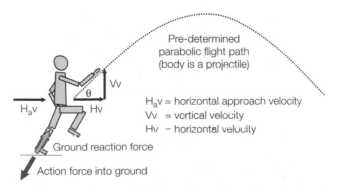

Fig. B2.6.   Newton's second law of motion (long jump take-off and flight)

At the point of take-off in the long jump we have seen that the athlete will have both vertical and horizontal velocity. In order to have created the changes in velocity (i.e., from horizontal during the run up to horizontal and vertical at take-off (i.e., change in motion or acceleration)) the athlete will have applied a force to the ground in order to drive himself or herself from the ground into the air. This force application is developed from the change in the athletes stride patterns at or just before the take-off point (athletes usually lower their center of gravity and lengthen the second to last and shorten the last stride into the take-off board). This adjustment of the body allows the athlete to be able to push his/her foot into the ground at the take-off board. This will create a resistive force from the ground that acts on the athlete. This ground reaction force will propel the athlete upwards and forwards. The resulting acceleration of the athlete (upwards and forwards) is demonstrated in horizontal and vertical velocity at the point of take-off (*Fig. B2.7*). Remembering that forces are vectors, it is possible to see that this propulsive force from the ground will have both vertical and horizontal components, and it is these two components that create the horizontal and vertical velocities used to determine the angle and resultant velocity of take-off.

Now let us consider the point of take-off and in particular look at only the vertical motion (change in velocity) of the long jumper (since we know from Newton's first law that the horizontal velocity is constant). As soon as the athlete leaves the ground, the force of gravity will try and pull the athlete back down to the ground. As soon as the athlete is airborne (at take-off) the only external force acting on him/her (neglecting air resistance) is the force of gravity (i.e., what causes the athlete to have weight). As the athlete travels upwards (remember we are only considering the vertical component of the parabolic flight path) the downward pull (acceleration) of gravity is immediately slowing the vertical ascent of the athlete. Even though the athlete is traveling upwards he/she is actually being slowed down (decelerated). As the athlete is slowed down (vertically) they will eventually come to a stop at the highest point in the flight path (*Fig. B2.3*). The acceleration of the athlete throughout his or her flight is **downward** even though it appears that they are going upward at the beginning (i.e., he/she is being slowed down or always being pulled downwards at a constant rate).

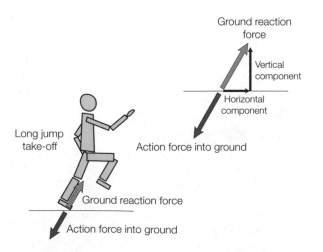

Fig. B2.7.   Ground reaction force during take-off in the long jump

Now, since we know that the weight (the force acting on the athlete because he/she has mass) of the athlete does not change and that the mass of the athlete remains the same, we can therefore say that the acceleration of the athlete is constant (derived from the previous equation $F = ma$). Next, we also know that because of the mass of the Earth and its gravitational pull on objects towards its center, this acceleration will be 9.81 m/s^2 (remembering that it does vary slightly between different positions on the Earth's surface – depending on how close or how far away you are from its center).

This means that as an athlete travels upwards at the beginning of the flight phase, they will be slowed down at a rate of 9.81 m/s for every second of movement (i.e., 9.81 m/s^2). Similarly, as the athlete begins to come back downward again, in the second half of the flight phase he/she will be accelerated downward (speeds up) at a rate of 9.81 m/s for every second of motion (i.e., constant acceleration).

This constant vertical acceleration acts on all objects and will accelerate a heavy object and a lighter object at exactly the same rate; which is why a heavy object and lighter object dropped from the same height will hit the ground at the same time (again obviously neglecting the affects of air resistance). In addition, this downward acceleration is totally independent of any horizontal motion (like in the case of the long jumper). It is unaffected by horizontal motion, nor does it have any affect upon horizontal motion (the other reason why the horizontal velocity of a long jumper in flight is constant – Newton's first law of motion). This can be demonstrated by placing a pen on a table and also at the same time holding another pen at the same height as the table. Next, get someone to push the first pen off the table with a large force (i.e., accelerate the pen rapidly off the table). At the same point as they push the pen off the table (at the same moment in time that it leaves the table) drop the pen that you are holding. The pen on the table (that has now been pushed off) will have horizontal (the push) and vertical (gravity) velocity and it will have projectile motion towards the floor. The pen you have dropped should only have vertical motion and should drop to the floor in virtually a straight line. However, both pens will hit the ground at the same time. Hence, horizontal motion (velocity) does not affect vertical motion (velocity). A further understanding of this constant vertical acceleration situation will be explained in more detail in the section B5 of this text.

In this example of the long jumper, Newton's second law of motion is used to identify and explain constant vertical acceleration and how and why a long jumper is pulled back towards the ground immediately after they have left it (jumped into the air). However, there are many other applications of Newton's second law that are applicable to the understanding of human motion but these will be discussed in more detail in section B3 entitled **The impulse–momentum relationship**.

3. **The law of reaction**: application to human movement.
This law states that for every action (a force) there will be an equal and opposite reaction (another force). In other words push on an object and you will feel the object push back on you with an equal and opposite force. For example, if you push on a wall you will feel an equal and opposite force that is coming from the wall and acting along your hands and arms. Similarly, if you stamp your foot against the ground you will feel a force through your leg that is exerted from the ground on your foot. *Fig. B2.8*, illustrates this in a number of examples within human movement.

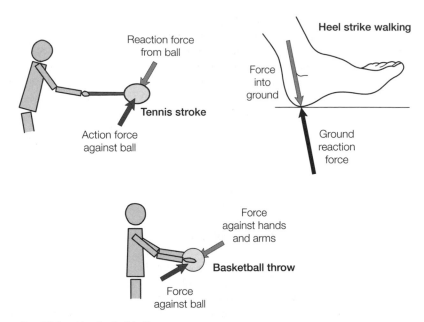

Fig. B2.8.   Newton's third law

However, when attempting to understand Newton's third law of motion it is important to be aware of two important facts: **First**, the effects of the forces are not canceled out (i.e., there is not a net effect of zero) because one acts on one object while the reaction (or other force) force acts on the other and, **second**, although the forces are both equal in magnitude and opposite in direction the effects of the forces are not the same. This can be seen when we consider *Fig. B2.7* (pushing into the ground) previously. In some cases the objects will accelerate (change their state of motion) and in other examples the objects will not accelerate (either move with constant velocity or remain stationary), although in each case there is an equal and opposite force acting (Newton's third law). This is explained from the consideration of the net force and the summation of the forces being balanced or unbalanced.

One of the most common applications of Newton's third law is seen in the consideration of the ground reaction force during walking. As you walk across the ground your foot exerts a force on the ground and consequently the ground exerts a force back on your foot. As your foot hits the ground at heel strike during walking it will do so with a force, which is derived from the mass of the foot and how it is accelerated into the ground at impact (i.e., $F = ma$). As contact is made and you drive your foot into the ground, an equal and opposite force will be exerted from the ground on your foot (i.e., you will feel the impact throughout your leg). However, the force from the ground (acting on your leg) will not cause you to move off into the air, and neither will it cause you to stop (although the effect will seem like a braking force on your body).

As you hit the ground at heel strike there are a number of forces that are acting in this example. For example, there will be the force from the mass of your foot and its acceleration into the ground once it has contacted the ground, there will also be the force of gravity (pulling your foot directly vertically downward), then there will be the ground reaction force from the ground on your foot (which will consist of the friction force between the foot and the ground (the anterior–posterior force), the normal reaction force and the medial–lateral force), and there will also be the force of the leg acting on the foot as it is driven over the foot during the stance phase (heel strike to toe off). All these forces act together and it is not simply a case of one force (**the action**) being opposed by an equal and opposite force (**the reaction**). These are all external forces and it is the net sum of all these external forces that will cause the body to accelerate or decelerate. In this context it is important to express that it is the external forces that will cause the internal forces within the joints. Hence, it is the external forces that cause the resulting change in motion (acceleration or deceleration).

Action – reaction forces   Although the terms action and reaction are widely used within biomechanics in the context of Newton's third law, there is a slight confusion when these forces are applied in sporting and/or other human movement situations. For example, it is difficult to determine which of the forces constitutes the **action** and which constitutes the **reaction**. In addition, there is a degree of confusion in that these forces (or terms action – reaction) when classified in this way could be misinterpreted to be movement rather than force. For example, when the racket hits the tennis ball during the ground stroke in the tennis game there will be a force exerted on the ball by the racket. There will also be a force exerted from the ball on to the racket. These forces are equal and opposite but it is the net effect of all the external forces that produces the change in movement (i.e., the acceleration or deceleration of one or both of the objects). The mass of the tennis ball is relatively small compared to the mass of the racket and once all the net forces are determined the net effect will be a force that causes the tennis ball to accelerate in the direction that the player intended to hit the ball. This equal and opposite force principle from Newton's third law (action – reaction) appears to falsely apply to movement as well as force. For example, if you are in the air during the flight phase of the long jump and you rapidly move your arms down (essentially by a muscle force within your body) towards your legs the reaction is that your legs will move upwards towards your arms (what appears to be an equal and opposite action – reaction (i.e., it appears as a movement rather than a force). However, this process is achieved by the equal and opposite torques (moments of force) that are applied to the body in order to cause this movement to happen. The torque that caused the trunk and arms of the athlete to move downwards caused an equal and opposite torque that caused the legs to move upward. *Fig. B2.9* illustrates this and shows how athletes during the flight phase of the long jump prepare themselves for a better position during landing (even though as we have seen they cannot change the pre-determined flight path).

This is the same principle used by a rocket and space shuttle to propel itself into space. Although the rocket does not have anything to push against the external vertical force that is being exerted downward (exerted by the jet engines) onto or into the air causes an opposite (reaction) force to be exerted on the rocket. The corresponding result (which is the net force) considering all the external forces (i.e., the vertical force acting downward from the rocket, the opposite force acting on the rocket upwards, the force of gravity pulling the rocket downward and possibly air resistance, friction and drag) accelerates the rocket vertically into the air and eventually into space. Similarly, the jet engines of a plane that are used to propel it horizontally through the air (although there are other forces such as aerodynamic lift and drag forces that also have a significant affect on the principle

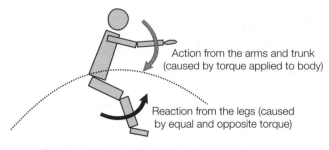

Action from the arms and trunk
(caused by torque applied to body)

Reaction from the legs (caused
by equal and opposite torque)

Creates a better position for landing

Fig. B2.9.   Newton's third law applied to the long jump flight phase

of flight), are acting backward to its direction of motion (which is forward). *Fig. B2.10* illustrates some more examples of Newton's third law of motion within human movement.

The application of Newton's laws of motion is seen in many examples of human movement and an understanding of these laws can be important with regard to both injury prevention and improvement in performance. For example, how can a person reduce the potentially damaging impact force that is created during the heel strike in running (i.e., the impact force that is experienced throughout the leg that can be between 2 and 5 times your body weight) or how can an athlete increase the ground reaction force acting on the athlete at the take-off in a high jump, so that they can potentially jump higher?

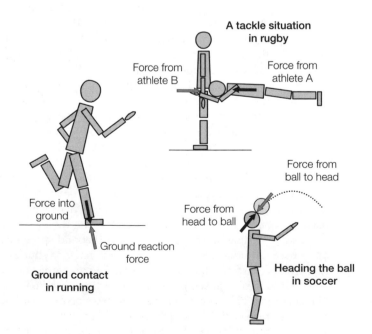

Fig. B2.10.   Newton's third law applied

B3 THE IMPULSE–MOMENTUM RELATIONSHIP

Key Notes

The impulse–momentum relationship ($Ft = m(v - u)$)	This develops from Newton's second law of motion ($\sum F = ma$) and is an important relationship within biomechanics.
Momentum	The linear momentum possessed by an object is a measure of the object's mass multiplied by the object's linear velocity. Since the mass of an object remains relatively constant the change in momentum experienced by an object represents a change in its velocity (increase or decrease).
Impulse	Impulse is defined as the force applied multiplied by the time of force application. It is equal to the change in momentum possessed by an object ($Ft = m (v - u)$). Impulse can be increased by either increasing the applied force or increasing the time of force application. In certain situations within human movement it is necessary to have a large force and small time of application and in other examples it is valuable to have the opposite situation.
Application	The shot putter in athletics applies a force to the shot for a long period of time in order to give the shot more impulse and hence a greater change in momentum (i.e., more velocity at release). The vertical (high) jumper applies a force to the ground in order to jump off the ground. The ground applies a reaction force to the jumper in order for them to be able to leave the ground. The net vertical impulse created during the preparation of the vertical jump will affect how high the athlete is able to jump. When catching a ball it is often necessary to increase the time of contact with the ball in order to reduce the force of impact (between ball and hand). This is achieved by following the ball's direction with your hands as you make the catch (as in the case of catching a cricket ball).

Impulse–momentum

The impulse–momentum relationship develops from Newton's second law of motion ($\sum F = ma$) and it allows us to apply this law to situations where forces are continually changing over time. For example, in many cases involving human motion forces will continuously change (i.e., they are applied over a period of time). Two rugby players who contact each other in a tackle situation will exert changing forces over time. Similarly when you run and jump on the ground you will apply forces that vary over time depending on a number of related variables: the speed of running, the surface of contact, the shoe type, the body position, and many other aspects.

In human movement it is usually the effect of these changing forces applied over time with which we are concerned. It is these forces and their effects that will be used to determine performance characteristics or injury potential (outcome measures). As we apply a force over the ground with our foot during the contact phase in running what will be the outcome of this application to our running speed? Does our speed increase or do the forces acting on our legs increase to a level that could develop a potential for injury? Similarly, in the athletic event of the shot putt how does the athlete apply enough force to project the 16 lb (7.27 kg) object through the air? *Fig. B3.1* illustrates some other examples of this varying force application over time within human movement.

Newton's second law of motion allows us to be able to understand this application of varying force over time (the impulse–momentum relationship) in more detail.

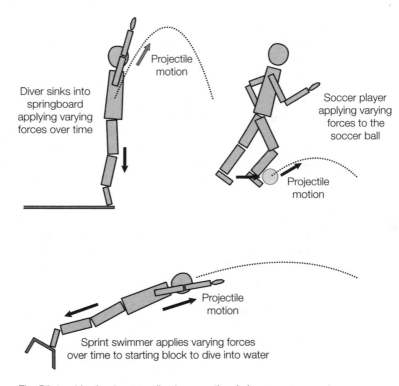

Fig. B3.1. Varying force application over time in human movement

From Newton's second law ($\sum F = ma$)

$$\sum F = ma$$

where
$\sum F$ = force (net force)
m = mass
a = acceleration

But linear acceleration (a) is also expressed as:

$$a = \frac{v - u}{t_2 - t_1}$$

where
a = linear acceleration (meters/second^2)
v = final velocity (meters/second) measured at position t_2
u = initial velocity (meters/second) measured at position t_1
t_2 = time at position t_2 (seconds) for final velocity (v)
t_1 = time at position t_1 (seconds) for initial velocity (u)

Now substitute for a into [$F = ma$] equation from Newton's second law

$$F = m \frac{(v - u)}{t_2 - t_1}$$

In order to cancel out the division component (on the right-hand side) we multiply both sides of the equation by ($t_2 - t_1$) or t (since ($t_2 - t_1$) will produce a single value for t (time)).

$$Ft = m (v - u)$$

Multiply out the brackets, and we have the equation for impulse:

$$Ft = mv - mu$$

where
Ft represents impulse measured in Newton second (Ns).
$mv - mu$ represents the change in momentum measured in kilograms. meters per second (kg.m/s).

Impulse

Impulse is defined as the force multiplied by the time (duration) for which the force acts. Impulse can be derived by using the average force acting over the same time period. **Linear (translational) momentum** is defined as the objects mass (kg) multiplied by the objects linear velocity (m/u). Hence, the faster an object is moving, or the more velocity it has, the greater will be the object's linear momentum. Similarly, if you could increase the object's mass you would also produce the same effect and increase the object's linear momentum.

In this example ($Ft = mv - mu$) we can see that the right-hand side of the equation ($mv - mu$) is referring to a **change in momentum**, which in the case of human movement is primarily brought about by a change in velocity (because the mass will remain relatively constant). Similarly, by considering the left-hand side of the equation (Ft) we see that this change in momentum can be affected by either increasing or decreasing either the force or the time for which the force acts (**either increase or decrease F or t**). If we increase the amount of force applied (say in the example from *Fig. B3.1* for the diver) the change of momentum would also increase (we would have a greater change in velocity). Similarly, if we increased the amount of time over which the force was applied we could also increase the amount of change in momentum and hence also increase the velocity (since the mass remains relatively constant). At this point it is also important to identify that by using this principle we can also decrease either of these components and thus cause a decrease in the change in momentum and hence a reduced velocity of movement. *Fig. B3.2* helps to illustrate this in more detail.

From the example in *Fig. B3.2*, where the soccer player applies a force to a soccer ball (with the foot) for a specific period of time (contact), we see that there are two components to this application. **For example, how would the soccer player either increase his/her force applied to the ball or how would they increase the contact time?** Both aspects are important in the understanding of the impulse–momentum relationship. Increasing the force applied is generally achieved by

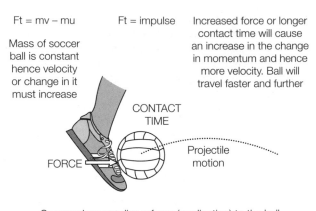

Ft = mv − mu

Mass of soccer ball is constant hence velocity or change in it must increase

Ft = impulse

Increased force or longer contact time will cause an increase in the change in momentum and hence more velocity. Ball will travel faster and further

CONTACT TIME

FORCE

Projectile motion

Soccer player applies a force (application) to the ball for a specific period of time (contact) (Ft = impulse). Measured in Newton second (Ns)

Fig. B3.2.   Soccer player applying a contact force to a ball to change its state of motion

either increasing the muscle strength of the player (and hence generally the mass), although this can also be achieved by moving the leg much faster ($\sum F = ma$). The player can try to move the leg faster towards the ball and thus impart a greater net kicking force applied to the ball at impact. **However, how is it possible to increase the contact time or time of application of this force?** The player would use technique and skill in attempting to increase contact time. For example, they may try to hit the ball in such a position that allows them to follow through the kick remaining in contact for longer – or indeed they may even intentionally impart a degree of spin and lateral contact with the soccer boot. In all these methods it is important to point out that it is not simply just a matter of being able to apply a large force for a long period of time. Anyone who has ever kicked a ball will understand that the contact time for the kicking of a ball is very short and it happens in a fraction of a second. Similarly, it is not possible to apply this large net kicking force all the time throughout this contact phase. The longer you try to stay in contact with the ball the smaller will be the average force applied and hence the smaller will be the impulse. As the foot makes contact with the ball an external force (from the foot) is applied to the ball. As the contact time increases the force applied is averaged over a longer period of time. Therefore, it becomes a careful, skilled compromise of force application and contact time to execute an efficient kick. As an example, try kicking a soccer ball with a pillow tied to your foot and you will see that you are unable to kick the ball with any great speed (velocity) or very far at all. You will have reduced the impact force (applied force) because the pillow is acting as a shock absorber for this force application and although you have probably increased the contact time you have much less average force and therefore much less impulse is acting on the ball. The result is that the ball has a much smaller change in momentum (less velocity) and will hence not travel as far or as quickly. Although this example is probably not very practical (i.e., it is not easy to fix a pillow to your foot) it does, however, clearly demonstrate the point. **The same effect can be achieved (although it is not as obvious) by changing your shoes: kick a ball with soccer boots or kick a ball with large novelty furry slippers and see what happens.**

Consider the example when you are performing a vertical jump from the ground (for maximum height jumped as in, for example, volleyball). As you prepare for the jump (from a stationary standing position) you will sink down into the ground while at the same time swinging your arms backwards. At the bottom of the sinking downward period you would then drive your arms forward and upward and push off with your legs, propelling yourself into the air vertically. *Fig. B3.3* illustrates this action in more detail.

In order to have achieved this action and jump into the air you will have applied a force over the ground for a period of time (contact with the ground). The ground reaction force (i.e., from the ground and acting on the person) would be the force that is used to determine the amount of impulse that is acting on the body (impulse = force × time). This impulse would provide a change in momentum (because the two are related by $Ft = mv - mu$). Now, since your mass is constant throughout this activity this change in momentum will result in a change in velocity. The greater the impulse (the more positive the net result) and the greater will be the change in velocity. Since at the beginning of the jump you are not moving (zero velocity – stationary) the more impulse you can generate the greater will be the take-off velocity in a vertical direction (since we are considering vertical impulse). The more take-off velocity you have, the higher you will jump, although, as we know, gravity, which is acting throughout this whole activity, will begin to slow you down at a constant rate as soon as you take-off. However, if you have more vertical velocity to begin with it will take longer for gravity to slow you down at a constant rate – hence you will jump higher.

Now let us use the equation ($Ft = mv - mu$) to look at this example in more detail. *Fig. B3.4* identifies this vertical jump example in a subject jumping from a force platform (in order that we can actually measure the amount of impulse that is created). In this figure it is important to identify that we are considering **vertical impulse**. As we have seen, gravity will affect the vertical components of movement. Throughout this exercise (vertical jump) gravity will continue to act on the person. At the beginning of the jump (as the person sinks down) the weight of the body is not being supported and the body will accelerate downward (this appears as a negative force effect as seen on the graph). At the point where the force trace returns to the body weight line the body will have maximum downward velocity.

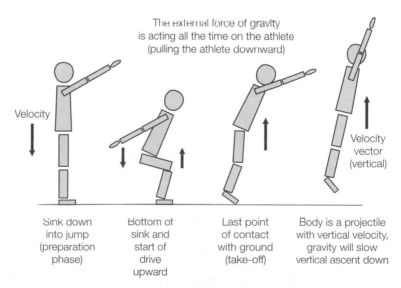

The external force of gravity is acting all the time on the athlete (pulling the athlete downward)

Velocity

Velocity vector (vertical)

| Sink down into jump (preparation phase) | Bottom of sink and start of drive upward | Last point of contact with ground (take-off) | Body is a projectile with vertical velocity, gravity will slow vertical ascent down |

Fig. B3.3.   *The vertical jump action*

Next, the deceleration (the stopping) of this downward velocity will occur in order to stop the body at the lowest point prior to the body beginning the drive phase (this is marked on the graph). In the case of the force trace shown in *Fig. B3.4*, the force from gravity (an acceleration downward) is inclusive in this net impulse calculation because the trace is presented about the subject's body weight ($W = m \times g$) where gravity is acting on the subject throughout the jump. The fact that the trace (vertical force trace) is plotted about the body weight line accounts for this effect from gravity (i.e., the negative part (under the body weight line) of the vertical force trace). Note: at this point it is important to clarify that in the consideration of **horizontal impulse** (that would also be created in a vertical jump) the effect of gravity is not considered as an external force (as it affects the vertical component).

From consideration of *Fig. B3.4* we can see that it is important to identify that in the impulse–momentum equation force is a vector quantity (i.e., it has magnitude and direction). An increase in impulse will cause a change in momentum in a specific direction (the direction of the force). For example, if you create a force downwards (which is necessary to initiate a vertical jump) the change in velocity (change in momentum but indicated as velocity because mass is constant) will also be in the downward direction. In the example of the vertical jump it is therefore possible to see that impulse will be created in both positive and negative parts (directions). In the case of you sinking down into the jump you are creating a negative impulse that is not contributing to the vertical component of the jump. However, it is necessary for you to be able to initiate the push-off propulsive phase. Hence, the downward (negative) impulse that is created is subtracted from the positive impulse and the result will be either a positive or negative net impulse. Considering *Fig. B3.4*, it is possible to calculate the vertical velocity of take-off from the net impulse that is produced.

Athlete mass $= 75\,\text{kg}$
Net impulse $= 352 - (18 + 10)\,\text{N.s}$
$ = [\text{B} - [\text{A} + \text{C}]]$ positive and negative components ($=$ net impulse)
$ = \textbf{324 N.s (positive impulse)}$

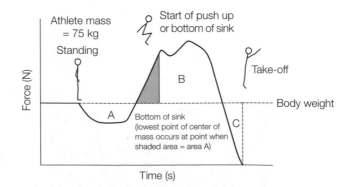

Area A = 18 Ns impulse (negative direction downward)
Area B = 352 Ns impulse (postive direction upward)
Area C = 10 Ns impulse (negative direction – athlete is leaving
        ground and is unable to maintain any propulsion)

Fig. B3.4.   Vertical force – time trace of a standing vertical jump measured using a force platform

Substitute this impulse value and the athlete's mass into the equation for impulse–momentum:

$$Ft = mv - mu$$
$$324 = 75 (v - u)$$

Considering that u (initial velocity) is zero (because you started from a standing stationary position) we can now see the following:

$$324 = 75 (v)$$

Divide both sides by 75 to get v (final velocity) on its own:

$$\frac{324}{75} = v$$
$$4.32 = v$$
$$\underline{4.32 \text{ m/s}} = v \quad \textbf{(vertical velocity at take-off)}$$

In this example (*Fig. B3.4*) the impulse derived was from the application of a **vertical** force (although there will also be other forces acting in different directions: anterior posterior forces, medial–lateral forces, and obviously gravity) acting over a period of time. In order to demonstrate the importance of this generation of impulse in human movement it is possible to adjust the values from *Fig. B3.4* to see what would happen if it were possible for us to create more positive impulse. This could be achieved by either by increasing the force applied or by changing our technique such that the application of the force was for a longer period of time (providing the average force was not significantly less). Alternatively, we could also have changed our technique such that we had less negative impulse (perhaps by modifying the descent phase). *Fig. B3.5* presents revised data for the vertical force–time trace in the vertical jump example.

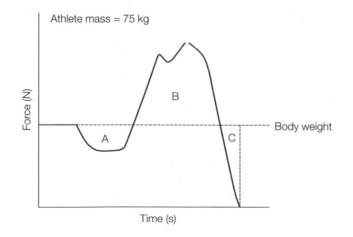

Impulse values (area under the curve)
Area A = 22 Ns
Area B = 400 Ns
Area C = 15 Ns

Fig. B3.5.   Vertical force – time trace of a standing vertical jump (modified values)

Considering *Fig. B3.5*, what is the effect on the vertical velocity if we create more positive vertical impulse?

$$\text{Net Impulse} = 400 - (22 + 15)$$
$$[B - \{A + C\}]$$
$$= 363 \text{ Ns}$$

Substitute this value and the athlete's mass into the impulse–momentum equation.

$$
\begin{aligned}
Ft &= mv - mu \\
363 &= 75\,(v - u) \\
363 &= 75\,(v) \qquad \text{since } u = 0 \text{ (stationary starting position)} \\
\frac{363}{75} &= v \\
\mathbf{4.84\ m/s} &= v \qquad \text{(an increase on the previous value of } \mathbf{4.32\ m/s)}
\end{aligned}
$$

Hence, in this application increase the amount of positive vertical impulse and you will increase the vertical take-off velocity and jump higher.

In human movement there are many other examples of where increasing the impulse will result in a greater change in momentum and hence greater velocity. In the case of the shot putt the athlete applies a larger force (by virtue of their strength (muscle size and muscle mass)) for a longer period of time (by virtue of their technique). The athlete would start by leaning over the back of the throwing circle and then by jumping backwards and rotating in the middle of circle finally to leaning over the front of the throwing circle. This would allow the athlete the time to apply a force to the shot for a much longer period and potentially (providing the average force was not substantially reduced) create more impulse (that is, acting on the shot). This would result in a greater change in momentum of the shot and hence more shot velocity (at release). The same applies in the sport of javelin, where the athlete would also try to apply a force to the javelin for a long period of time by leaning back into the run up to rotating and leaning forward into the delivery phase (thus creating more impulse).

However, within human movement it is not always desirable to create large amounts of impulse and it is sometimes the case that the net force needs to be reduced (or averaged over time) in order to minimize the potential for injury. Imagine trying to catch a cricket ball that is thrown at you. If you stand still and hold your arms outstretched (and rigid) you will feel a large force acting on your hands and arms as you catch the ball. Why? As it contacts your hand, the ball will require an impulse that is applied to the ball in order for it to be stopped (i.e., to change its momentum). The amount of impulse that will be required to change the momentum possessed by the ball (i.e., it will go from traveling quickly to almost a sudden stop) will be large, depending of course on its mass and velocity (momentum) before impact. Hence, the reason you feel a large force is that you have allowed the contact period (between your hands and the ball) to be a very small period in time (by holding out your hands rigidly the ball will just hit your hands and stop suddenly). The force that is applied to the ball in order to stop it (and consequently to your hand – Newton's third law) is high because it is acting over a short period of time (contact time). Therefore, if you now try to catch the cricket ball by moving your hands in the direction the ball is traveling, this time as you begin to catch it you will increase the contact time and thus average out the force (i.e., it will feel much easier to catch the ball this way).

As an example, an object with 50 units of momentum (say the ball) must experience 50 units of impulse (from the hands) in order for it to come to a stop

$(Ft = m(v - u))$. Any combination of force and time could be used to provide the 50 units of impulse needed to stop the ball. In this case if the contact time was 2 units, the force would need to be 25 units. Similarly, if the contact time was increased to 4 units, the force would be reduced to 12.5 units. The same principle applies for many other situations in human movement where it is important to increase contact time to reduce potentially damaging impact/contact forces. *Fig. B3.6* illustrates some of these examples in human movement.

The impulse–momentum equation is one of the most important principles in biomechanics and it provides a method for understanding both improvement in performance and injury prevention within human movement. Hence, it is critical that the student should have a good working knowledge of this topic.

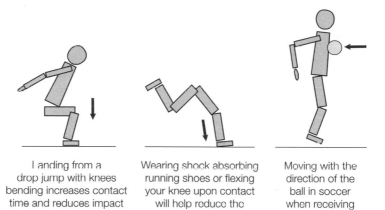

| Landing from a drop jump with knees bending increases contact time and reduces impact force (shock absorbing) | Wearing shock absorbing running shoes or flexing your knee upon contact will help reduce the impact forces | Moving with the direction of the ball in soccer when receiving a pass on the chest |

Fig. B3.6.   Impulse–momentum within sport (increasing contact time to reduce impact force)

B4 CONSERVATION OF LINEAR MOMENTUM

Key Notes

Linear momentum

Linear momentum is the product of the mass of the body multiplied by its linear velocity. To increase the linear momentum possessed by an object we could either increase its mass or increase its velocity. Generally, within human movement the component of mass remains constant and therefore it is the change in velocity that is used to change the momentum possessed by a body.

Principle of conservation of linear momentum

This principle states that in any system where bodies (or objects) collide or exert a force upon each other, the total momentum in any direction remains constant unless an external force acts on the system. The term system is used to describe two or more bodies that are in motion and that exert a force on each other. In determining linear momentum it is therefore important to specify the direction in which the momentum is considered (i.e., consider all the forces that are acting in that direction: vertically, anteriorly, and posteriorly, or medially and laterally (horizontal)). The principle of conservation of linear momentum is only valid when: 1) there is no external impulse acting on the system (i.e., no external force) and 2) the total mass of the system remains constant (before and after collision).

Application

When a goalkeeper in soccer catches a ball in the air the momentum in the system before the collision (where collision equals contact of goalkeeper and ball) is equal to the momentum in the system after the contact or collision (i.e., when goalkeeper and ball are together). The example is given as "in the air" because when the goalkeeper is in contact with the ground there would be other external forces acting on the system.

Linear momentum

Linear momentum is defined as the product of the mass of the body multiplied by its linear velocity:

$$\text{Linear momentum} = \text{mass} \times \text{linear velocity}$$
$$(\text{kg.m/s}) = (\text{kg}) \times (\text{m/s})$$

In order to increase the linear momentum possessed by a body it would be necessary either to increase its mass or increase its linear velocity. Generally, within human motion it is difficult to increase the mass of the body (as this remains relatively constant), so in order to increase momentum we would therefore increase the object's linear velocity.

In human movement there are many situations where collisions between objects or bodies occur. For example, two rugby players (or American football or

Australian rules players) collide with each other in a tackle situation and the soccer player would collide with the ball on numerous occasions throughout a game: such as in the case of receiving a chest pass or heading the ball. *Fig. B4.1* illustrates some other examples of collisions with human movement.

Consider Newton's first law of motion, the law of inertia where a body will remain in a state of rest or constant velocity unless it is acted upon by an external force. We can now extend this law for examples involving collisions to explain the **principle of conservation of linear momentum**.

Principle of conservation of linear momentum

The **principle of conservation of linear momentum** states that in any system where bodies collide (and there can be more than two bodies) or exert a force upon each other, the total momentum in any direction remains constant unless some external force acts on the system in that direction. In this context the term system can be used to describe two or more bodies in motion that exert forces on each other. *Fig. B4.2* illustrates this in the catching of a ball during goalkeeping in soccer (the goalkeeper is shown catching the ball in the air because of the external forces that would need to be considered if the goalkeeper was on the ground).

As we can see from *Fig. B4.2*, the momentum of the ball and the goalkeeper (the system) before the collision (the catching of the ball) is equal to the momentum of the system (the ball and the goalkeeper together) after the collision (the catch). For simplicity the ball and the goalkeeper before contact could be termed system-1 and the ball and goalkeeper together after contact (when the goalkeeper holds the ball) termed system-2.

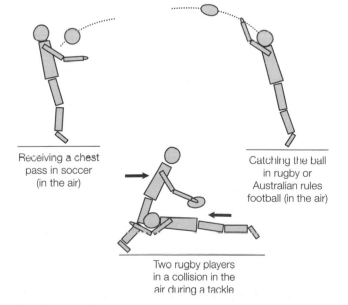

Receiving a chest
pass in soccer
(in the air)

Catching the ball
in rugby or
Australian rules
football (in the air)

Two rugby players
in a collision in the
air during a tackle

Fig. B4.1. Collisions between bodies within human movement

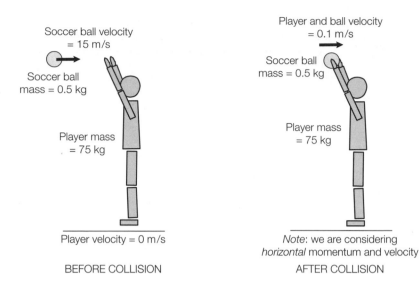

Momentum in system before impact = momentum in system after impact

Fig. B4.2.   A system of forces before and after collision: goalkeeper catching a ball in soccer (in the air)

Before collision (momentum in system-1 equals)

Momentum of the ball + Momentum of the goalkeeper
[mass of ball × velocity (horizontal) of ball] + [mass of goalkeeper × velocity (horizontal) of goalkeeper]

Note: it is important to point out that we are **considering linear momentum in a horizontal direction** and, as we have seen previously, in Section B2 (for horizontal motion), we can neglect the effects of gravity (an external force) as we are only considering the momentum in this direction in this example.

After collision (momentum in system-2 equals)

Momentum of the ball + goalkeeper combined
[mass of ball and goalkeeper × velocity (horizontal) of ball and goalkeeper combined]

Now according to the principle of conservation of linear momentum, the momentum possessed by the system before the collision equals the momentum possessed by the system after the collision (the amount of momentum is constant – it is conserved). In order to prove this we can use the values for mass and velocity as shown in *Fig. B4.2*.

Momentum before collision = Momentum after collision
$$(0.5 \times 15) + (75 \times 0) = (75.5 \times 0.1)$$
$$7.5 + 0 \text{ kg.m/s} = 7.5 \text{ kg.m/s}$$

Again, it is important to note that this is **horizontal** linear momentum that we have determined. In addition, it is also possible to see that if we did not already know the velocity for the ball and the goalkeeper after collision we could use this equation to calculate the combined velocity.

Application   It is important to remember that linear momentum possessed by a system will
remain constant in both magnitude and direction, and that the **principle of conservation of linear momentum is valid only if the following conditions are met**:

1. There is no external impulse (since as we have seen impulse = force × time and
    it is related to a change in momentum) in other words, no external force.
2. The total mass of the system (bodies that are colliding) remains constant.

To illustrate this principle in a more simplified form (i.e., not involving human
bodies or projectile objects such as soccer balls), *Fig. B4.3* identifies a more
practical example.

Considering *Fig. B4.3*, we can see that ball A has a mass of 2 kg and is moving
towards ball B with a horizontal velocity of 8 m/s. Ball B is also moving in the
same direction away from ball A but with a velocity (horizontal) of 2 m/s. Ball B
has a mass of 1 kg.

As the balls collide, there will be an impulse exerted by one ball on the other
(i.e., a force applied for a period of time). In this case the contact time is expected
to be small and therefore it is likely that the force will be high. There will be a
change in momentum brought about by the impulse but the total amount of
momentum (before and after collision) will remain constant. Ball A will experience
an impulse in the direction from RIGHT to LEFT (as it will experience a force from
ball B (the action–reaction law)) whereas ball B will experience an impulse in the
direction from LEFT to RIGHT because it is ball A that is making the contact (the
action). The two balls (A and B) will experience a change in momentum that is
equal to the amount of impulse that is created and this change in momentum as we
have seen is dependent upon the force and the amount of time that it is applied for.
Similarly, we have seen that momentum is related to mass and velocity and in the
case of the ball with more mass (2 kg) there will be less change in velocity for a
given momentum. Similarly, for the ball with a smaller mass (1 kg) there would be
a greater change in velocity in a given direction. It is important to point out that
this change in momentum (or velocity in each case since the mass is constant) will
take place in the direction of the impulse (the applied force). For example, for ball
A there will be a change in momentum in the direction from RIGHT to LEFT

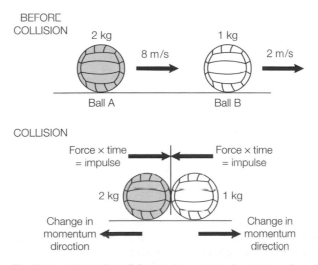

Fig. B4.3.   Two balls colliding to demonstrate the conservation of linear momentum principle

whereas for ball B there will be a change in momentum in the direction from LEFT to RIGHT. Again, it is important to reiterate that we are referring to **horizontal linear momentum** in this application.

The impulse on ball A (from RIGHT to LEFT – is given a **negative** sign since it is important to identify direction in this application) is equal to the change in momentum of ball A (the difference between momentum before and after collision) and this can be expressed as follows:

$$-Ft = m_A v_A - m_A u_A$$

where

$-Ft$ = impulse in the RIGHT–LEFT direction
            (given a negative value to indicate direction)
m_A = mass of ball A
v_A = final velocity of ball A (after collision)
u_A = initial velocity of ball A (before collision)

The impulse on ball B (from LEFT to RIGHT – is given a **positive** sign since again it is important to identify direction in this application) is equal to the change in momentum of ball B (the difference between momentum before and after collision) and this can be expressed as follows:

$$+Ft = m_B v_B - m_B u_B$$

where

$+Ft$ = impulse in the LEFT–RIGHT direction
            (given a positive value to indicate direction – but there is no need to express the + sign)
m_B = mass of ball B
v_B = final velocity of ball B (after collision)
u_B = initial velocity of ball B (before collision)

Now, considering that the impulses acting on the two balls are of **equal** magnitude (i.e., the forces acting on each are the same (action–reaction) and the contact time is the same for both balls) we can now express the equation to demonstrate the conservation of linear momentum principle:

$$Ft = - (m_A v_A - m_A u_A) = (m_B v_B - m_B u_B)$$

(minus sign to indicate direction of momentum change)
Impulse = change in momentum of ball A = change in momentum of ball B

Rearrange this equation and we have:

$$m_A u_A + m_B u_B = m_A v_A + m_B v_B$$
Momentum before collision = Momentum after collision

which confirms the conservation of linear momentum principle that the momentum in the system before collision or impact equals the momentum in the system after the collision. *Fig. B4.4* shows that the two balls (A and B) have continued to move forwards but with different velocities.

Although each body will undergo a change in momentum **separately** (even though they will experience a different change in velocity because their respective masses are different (A = 2 kg and B = 1 kg)) this change in momentum will be equal and in opposite directions. The conservation of linear momentum equation in this example (where both balls continue but with different velocities) therefore leaves us with one equation to find two unknown quantities (i.e., the two final

AFTER COLLISION

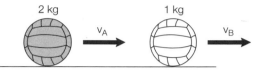

$$m_A u_A + m_B u_B = m_A v_A + m_B v_B$$
conservation of linear momentum

Momentum before collision = momentum after collision

Fig. B4.4. The two balls after collision indicating the conservation of linear momentum principle

velocities (v) of ball A and ball B). Therefore, we need one other equation to solve this problem and hence we must know either of the following:

1. Either of the final velocities v_A or v_B
2. That there is no rebound and the two balls will continue forward coupled together with a common velocity ($v_A = v_B = v$)

We can now rewrite the conservation of linear momentum equation for this specific case of two bodies that continue to travel, coupled together, with a common velocity (which is often the case in many human movement applications).

Conservation of linear momentum (**no rebound**)
$$(m_A \times u_A) + (m_B \times u_B) = (m_A + m_B)\, v$$
Momentum before impact = Momentum after impact
(for cases where bodies continue coupled together with a common velocity (v))

Fig. B4.5 illustrates the many different collision situations that can occur within human movement: objects travel on with different velocities, objects rebound and objects or bodies travel on with a velocity that is combined.

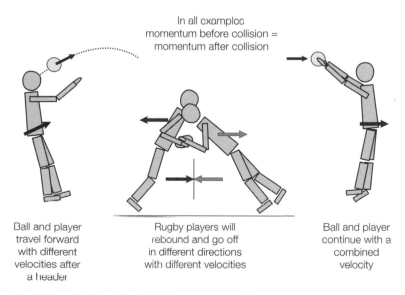

In all examples
momentum before collision =
momentum after collision

| Ball and player travel forward with different velocities after a header | Rugby players will rebound and go off in different directions with different velocities | Ball and player continue with a combined velocity |

Fig. B4.5. Collision of bodies in human movement

We can now see that it is possible to use these equations, which have developed from Newton's first law and the conservation of linear momentum principle, in many applications in human movement. For example, it is possible to work out the momentum or change in momentum experienced by bodies that collide or indeed the forces that are exerted due to the impulses that are created in such collisions.

B5 GRAVITY, WEIGHT, AND VERTICAL PROJECTION

Key Notes

Newton's law of gravitation

This law states that any two objects that have mass exert an attractive force on each other. This force is directly proportional to the mass of the objects and inversely proportional to the distance between the objects.

The force of gravity

The pages of this book and the person who is reading it will be exerting an attractive force on each other. However, because of the relatively small masses involved (i.e., the pages of the book and the human body) the force will be very small and it will not be possible to observe its effects. The planet Earth (due to its very large mass) exerts a significant force on the human body. This force produces the weight of an individual or object. The attractive force of the Moon on your body is less than the attractive force you experience on the planet Earth (because the Moon has much less mass than the Earth). Hence you will weigh less on the Moon although your mass will be exactly the same. This is the reason that astronauts are able to jump large distances when they are on the surface of the Moon. The force of gravity is an external force that acts on all bodies.

The force of gravity is constantly acting on all bodies. The effect of the force of gravity (in the balance of all external forces that are acting, i.e., the net force) results in an acceleration of the body or object. The acceleration on the planet Earth is given as -9.81 m/s^2 (presented as a minus value because the force of gravity from the Earth acting on the body will be trying to pull the body downward toward the center of mass of the Earth).

Acceleration caused by the force of gravity

On the planet Earth the effects of the force of gravity due to the mass of the planet act on all objects that have mass. However, this effect will only act on the vertical component of any movement. Any horizontal component of movement will be independent of the external force of gravity. For objects that are at or close to the surface of the Earth the acceleration of -9.81 m/s^2 is considered to be constant. This acceleration (because of the position of the object in relation to the center of the Earth and because of the Earth's relative large mass compared to the object's small mass) will act on all objects with the same rate regardless of their mass. Hence, dropping a hammer and a pen from the same height while on or at the surface of the Earth will result in both objects hitting the floor at the same time (neglecting air resistance).

Air resistance

In some situations, within human movement and sport, the effects of air resistance (as an external force) are **not** negligible. Air resistance will affect the trajectory of a golf ball and the trajectory of a javelin during flight. Often long jumpers who have a strong "tail wind" during their jump are not allowed the distance that they have achieved because of the contribution of this external force (and often the jump is disallowed in competition).

Newton's universal law of gravitation

In addition to developing the three laws of motion that we are familiar with, Isaac Newton also formulated the universal law of gravitation. This law states the following.

Any two objects exert a gravitational force of attraction on each other. The magnitude of this force is proportional to the masses of the two objects and inversely proportional to the square of the distance between them.

Numerically, this attractive force that each mass exerts on the other can be expressed by the following:

$$F = \frac{G\,M\,m}{r^2}$$

where

G = the Newtonian gravitational constant (6.67×10^{-11} Nm^2/kg^2)
M = mass 1 (measured in kg)
m = mass 2 (measured in kg)
r = the distance between the centers of the two masses
(measured in meters (m))

It is important to point out that we have seen that the inertia of an object (by virtue of its mass) determines the force needed to produce a given acceleration of the object. Gravitational mass determines the force of attraction between two bodies. In mechanics the value referred to as the Newtonian gravitational constant is the gravitational force that exists between two 1 kg objects separated by a distance of 1 m. Hence 6.67×10^{-11} Newtons of force is an exceptionally small attractive force.

This equation for the force of gravity can be further developed to express the value for the acceleration (a) of mass 2 (m) as it is pulled towards mass 1 (M):

$$a = \frac{G\,M}{r^2}$$

This law is stating that **any two objects that have mass** will exert an attractive gravitational force on each other. Although it is hard to imagine this being the case, you will be exerting an attractive force on the Earth as well as the one you can feel from the Earth acting on you. Similarly, you will (because you have mass) be exerting an attractive force on this book or computer as you read this text. The book or computer will also exert an attractive gravitational force on you. All objects that have mass will obey this universal law. *Fig. B5.1* helps to illustrate this attractive gravitational force with examples from human motion.

The force of gravity

As we stand and move about on the planet Earth we can experience the attractive force of gravity quite regularly and very obviously. As we get up from a chair it requires an effort because the force of gravity from the planet Earth is pulling us downward (although, as we are clearly aware, gravity will be acting on us all the time, even when we are just sitting in the chair and not moving). As we walk gravity holds us to the Earth's surface so that we are able to generate forces to overcome external forces (such as friction) and move forward. As we throw a ball in the air and then try to catch it again we experience gravity: first, in holding the ball in our hand stationary, then in trying to get it in the air, then by trying to catch it and finally by holding it in our hand again (i.e., gravity is acting all the time). The gravitational attractive force of the Earth on our body will affect all activities we perform on this planet or on any objects we choose to use while we are on it. The reason

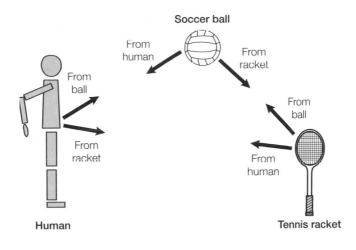

All these bodies will exert an attractive force on each other
that is proportional to their mass and inversely proportional
to the distance between their mass centers

Fig. B5.1. Gravitational attractive forces between masses

we can feel this force so obviously is that the Earth's mass (current estimate is 5.9725 billion trillion tonnes (or 5.9725×10^{24}) – where 1 metric tonne is 1000 kg) is very large in relation to the mass of our bodies (or indeed all other masses on its surface or close to its surface). Similarly, the reason we do not feel the pages of this text pulling us toward them is because the two masses (the pages of the book and our body) involved are relatively small (compared with the mass of the Earth).

Although this **gravitational attraction** between two bodies is **a force**, the effects of this force (actually the effects of the net force acting on the bodies) is usually expressed in the form of acceleration: **for example, how much is one object accelerating (or pulling it) the other one towards it?**

Acceleration caused by gravity

At the surface of the Earth the gravitational acceleration (the pull towards its center) on our bodies and all other objects that have mass (pens, books, soccer balls, tennis balls, javelins, and so on) is given as close to 10 m/s². In England the acceleration due to the pull from the mass of the Earth is said to be 9.81 m/s² (because of the relative large mass of the Earth when compared to the mass of an object). This means that any object that is dropped towards the center of the Earth (or towards the ground in our case) will increase its velocity by 9.81 m/s for every second of motion. On the surface of the planet Earth this acceleration is considered to be **constant** and it is the same for an object that is 100 kg mass as it is for an object that is 0.5 kg in mass. **To illustrate this, try dropping two different objects of obviously different masses from the same height and see which one hits the floor first?** The gravitational force from the Earth acting on all objects is directed vertically downward (or to be more correct is directed towards the Earth's mass center) and as we sit or stand on the Earth we are (in the balance of net external forces) being accelerated vertically downward at 9.81 m/s² (and again to be technically precise this amount (9.81 m/s²) actually depends on where exactly we are on its surface and on the magnitude of the two masses involved). However, the reason we do not continue downward is because we have the ground to stand on and the ground will be exerting an equal and opposite reaction force on us (upward). If there was no surface for the planet (the ground) and there was a large hole towards its center we would continue to accelerate down at this rate until we eventually came to a stop at its center (because at its center there will be no more gravitational force pulling you downward as its mass is equally distributed all around you).

The gravitational acceleration from the Earth will vary slightly depending on the masses involved and on where you are on the planet's surface. As we have seen already, in England it is specifically 9.81 m/s² and because the Earth varies from its width at the equator to its width at the poles (the Earth is approximately 43 kilometers wider at the equator than it is at the poles) the gravitational acceleration will also vary (because we will be nearer or further away from its center, which is where the Earth's mass is primarily concentrated). For example, it varies by 1 part in 200 from the equator to the poles (i.e., 0.5%). Similarly, it will also vary if you are either at sea level or if you are standing on the top of a mountain (because on the top of the mountain you should be further away from its center and the gravitational force (or effect (acceleration)) should be slightly less). However, such variation (due to being at sea level or on a mountain at altitude) is even smaller than that described previously because of the shape of the Earth. This variation is said to be not more than a maximum of 0.001 m/s². For example, in human movement it is often argued that it is easier to jump higher at altitude than it is to jump for height at sea level (i.e., many Olympic records in athletics have been set at high altitude meetings such as in Mexico in 1968). While it is true that the gravitational effects will be less at altitude it is unlikely that this small difference (variation) will have any effect on your ability to jump higher into the air. However, if you were on the surface of the Moon (which has only 1.23% of the Earth's mass) you would be able to jump much higher into the air because you would be far enough away from the Earth (and much nearer the Moon's center of mass) for it not to significantly affect your jump (although the Earth will still be pulling both you and the Moon towards it and you and the Moon will also be pulling the Earth toward you). The ability to be able jump higher on the surface of the Moon is because the Moon is unable to cause a gravitational acceleration like that experienced on the surface of the Earth (having only 1.23% of the mass of the Earth the Moon causes a gravitational acceleration of objects of only 1.6 m/s² (about 1/6th the gravitational acceleration of the Earth)).

So as we can see on the surface of the Moon you would have the same mass (because this is a measure of the quantity of matter in your body, i.e., the number of atoms and molecules in your body) and you would also have the same strength (related to muscle mass, size, and girth) but you would weigh much less (because of the reduced downward gravitational pull from the Moon on your body).

Weight

As we have already seen from section B1, the weight of a body is defined as the gravitational force acting on your body. Since this force is expressed as an acceleration value we can use this to calculate our weight.

Using the equation proposed by Newton to demonstrate this force of gravity we can see the following:

$$F = \frac{G\,M\,m}{r^2}$$

If this is considered in the context of human movement where we are concerned with the effects of the Earth's gravity on our bodies or on the movement of our bodies we can observe the following:

F = the gravitational force acting on us because of the Earth's mass and that we move on or near to the surface of it (i.e., our weight)

G = the gravitational constant provided by Newtonian mechanics

m = the mass of our body

M = the mass of the Earth (a constant value)

r = the distance between the center of mass of our body and the center of mass of the planet Earth. This value will also remain relatively constant even if we jump into the air or are on the top of a mountain – as we have seen the variation on gravity by virtue of position on the Earth is no more than 0.5% depending on where you are on its surface or 0.001 m/s² depending on how far away you are from its center of mass (i.e., at sea level or on top of a mountain)

Since we have several constants in this equation we can now use this knowledge to develop the equation to calculate our weight (or the force acting on our body due to the gravity of the Earth), which is more relevant to our studying human motion.

$$F = \frac{G\,M\,m}{r^2}$$

Since G = constant value, M = constant value which creates an acceleration at the surface of the Earth of 9.81 m/s², r = constant value we can rearrange this equation to represent our weight on the surface of this planet.

Weight at the surface of the planet Earth.

$$W = m \times g$$

where

Weight (gravitational force) = mass × acceleration due to gravity
(Newton's (N))   = kg × m/s²

For a 75 kg person standing on the surface of the planet Earth in England their weight would be calculated as follows:

$W = m \times g$
$W = 75 \times 9.81 \text{ m/s}^2$
W = 735.58 Newtons (to two decimal places)

As an example it is also possible to calculate (in order to illustrate how weight changes because of different gravitational forces) the weight of the same person standing on the surface of the Moon. In this case their weight would be calculated as:

$W = m \times g$
$W = 75 \times 1.6 \text{ m/s}^2$ (the Moon's gravitational acceleration)
W = 120 Newtons

In both cases the subject's mass would be exactly the same (75 kg) and the number of atoms and molecules that make up the person (the measure of their mass) would also be exactly the same. However, this is a clear illustration of why it is easier for astronauts to jump higher while they are on the surface of the Moon (i.e., the reason why you see them able to take large leaps and bounds while Moon walking). However, for the purpose of studying biomechanics the value for the Earth's gravitational acceleration should be considered as 9.81 m/s².

Vertical projection Gravity, as we have seen previously, is an **external force** that affects only the vertical component of projectile motion. In previous sections within this text we have seen that gravity does **not** affect the horizontal component of projectile motion. The effect of the force of gravity in the balance of the net forces acting is often expressed as an acceleration value (9.81 m/s^2) and in the understanding of vertical projection it is important to represent velocities and accelerations with directional components (as they are vector quantities that have both magnitude and direction).

If we throw a ball into the air, and we were able to throw this ball perfectly vertically upwards (although in practice this is not so easy to achieve) gravity would be acting on the ball (actually gravity is acting on both us and the ball all the time). The acceleration due to gravity in this case would be expressed as –9.81 m/s^2. The minus sign would denote that gravity is acting vertically downward (i.e., trying to pull the ball downward towards the Earth's mass center or trying to slow down its vertical ascent when we throw it into the air). *Fig. B5.2* helps to illustrate this exercise in more detail.

In *Fig. B5.2* the ball leaves our hand with a specific amount of upward vertical (+ve) velocity. This is created from how much net force was eventually applied to the ball and for how long it was applied (i.e., net vertical impulse = force × time = change in momentum (vertical momentum)). The amount of this vertical velocity will determine how high the ball will travel (since the acceleration caused by gravity is considered **constant** at or near to the surface of the Earth regardless of the mass). Hence, the ball with the largest vertical velocity at the point of release from the hand will travel to a higher point in its flight path vertically upwards.

As the ball leaves the hand, the force applied to the ball to make it leave the hand becomes zero and gravity will be the only force still acting on the ball (ignoring air resistance). Although the ball will still travel upwards, gravity will be acting by slowing down its vertical ascent (i.e., pulling it back downward). Eventually, gravity will bring this ball's vertical movement to a stop (it will have slowed it down such that there will be no more positive vertical velocity upward) and its vertical velocity at this point will become zero and it will instantly change its direction of motion (zero positive vertical velocity at the peak height of the flight path upwards). The ball will now start to move downward and it will do so at an acceleration rate of –9.81 m/s^2, although it is important to remember that throughout this action (flight) it has always been accelerating downwards (i.e., from when it left our hand). Similarly, it will also have an accumulating negative vertical velocity downward (negative indicating it is moving downwards). If you manage to catch the ball at the exact same height as that at which you released it,

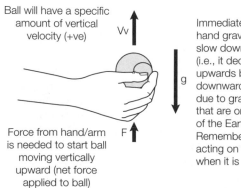

Ball will have a specific amount of vertical velocity (+ve) Vv

Force from hand/arm F is needed to start ball moving vertically upward (net force applied to ball)

Immediately the ball leaves our hand gravity (–9.81 m/s^2) will slow down its vertical ascent (i.e., it decelerates its movement upwards by pulling it downward). This acceleration due to gravity acts on all masses that are on or near to the surface of the Earth with an equal amount (g). Remember the force of gravity is acting on the ball at the time even when it is stationary in our hand

Fig. B5.2.   Gravitational acceleration from the Earth

you will find that at this point the ball will have the **same** vertical velocity as it had when it left your hand (although it will now have a negative sign indicating downward movement). *Fig. B5.3* identifies this in more detail.

If the ball is not caught and it is allowed to continue until it hits the ground, it will continue to accelerate at a rate of –9.81 m/s² in this direction. The ball will accelerate at –9.81 m/s² until it is acted upon by some external force (i.e., it is stopped by the force of contact with the ground (the force from the ground on the ball) or by contact with any other object).

Now, if we take the same ball and this time throw it with both a vertical velocity (the same as in our previous experiment) and a horizontal velocity (i.e., it would now project at an angle) we could demonstrate exactly the same effect from the force of gravity. *Fig. B5.4* illustrates this in more detail.

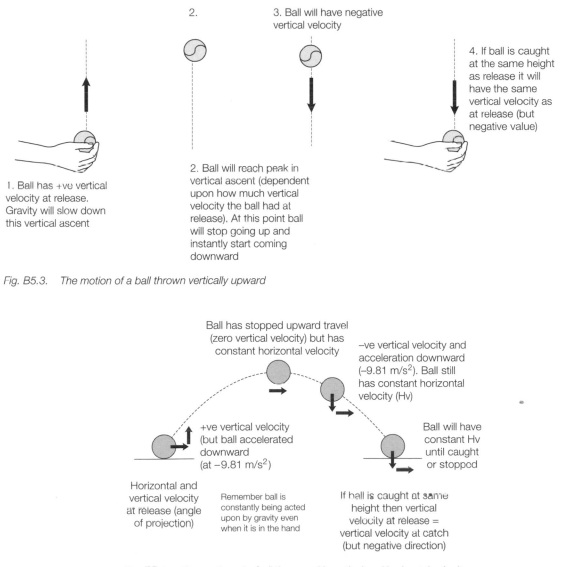

2.

3. Ball will have negative vertical velocity

4. If ball is caught at the same height as release it will have the same vertical velocity as at release (but negative value)

2. Ball will reach peak in vertical ascent (dependent upon how much vertical velocity the ball had at release). At this point ball will stop going up and instantly start coming downward

1. Ball has +ve vertical velocity at release. Gravity will slow down this vertical ascent

Fig. B5.3.   The motion of a ball thrown vertically upward

Ball has stopped upward travel (zero vertical velocity) but has constant horizontal velocity

–ve vertical velocity and acceleration downward (–9.81 m/s²). Ball still has constant horizontal velocity (Hv)

+ve vertical velocity (but ball accelerated downward (at –9.81 m/s²)

Ball will have constant Hv until caught or stopped

Horizontal and vertical velocity at release (angle of projection)

Remember ball is constantly being acted upon by gravity even when it is in the hand

If ball is caught at same height then vertical velocity at release = vertical velocity at catch (but negative direction)

Fig. B5.4.   The motion of a ball thrown with vertical and horizontal velocity

The ball will travel to the same height as it did in our experiment where we just threw it vertically upward, but because it also has a component of horizontal velocity it will travel in a **parabolic flight path** (forward). If the ball is caught at the same height it was released we know that it will have the same vertical velocity as when it was released (even though it is now traveling in a parabolic flight path with horizontal displacement). We have also seen that from Newton's first law of motion this ball will travel forwards with constant (no acceleration) horizontal velocity (in the absence of an external force and remember gravity is not considered to act as an external force on the horizontal component of motion) until it hits the ground or any other object in its flight path. This is why the space shuttle or satellites continue to orbit the Earth, that is, although they are constantly being pulled downward to Earth, because the direction of the gravitational effect from the Earth is changing (i.e., the Earth is rotating and they are also moving around the Earth) they continue to orbit the Earth in a circular path. The horizontal component of its motion is completely independent of the vertical component of its motion. Graphically for the ball experiment this can be shown in *Fig. B5.5*.

Considering *Fig. B5.5* in terms of the vertical component of the ball's motion, we can see that it travels upward and downward (displacement/time graph; *Fig. B5.5*: graph 1) with a decreasing vertical velocity (positive value) as it travels upward. The ball then reaches the peak height of the flight path and the velocity changes direction (i.e., it stops going upward and instantly starts coming downward) and throughout this action it has been accelerating at a constant rate (-9.81 m/s^2) with a decreasing positive vertical velocity and an increasing negative vertical velocity (graphs 2 and 3). This is exactly the same as when the ball that was thrown perfectly vertically (providing the vertical release velocity was the same in both experiments). Horizontally, the ball will be displaced as shown *Fig. B5.5*: graph 4. It will travel forwards with constant horizontal velocity (graph 5) in accordance with Newton's first law and it will do so with zero horizontal acceleration (constant velocity horizontally as in graphs 5 and 6). Hence, vertical and horizontal motions during projectile flight are independent of each other and gravity affects the vertical component only.

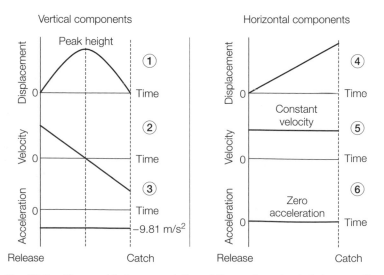

Fig. B5.5.   The graphical representation of the motion of a ball thrown with vertical and horizontal velocity

Air resistance

In the understanding of vertical projection it is worth making a comment about the effects of air resistance. Normally, in human motion we consider the effects of air resistance to be negligible (particularly on the human body as it travels as a projectile through the air). However, in certain applications the effects of **air resistance** will not be negligible and will be considered as an **external force** that affects motion. For example, in the case of dropping objects vertically, we know from Newton's law of gravitation that any object near to or on its surface regardless of its mass will accelerate toward the ground at a constant rate (i.e., two objects of different masses when dropped at the same height will both hit the ground at the same time). However, if you take the case of dropping a piece of paper and a golf ball you will see that the golf ball will hit the ground first. In this case air resistance will affect the piece of paper by a significant amount such that its descent towards the Earth will be slowed down (air resistance becomes an external force). Similarly, in sports such as javelin, hammer throwing, and discus, and even to an extent in long jumping when there are "head and tail" winds air resistance will have an effect. Often long jumps that are wind assisted are not legitimate jumps (in this case the tail wind would be an external force of assistance). Hence, in certain sports and movements it may be the case that the air resistance effects should be considered to be more than negligible. **Experiment with dropping different objects from the same height to see if you can demonstrate the effects of air resistance on the vertical downward acceleration of objects caused by the force of gravity.**

B6 FRICTION

Key Notes

Friction forces	Friction forces act between any two surfaces in contact. This friction force opposes the motion or sliding between the two objects. The frictional force that exists between objects is an essential necessity for human movement. Imagine trying to walk over the ground without the frictional force that exists between the foot and the ground. As an example, when walking on ice the frictional force between the foot and the ground is reduced and the result is often the foot slipping or sliding across the ice.
The coefficient of friction	The relationship that exists between the two surfaces in contact that gives rise to the frictional force can be described by what is termed the coefficient of friction. The symbol μ (mu) is used to denote the coefficient of friction between two surfaces in contact ($\mu = \mathrm{Tan}\ \theta$). Increase the coefficient of friction value between the two objects in contact and there will be an increase in the maximum frictional force. Similarly, decrease the coefficient of friction and the maximum frictional force is reduced.
Maximum frictional force	The maximum frictional force (Fmax) that exists between two surfaces in contact is the maximum force offered by friction in resistance to motion of the body. Hence in order to move the body or object (i.e., slide one object over another) the maximum frictional force must be overcome.
Types of frictional force	Friction can be classed as dry friction or fluid friction. Dry friction exists between two surfaces that are not lubricated. Fluid friction exists between two layers of fluid (i.e., water on water or air on water). Dry friction can be both static and dynamic. Static dry friction is when the objects in contact are not moving, and dynamic dry friction is when one or both of the objects in contact are in motion. The frictional force, whether it is static or dynamic, depends on the type and nature of surfaces in contact (i.e., types of materials, smoothness or roughness of their surfaces). The frictional force that exists between two surfaces in contact is, however, independent of the area of contact between the two surfaces. The maximal friction force that exists between a book and a table will be the same if the book is closed or open (providing it is placed on the table with its outside cover contacting the table in both applications).
Frictional force and the normal reaction force	The normal reaction force (N), which acts at 90° to the surface of contact, increases when the mass of one of the objects increases. The normal reaction force is proportional to the frictional force. Hence the frictional force increases when the mass of one of the objects in contact increases.
Application	Within human movement, athletes have a need to both increase and decrease the frictional force that exists between two surfaces in contact. In running, the grip between the running shoe and the ground is essential. Whereas in swimming, the one-piece fast skin swimming suits are designed to reduce the friction between the swimmer and the water.

Friction forces

As we know, biomechanics is concerned with the study of forces and the effects of these forces on living things. Most of the forces with which we are concerned in biomechanics tend to be external forces that are acting on the body or object of interest (the forces that cause the body to move). External forces are outside of the body (external) and these can be both contact and non-contact type of forces (gravity could be described as a non-contact external force). Internal forces are forces that are within the body (internally) and these are usually forces that result from the net effect of the external forces. The net force on the player's foot as he/she kicks a soccer ball would be an external force whereas the force on the anterior cruciate ligament in the knee caused by the kicking action would be an internal force. In mechanics (and biomechanics) it is important to distinguish between these types of forces. For example, a force applied at part D in a body or object will tend to distort some other part of the body (i.e., part E). The forces between the two parts of the body (D and E) are called internal forces. If the body is in equilibrium (when the algebraic sum of the all the forces or moments acting is zero) under the action of external forces both the external and internal force systems are separately in equilibrium.

Forces can be resolved into individual component parts, such as vertical and horizontal forces. *Fig. B6.1* shows the contact forces that exist between the foot and the ground at heel strike during running (sagittal plane only).

The ground reaction force (GRF) that exists as a result of the foot contacting the ground at heel strike in walking is the result of all the reaction forces acting between the foot and the ground during this contact (i.e., in three dimensions). This GRF, which is only shown in the sagittal plane (two dimensions) in *Fig. B6.1*, can be resolved into two components which are shown as a vertical and horizontal component. In this case it is again important to point out that we are only considering this (*Fig. B6.1*) in two dimensions (about a single sagittal plane). The actual forces acting in this case will be in three dimensions and there will be a medial–lateral (side to side) force that will also be a component of the ground reaction force.

The force that is perpendicular to the surface (vertical) is called the **normal force** and this always acts at 90° to the contact surface. The force that is acting parallel to the surface of contact (horizontal) is termed the **friction force**.

Friction forces act between any two surfaces that are in contact and the friction force opposes motion or sliding between the two objects. *Fig. B6.2* shows other examples of contact forces and demonstrates that frictional forces would be present in all these examples.

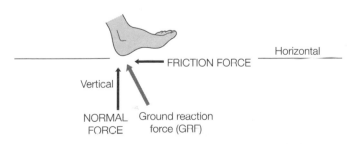

Fig. B6.1. *Normal and frictional forces at heel strike during walking (sagittal plane components only are shown (two-dimensional))*

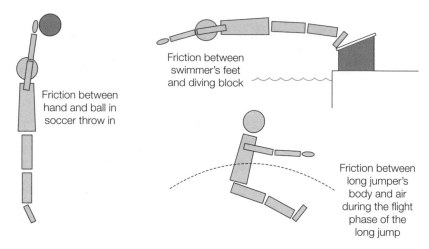

Friction between swimmer's feet and diving block

Friction between hand and ball in soccer throw in

Friction between long jumper's body and air during the flight phase of the long jump

Fig. B6.2.   Contact forces within sport

Coefficient of friction

The friction force is an essential necessity of human movement and locomotion, and without frictional forces between two objects it would be very difficult to initiate and maintain movement. For example, imagine trying to run across an ice rink in normal shoes. The frictional force between the ice and the shoe is very small and the result is a slipping of the foot during locomotion. The relationship between the two surfaces in contact that gives rise to friction can be described by what is termed the **coefficient of friction.** This is represented by the **symbol μ** (the Greek letter mu). *Fig. B6.3* helps to define what is understood by the term coefficient of friction (μ).

In *Fig. B6.3* the diagram (left) shows that if you place a brick on a surface and try to apply a force (Q) to slide the brick across the surface the frictional force (F) will resist the pushing of the brick. Hence, the brick will not move until you have exerted enough force (Q) to overcome the maximum frictional force (Fmax) created between the two surfaces. The coefficient of friction that describes the friction between the two surfaces is determined by imagining that you are able to tilt the surface on an angle (as shown in the diagram on the

Friction force

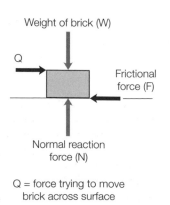

Weight of brick (W)

Q

Frictional force (F)

Normal reaction force (N)

Q = force trying to move brick across surface

Coefficient of friction

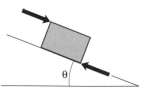

θ

θ = angle of inclination that causes horizontal component of gravitational force to cause block to slide

μ = coefficient of friction (mu)
μ = tan θ
Fmax = μ × N

Fig. B6.3.   Coefficient of friction

right of *Fig. B6.3*). As the surface is tilted upward there will be a point (an angle of inclination) where the block (brick) will start to slide down the slope. At this point the Fmax force between the two surfaces will be overcome (i.e., by the force of gravity and in particular the component of this force that is parallel to the surface of the ramp) and the block will slide down the slope. The tangent (opposite divided by adjacent in right-angled triangles) of the angle that is created when the block begins to slide is the measure entitled **coefficient of friction (μ (mu) = tan θ)**.

Example

The angle of inclination required to start a 20 kg mass sliding down a plastic covered surface is 35°. Calculate the coefficient of friction (μ) and the maximum frictional force (Fmax) which exists between the two surfaces in contact (the 20 kg mass and the slope).

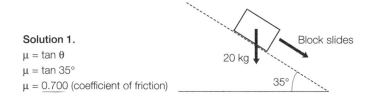

Solution 1.

$\mu = \tan \theta$

$\mu = \tan 35°$

$\mu = \underline{0.700}$ (coefficient of friction)

In order to calculate the maximum frictional force (Fmax) we use the formula that was developed in *Fig, B6.3* (Fmax = $\mu \times$ N) but first we need to establish the normal (N) reaction force acting between the two surfaces.

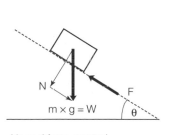

Normal force created
from ramp acting on
block (upwards to right)

Solution 2.

Normal reaction force (N) = cos $\theta \times$ W

$F_{max} = \mu \times N$

N = W cos θ

where

  N = normal reaction force

  θ = angle of inclination

  W = weight of block (force
      due to gravity)

  g = acceleration due to gravity
      9.81 m/s^2

  m = mass of block

We can now use the equation **N = W cos θ** to solve the problem for the **maximum frictional force (Fmax)** that exists between the two surfaces in contact.

Friction can be classed as being either **dry friction** or **fluid friction**. Dry friction is the force that exists between the surfaces of two objects in contact that are not lubricated (i.e., they are dry). Fluid friction exists between two layers of fluid, such as air and water, or water and water. This type of frictional force does not occur frequently in sport or human movement and the mechanics involved in the understanding of fluid friction are complex and are beyond the scope of this text in biomechanics.

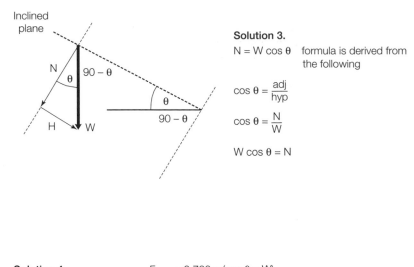

Inclined plane

Solution 3.

$N = W \cos \theta$ formula is derived from the following

$$\cos \theta = \frac{adj}{hyp}$$

$$\cos \theta = \frac{N}{W}$$

$$W \cos \theta = N$$

Solution 4.

Using

$F_{max} = \mu \times N$

$N = W \cos \theta$

$$F_{max} = 0.700 \times (\cos \theta \times W)$$
$$= 0.700 \times (\cos 35° \times (20 \times 9.81))$$
$$= 0.700 \times (0.819 \times 196.2)$$
$$= 0.700 \times 160.68$$
$$= \textbf{112.48 N}$$

This is the maximum frictional force that exists between the two surfaces in contact. This force would need to be overcome before one object could be pushed across the surface of the other object. Hence you would need to exert a force *of more than* **112.48 N** to start the block sliding (in the flat condition)

Dry friction can be **static dry friction** (when objects are not moving) or **dynamic dry friction** (when one or both of the objects in contact are in motion). The friction force, whether in the static or dynamic situation, depends on the type and nature of each surface in contact. For example, different surfaces in contact will have different coefficients of friction. Similarly, different roughness of surfaces in contact will also have different frictional properties: steel and plastic (as used in artificial hip joint replacements) have very low coefficients of friction and move easily over each other; a rough surface acting on another rough surface will have frictional properties different from two smooth surfaces acting together and it should be easier to slide or move the smooth surfaces across each other. Many of these examples can be seen throughout sport and human movement, for example the type of grip on the javelin; the chalk used by weightlifters or gymnasts for better grip; the table tennis bats with rough and smooth surfaces; and even soccer boots with modified uppers for better contact and control of the ball.

The frictional force that is created between the contact of two objects is independent (not connected with) of the surface area of contact. For example, place a book on a table and try to push it. Now open the outside covers of the book place it flat on the table and try to push it again. The book with its covers closed will create the same frictional force as the book open with both its outside covers in contact (effectively doubling its contact area). The reason for this is that although you have increased the surface area of contact (i.e., when you opened the book) you have also distributed the same mass over a larger area of contact, and have thus created a smaller average force because it is spread over a larger area (the net result of both conditions is the same because you have not changed the mass of the book). In other words, you have maintained the mass of the book but spread it over a larger area thus making each small contact force less – because you have spread the initial load over double the surface area.

Friction force and normal reaction force

From consideration of the equation in *Fig. B6.3* we can see that the frictional force is proportional to the normal reaction force. Hence, if you increase the normal reaction force you will increase the frictional force between the two objects. In the case of the open and closed book you did not increase the normal reaction force (you spread the same force over a larger surface area). The frictional force between the two objects remains the same in both the open and closed book situations because the mass and the normal reaction force also remain the same. As a further example of this you will see that by adding another book on top of the initial closed book you increase the frictional force and it will be harder to push or slide the two books across the table (i.e., you have increased the mass, the normal reaction force, and also the frictional force between the two books and the table). Since the frictional force resists motion between two objects it will be harder to push the two books than it is to push one. *Fig. B6.4* illustrates this in more detail.

Another example of this can be seen by placing your hand flat on a table and then see how easy it is to initiate movement (i.e., slide it across the table). Next, repeat the same experiment but this time press hard down onto the table. In the latter example it will be more difficult to slide your hand across the table because you have increased the normal reaction force and thus the frictional force existing between the two surfaces in contact (i.e., the hand and the table).

The normal reaction force (N) is proportional to the frictional force (Fmax) as we can see from the equation **Fmax = μ × N.** The normal reaction force increases when the mass of one of the objects in contact is increased. Hence, in this case (increasing the mass of the object resting on another object or surface) the

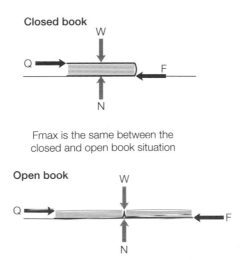

Fmax is the same between the closed and open book situation

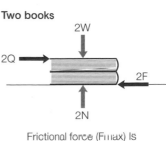

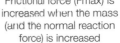

Frictional force (Fmax) is increased when the mass (and the normal reaction force) is increased

Fig. B6.4. Area of contact between two surfaces

frictional force (Fmax) increases and it is harder to slide the object across the surface of the other. In both static and dynamic friction situations the frictional force between two objects is not affected by the surface area of contact. In addition, it is more difficult to start an object moving than it is to keep it moving. Hence, the static friction between two objects is greater than the dynamic friction between two objects.

Application

The coefficient of friction describes the relationship between the two surfaces in contact. Increase the coefficient of friction value and the frictional force between the two objects will also increase (Fmax = $\mu \times$ N). Similarly, decrease the coefficient of friction between the two objects and they will slide across each other more easily. Within sport and exercise there are many examples where it is desirable both to increase and decrease the coefficient of friction between two objects.

Running shoes with rubber soles are designed to grip the floor so that the athlete can push into the ground with a large enough force to cause a reaction force from the ground which results in a net force applied to move the body forward with speed. Artificial joints, such as knees and hips, within the human body require low coefficients of friction so that they are made to last for a number of years before they begin to wear down. *Table B6.1* presents some of the co-efficient of friction values for a number of different surfaces in contact.

Table B6.1.

Surfaces		Coefficient of friction (μ)
Rubber on concrete	(dry)	0.60–0.85
Rubber on concrete	(wet)	0.45–0.75
Polystyrene (plastic) on steel	(dry)	0.35–0.5
Wood on wood	(wet)	0.20

The force that opposes the motion of one solid surface of an object sliding over another is termed **kinetic** or **dynamic friction**. The force that opposes the initial movement of the object is slightly greater and is called **static** or **limiting friction**.

Kinetic friction occurs when two objects are moving relative to each other and they rub together. The **coefficient of kinetic friction** for two objects is usually **less than** the **coefficient of static friction**. The drag of air particles acting on a javelin or the water particles acting on the swimmer are two examples of **kinetic friction**. For a car tyre the coefficient of dynamic friction is much less than the coefficient of static friction. The tyre provides the best traction with the road when it is not sliding. However, in the case of the car "skidding" the tyres become less effective because of the reduced sliding coefficient of dynamic friction. The coefficient of kinetic friction for metal on metal (same type) can be as low as 0.15, which as you can see is lower than any of the values presented for limiting or static coefficient of friction shown in *Table B6.1*.

The frictional force between two objects is essential for initiation and main-tenance of human motion. If the applied force equals the frictional force (Fmax) then the objects in contact will not move over each other. It takes more force to start an object moving over another than to keep an object moving in this way. As long as the two surfaces in contact are at rest the coefficient of friction between the objects remains constant. However, once motion begins the coefficient of friction between the two objects has a lower value and hence it is easier to keep the object/s moving. The **frictional force** can be any value from **zero to Fmax** depending upon how much force is being applied to move the two (or one of the two objects in contact) objects. The direction of the frictional force is always opposite to the intended direction of motion of one or both of the two objects.

Example

A force of any more than 100 N is required to start a 70 kg mass sliding across a wooden floor. Calculate the coefficient of friction between the mass and the wooden floor.

Solution 5.

$F_{max} = \mu \times N$

$100\,N = \mu \times (70 \times 9.81)$

$\dfrac{100}{686.7} = \mu$

$0.146 = \mu$

Heat is often generated at sites of friction between two objects in contact. Within the human body this heat can cause damage to the soft tissue structures. Blisters would be an example of excessive amounts of friction between two surfaces in contact in the human body. The body would respond by producing a layer of fluid between the superficial and deep layers of skin, thus trying to protect the deeper layers. In the long term often the superficial layer of skin is thickened, as in the case of the skin on the ball of the foot. Lubrication of the surfaces in contact helps reduce the amount of friction between objects in the dry condition. Articulating joints within the body that are lubricated with synovial fluid can produce a sliding system that is five times as slippery as ice on ice. As a result human joints can last for well in excess of 70 years before significant wear and tear issues occur (such as arthritis and joint degeneration). Finally, the one-piece fast skin swimming suits, seen at many Olympic Games, are designed to create a layer of water around the suit (eddy currents) that acts against the water in the pool creating water on water friction situation. These allow the athlete to slip and glide through the water much easier. Friction in sport and exercise is essential and there, are many examples when it should be both increased and decreased in order to perform more efficiently and effectively.

C1 TORQUE AND THE MOMENT OF FORCE

Key Notes

Torque	A torque is a twisting or turning moment that is calculated by multiplying the force applied by the perpendicular distance (from the axis of rotation) at which the force acts (the moment arm). Torques cause angular accelerations that result in rotational movement of limbs/segments.
Clockwise and anti-clockwise rotation	Clockwise rotation is the rotary movement of a limb/lever/segment in a clockwise direction (–ve). Clockwise is referring, in this case, to the hands of a clock or watch. Anti-clockwise rotation is rotary movement in the opposite direction (+ve).
Force couple	A force couple is a pair of equal and opposite parallel forces acting on a system.
Equilibrium	This is a situation in which all the forces and moments acting are balanced, and which results in no rotational acceleration (i.e., a constant velocity situation).
Second condition of equilibrium	This states that the sum of all the torques acting on an object is zero and the object does not change its rotational velocity. Re-written, this condition can be expressed as the sum of the anti-clockwise and clockwise moments acting on a system is equal to zero ($\sum$ACWM + $\sum$CWM = 0).
Application	Swimmers are now utilizing a pronounced bent elbow underwater pull pattern during the freestyle arm action. This recent technique change allows the swimmer to acquire more propulsive force and yet prevent excessive torques being applied to the shoulder joint (which were previously caused by a long arm pull underwater pattern). Large torques are needed at the hip joint (hip extensor and flexor muscles) to create the acceleration of the limbs needed to kick a soccer ball.

Torque

A **torque** is defined as a twisting or turning moment. The term **moment** is the force acting at a distance from an axis of rotation. Torque can therefore be calculated by multiplying the **force** applied by the **perpendicular distance** at which the force acts from the axis of rotation. Often the term torque is referred to as the **moment of force**. The moment of force is the tendency of a force to cause rotation about an axis. Torque is a **vector quantity** and as such it is expressed with both **magnitude and direction**. Within human movement or exercise science **torques cause angular acceleration that result in the rotational movements of the limbs and segments**. These rotational movements take place about axes of rotation. For example, the rotational movements created in the leg while kicking a soccer ball would occur about the ankle (the foot segment), the knee (lower leg segment) and the hip (upper leg segment) joints or axes of rotation. If an object is pushed with a force through its center of mass it will move in a straight line (linear motion) in the same direction as the applied force. However, if an object is pushed with a force at a perpendicular distance away from its center of mass it will both rotate (about an axis of rotation) and its center of mass will translate (move in a straight line). *Figs C1.1, C1.2* and *C1.3* illustrates this concept of **torque** in more detail.

Clockwise and anti-clockwise rotation

In *Fig. C1.3* it is possible to see that when a **force** is applied at a **perpendicular distance** from the center of mass (which in this example is considered to be the axis of rotation), the box (object) will both rotate and move forwards. The torque that is created as a result of applying this force at a perpendicular distance will cause the box to rotate about its **axis of rotation**. However, the box will also move forwards (translate) as the force is applied in a horizontal direction. Although the

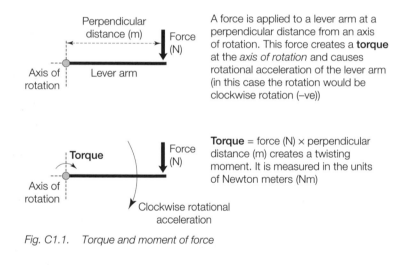

Fig. C1.1.   *Torque and moment of force*

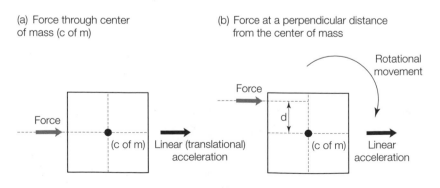

Center of mass is the axis of rotation in this example and we
are ignoring the external force of gravity

Fig. C1.2.   *Torque and moment of force*

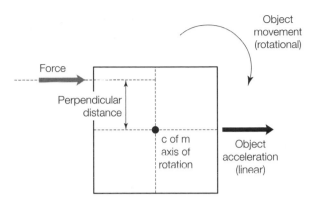

Torque = force × perpendicular distance

Fig. C1.3.   Torque and moment of force

force is acting at a perpendicular distance from an axis of rotation and it will create a torque or twisting moment, it will also have a horizontal component of force acting on the box (because the force is being applied horizontally). When we apply a **force** at a **perpendicular distance** to an **axis of rotation** we have seen that we create a torque or twisting moment (a tendency to rotate). The perpendicular distance from the center of rotation is called the **moment arm**. The torque that is created causes a **potential** for the **rotational acceleration** and thus the resulting rotation of the limb, lever or segment on which it is being applied. This rotation can be described as being either **clockwise rotation** or **anti-clockwise rotation** (described by reference to the direction of the movement of the "hands" on a clock or watch). Within biomechanics **clockwise rotation** is usually given the **negative symbol (–ve)** whereas **anti-clockwise** rotation is given the **positive symbol (+ve)**. In many situations within biomechanics it is often the case that pairs of forces act about a segment and about an axis of rotation. Two equal and opposite forces that are acting on a system create what is termed a **force couple**. The term **couple** is therefore defined as a pair of **equal** and **opposite parallel forces**. *Figs C1.4* and *C1.5* are useful in clarifying these terms in more detail.

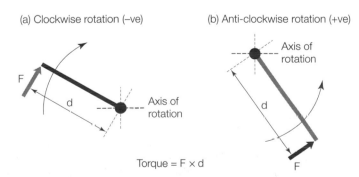

Fig. C1.4.   Clockwise and anti-clockwise rotation

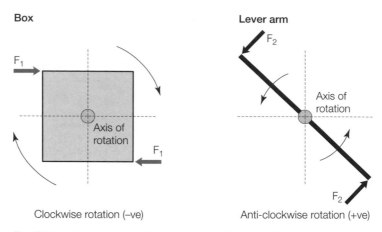

Box   Lever arm

Clockwise rotation (–ve)   Anti-clockwise rotation (+ve)

Fig. C1.5.   A couple: pair of equal and opposite parallel forces

Force couples

In *Fig. C1.5* it is possible to see the effect of a couple on two objects (a box and a lever arm). In each example in *Fig. C1.5* a couple (a pair of equal and opposite parallel forces) is seen applied to the objects. In both cases the objects (a box and a lever arm) rotate (and in these cases the objects only rotate because the only forces acting on them are equal and opposite – obviously in this example we have ignored the external force of gravity) in both a **clockwise** (the box) and **anti-clockwise** (the lever arm) direction. The couples create rotation about the axes of rotation. In these examples in *Fig. C1.5* there is no translation (linear motion) because the total net force on the systems is zero (i.e., the forces are equal and opposite). According to Newton's first law of motion an object will remain at rest or continue with uniform linear motion unless it is acted upon by an external force. In both these cases the net linear force on the objects from the couples would be zero and the objects would remain in the same positions (i.e., they would not move linearly). However, they would rotate about their respective axes of rotation because the couples cause torques and hence a combined rotational effect.

Since torque is expressed as force multiplied by perpendicular distance from an axis of rotation it can be expressed mathematically as follows:

$$\text{Torque} = \text{force} \times \text{perpendicular distance}$$
$$T \qquad = F \times d$$

where
F = force (measured in Newtons (N))
d = perpendicular distance from axis of rotation (measured in meters (m))
T = torque (measured in Newton meters (Nm))

Considering this equation, it is possible to see that in order to **increase** the **torque**, and hence increase the turning moment (or rotational effect) applied to an object, we can either **increase** the **force** applied or **increase** the **distance** from which the force is applied (i.e., from the axis of rotation).

Force applied = 20 N and the distance is 0.3 m

$T = F \times d$

$F = 20$ N

$d = 0.3$ m

$T = 20 \times 0.3$

T = 6 Nm

Force applied = 35 N and the distance is 0.3 m

$T = F \times d$

$F = 35$ N

$d = 0.3$ m

$T = 35 \times 0.3$

T = 10.5 Nm

Similarly, in order to **decrease** the amount of **torque** applied to a system we can either **reduce** the **force** applied or **shorten** the **distance** from the axis of rotation from which the force is applied. Both of these applications have particularly important implications within Human Movement and Exercise Science.

Application

There are many applications within human movement where is it beneficial both to increase and decrease the amount of torque or twisting moment acting on an object (or within a system where a system could be a group of leg segments). For example, in the case of swimming freestyle you may notice that many Olympic level swimmers now use a pronounced bent elbow action during the pull phase of the freestyle stroke. Indeed, after the swimmer's hand has entered the water, the swimmer will immediately bend the elbow and pull through the stroke almost entirely in this bent elbow position. This technique is designed to allow the athlete to be able to create as much propulsion as possible and yet at the same time protect the shoulder joint from excessive torque and loading (which was the case when previously using an extended arm to pull in freestyle (i.e., a long lever arm)). In the Athens Olympics in 2004 you may have seen the Australian 1500 m freestyle swimmer Grant Hackett clearly adopt this bent elbow technique.

Similarly, within the exercise of weightlifting, using the action of "arm curling" with weights requires the consideration of torque. The arm curl is where the elbow is flexed and extended while the athlete holds a weight in the hands. This action requires the biceps brachii muscle in the arm to create a torque (or turning moment) in order to resist the turning effect (or torque) created by holding the weight at a distance from the elbow (i.e., in the hand). In order to perform the action the athlete must exert a force in this muscle that creates a turning moment that overcomes the turning effect (the torque) created by the weight at the hands (due to its position in the hand from the axis of rotation (i.e., the elbow joint)). *Figs C1.6* and *C1.7* help to illustrate this concept in more detail.

As with forces, torques are vector quantities and their properties have both magnitude and direction (i.e., clockwise or anti-clockwise moments) and they can be summated and resolved. In the example within *Fig. C1.7* it is possible to see that the system (the arm curling example with weights) has two moments acting (in the sagittal plane and about transverse axis). A clockwise moment (–ve) which is created by the weight and the distance this weight acts from the elbow axis of rotation and an anti-clockwise moment (+ve) which is created by the muscle force (biceps brachii) and the distance that this force acts from the elbow joint axis of rotation. Note that it is important to point out that both moments are expressed with reference to the same axis of rotation (i.e., the elbow joint).

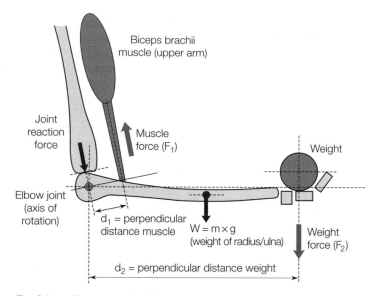

Fig. C1.6.   Torques at the elbow joint during an arm curl with weights

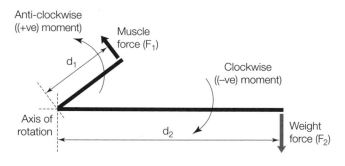

Ignoring joint reaction forces and effects of gravity on radius/ulna

Fig. C1.7.   Torques at the elbow joint during an arm curl with weights

If the turning moment created by the muscle force equals the turning moment created by the weight then the system will be in what is called **equilibrium** (balanced forces and moments – resulting in no rotational acceleration – a constant or no velocity situation). However, if the muscle force creates a larger torque (turning moment) then the result is that the weight will be lifted and the arm will move in flexion. Alternatively, if the weight creates the greater turning moment then the arm will drop and the elbow will extend. This latter case happens when athletes lower weights down in a controled manner (i.e., meaning at a constant velocity (no acceleration) where the muscle is creating a moment equal to the moment created by the weight). Throughout the action of arm curling with weights, the muscle force exerted by the biceps brachii will need to continually change. The reason for this change is due to the position of the arm at various points in the flexion–extension movement. In this case both the moment arm of the weight and the moment arm of the muscle force will also continually change. Since the "weight" has a constant mass and therefore a constant weight (force =

mass × acceleration due to gravity), the muscle force will need to change to accommodate the different torque or twisting moment created by the differing moment arms. *Fig. C1.8* shows two positions of the arm in this example which illustrate the changing lengths of the respective moment arms.

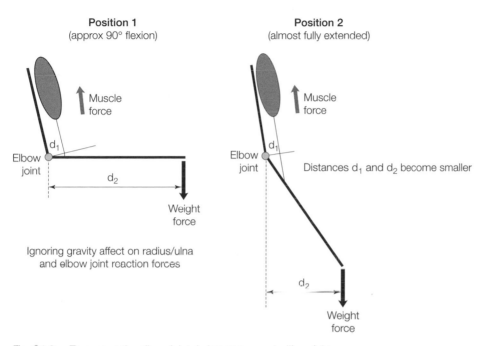

Fig. C1.8. Torques at the elbow joint during an arm curl with weights

The second condition of equilibrium

The first condition of equilibrium that is derived from Newton's first law of motion states that an object in a state of equilibrium does not accelerate (i.e., change its state of motion). In this case the sum of all the forces acting on the object is zero and the object is in a state of balance (i.e., it does not accelerate linearly in any direction).

Similarly, an object that is in a state of equilibrium does not change rotational velocity and the second condition of equilibrium states that the sum of all the torques acting on an object is also zero. Since the second condition of equilibrium is related to torque or turning moments we can write the second condition of equilibrium as:

The sum of the anti-clockwise moments and clockwise moments about a point (rotational point) is equal to zero.

Since we know that torques cause rotation and they are vector quantities (with magnitude and direction) we can use this knowledge to expand the second condition of equilibrium as follows:

$$\sum ACWM + \sum CWM = 0$$

where

$$\sum \qquad = \text{the sum of}$$
$$\text{ACWM} \quad = \text{anti-clockwise moments (+ve)}$$

$$\text{CWM} \qquad = \text{clockwise moments (–ve)}$$

In this context, it is important to point out that this is referring to a system of torques about a common axis of rotation. Hence, it is important to use the same origin (point of rotation) for anti-clockwise and clockwise moments in the calculation of the second condition of equilibrium. In the example in *Fig. C1.7* we had a system with two moments that were acting about a single axis of rotation (a common axis or origin) which was the elbow joint. If we now add some values to this example it would be possible to work out the muscle force that would be needed in order to balance the system (i.e., hold the weight in the hand in a stationary position).

Consider the example in *Fig. C1.7*, with the arm held in the 90° flexion position. The free body diagram would be represented as follows.

Free body diagram

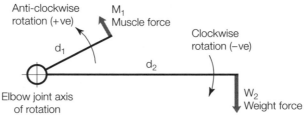

d_1 = perpendicular distance of muscle force

where
d_1 = the perpendicular distance to the muscle force (**0.05 m**)
d_2 = the perpendicular distance to the weight force (**0.45 m**)
M_1 = the muscle force (**unknown**)
W_2 = the weight force. In this case the weight is a 5 kg dumbbell. Hence the force acting would be determined by the mass multiplied by the acceleration due to gravity ($F = m \times g$). For the weight this would be equal to $5 \text{ kg} \times 9.81 \text{ m/s}^2$ = **49.05 N.**

Hence we can now use the formula for torque ($T = f \times d$) and the second condition of equilibrium to solve the problem for the muscle force needed to hold this weight stationary.

$$\text{Clockwise moment or torque (–ve)} = F \times d$$
$$= W_2 \times d_2$$
$$= 49.05 \times 0.45$$
$$= \underline{\textbf{–22.07 Nm}}$$
$$\text{Anti-clockwise moment (+ve)} \qquad = F \times d$$
$$= M_1 \times d_1$$
$$= \underline{\textbf{+M}_1 \times \textbf{0.05}}$$

Now substitute these into the second condition of equilibrium equation:

$$\sum \textbf{ACWM} + \sum \textbf{CWM} = 0$$
$$\textbf{M}_1 \times 0.05 + (-22.07) = 0$$

(Note: the clockwise moment is given a negative sign.)

Now rearrange the equation to get M_1

$$M_1 = \frac{22.07}{0.05}$$

(Note: 22.07 has now become positive because we have moved it to the other side of the equation and hence changed its sign.)

$$M_1 = \underline{\textbf{441.4 N}} \qquad \text{(Muscle force)}$$

Therefore, it is clear that we have to exert 441.4 N of force in the biceps brachii muscle in order to hold the weight in this static 90° flexion position. In order to overcome this weight and flex the joint further we would have to exert more force than this and thus create a larger turning moment (anti-clockwise). In this example it is useful to note that because the moment arm of the muscle force is small (0.05 m) we have to exert a large force in the muscle to balance the effect from the weight because it is acting at a much longer moment arm (0.45 m).

Probably the simplest way to understand clockwise and anti-clockwise moments in action and their relationship with equilibrium is to imagine the seesaw you probably sat on as a child. In order to balance the seesaw it was necessary to move either person further in or further out from the center of the device (the fulcrum or pivot point). The central point of the seesaw in this case would be the axis of rotation. The weight of each person sitting on the seesaw created the **forces** and the distances from the center of the seesaw at which each person sat created the **moment arm**. One person would create a clockwise rotation (moment) of the seesaw and the other would create an anti-clockwise rotation (moment). In order to move the seesaw or balance the seesaw you either had to move in, move out or push off the ground with a force (hence changing the moment or torque created and moving the seesaw up or down).

Throughout human movement and exercise science there are numerous situations of the use and application of torques and levers. For example, the torques created at the joints during the pull phase in swimming; the torques on the lower back during the golf swing; the levers and torques created by the canoeist and paddle in white water slalom; and the torques needed in the limbs of the soccer players effectively to kick a soccer ball with both speed and accuracy. In many of these examples it is often desirable both to increase and decrease the torques that are created. The use and application of levers is one example where the consideration of torque is clearly applied and within human movement there are many applications of different types of levers. These will be considered more carefully in section C6.

C2 NEWTON'S LAWS OF MOTION – ANGULAR MOTION

Key Notes

Newton's first law of motion (angular analog)

The angular momentum of a body remains constant unless a net external torque is exerted on the body. The angular momentum (L) of a body can be determined by the moment of inertia (I) multiplied by its angular velocity (ω). The moment of inertia of a body is described as the reluctance of an object to start or stop rotating or change its state of motion. Moment of inertia of a body is calculated from the distribution of mass (m) about an axis of rotation (r). Moment of inertia = mass (m) × radius^2 (r^2). The further away from the axis of rotation a mass is distributed the larger will be the moment of inertia.

Application

The ice skater, in a jump, holds the arms close to the body during a pirouette move (rotation about the longitudinal axis) which reduces the moment of inertia of the body about this axis. This offers less resistance to a change in its state of rotational motion (about this axis). If the skater had created an amount of angular momentum before he/she left the ice this angular momentum (in the absence of an external torque or force) would remain constant. Since angular momentum = moment of inertia × angular velocity a reduced moment of inertia would result in an increased angular velocity (i.e., more rotations in a given time).

Newton's second law of motion (angular analog)

When a torque acts on an object the change in angular motion (angular momentum) experienced by the object takes place in the direction of the torque and this is proportional to the size of the torque and inversely proportional to the moment of inertia of the object. Algebraically this is expressed as $T = I \times \alpha$ (where T = torque, I = moment of inertia and α = angular acceleration).

Application

In the case of arm curling with weights, the biceps brachii muscle applies a torque to the lower arm (the forearm). Depending on the moment of inertia of the arm and the weights this torque may cause an acceleration of the arm (anti-clockwise). The amount of this acceleration is dependent on the moment of inertia offered in resistance to this movement. The smaller the moment of inertia the greater will be the acceleration for a given applied net torque.

Newton's third law of motion (angular analog)

Whenever an object exerts a torque on another there will be an equal and opposite torque exerted by the second object on the first.

Application

The torque created on the upper leg by the hip flexors during the kicking action in soccer will create an equal and opposite torque that is exerted on the pelvis. This has important implications for hamstring injury. The torque that is created on the body by the shoulders and hips during the backswing in golf will create a reaction torque acting in the lower back (lower back injury implications).

As we have seen in section B2, Newton's first law of motion relates to situations where forces are balanced and the net effect of external forces acting on an object is zero. As a reminder, Newton's first law, which related to linear motion, states the following.

Newton's first law of motion

Every object will remain at rest or continue with uniform motion unless it is acted upon by an unbalanced force.

This law can also be applied to the linear momentum of a body in that it is also true that the momentum (mass × velocity) possessed by a body is constant in the absence of any external force. This law is saying that a body will either stay at rest (with no momentum) or keep moving (with a constant momentum) unless it is acted upon by an external force. An object in motion that is not affected by a net external force will have a constant linear momentum because it has a constant linear velocity and a fixed mass.

In angular terms this law can be reworded as follows and can be made applicable to angular motion:

The angular momentum of a body remains constant unless a net external torque is exerted upon the body.

In this application we are dealing with rotation and rotational effects and the term **force** has therefore been replaced by the term **torque** (a twisting or turning moment). In this context, it is worth identifying that in the application of this law to rotation it has purposefully **not** been stated in the context of **constant angular velocity** (like in the case of the linear analog) because as we will see in the rotational application for the human body (which is not a rigid body) this is not necessarily the case.

The **angular momentum** (usually represented by the **symbol L**) of a body or object can be expressed by the **moment of inertia** (the reluctance of the body to start, stop, or change its rotational state) of the body multiplied by its **angular velocity**.

$$\text{Angular momentum} = \text{Moment of inertia} \times \text{Angular velocity}$$
$$L = I \times \omega \; (\text{kg.m}^2/\text{s})$$
$$L = I\omega$$

Angular momentum is measured in the units of **kilogramme meter squared per second (kg.m²/s)**. The moment of inertia of an object is the reluctance of the object to start or stop rotating, or to change its state of rotation. **Moment of inertia** is measured in the units of **kilogram meter squared (kg.m²)**. The concept of moment of inertia will be covered in more detail in section C3 but for the purpose of understanding the angular analogs of Newton's laws it is necessary to provide a brief description here.

The moment of inertia of an object refers to the object's ability to resist rotation. The larger the moment of inertia the more the object will resist rotation. Similarly, the smaller the moment of inertia of an object the less will be its resistance to start, stop, or change its rotational state. The **moment of inertia** is calculated from the distribution of **mass (m)** about an axis of **rotation (r)**. It can be expressed mathematically as: $I = mr^2$. The moment of inertia of a body is related to a specific axis of rotation and there will be different moment of inertia values for each axis that the body is rotating about. For example, there may be a moment of inertia of the whole body about a longitudinal axis or about an anterior–posterior axis. Also, moment of inertia can be expressed for individual parts or individual segments of a body (i.e., the upper leg can have a moment of inertia about the hip joint axis of rotation or the lower leg a moment of inertia about the knee joint axis of rotation).

In order to understand moment of inertia in a more detail, *Figs C2.1* and *C2.2* show moment of inertia conditions for different positions of the body and different axes of rotation.

In *Figs C2.1* and *C2.2* it can be seen that when the mass is distributed much closer to the axis of rotation the moment of inertia is reduced. In *Fig. C2.2* (Example A) the arms are held close in tightly around the body. This has the effect of distributing the mass (i.e., the mass of the arms) much closer to the axis of rotation (the longitudinal axis). This reduces the moment of inertia of the body about this axis of rotation. In the case of Example B the arms are held outward

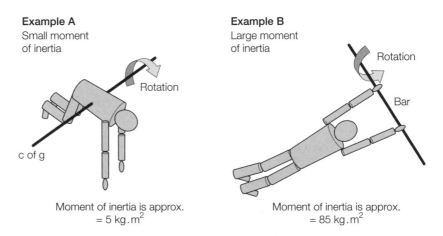

Example A
Small moment
of inertia

Example B
Large moment
of inertia

Moment of inertia is approx.
$= 5 \ \mathrm{kg.m}^2$

Moment of inertia is approx.
$= 85 \ \mathrm{kg.m}^2$

Note: moment of inertia (I) is the distribution of mass about the axis of rotation. In Example A, which has a small moment of inertia, the mass is tightly collected around the axis of rotation (which is through the center of gravity). However, in Example B the mass is distributed away from the axis of rotation (which is the bar) and the moment of inertia is much larger

Fig. C2.1.   Moment of inertia (transverse axis of rotation)

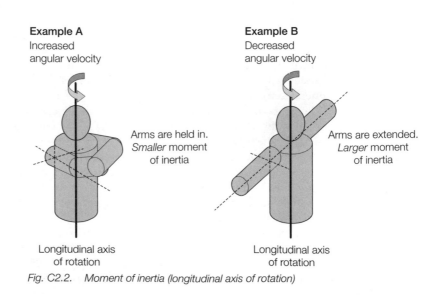

Example A
Increased
angular velocity

Example B
Decreased
angular velocity

Arms are held in.
Smaller moment
of inertia

Arms are extended.
Larger moment
of inertia

Longitudinal axis
of rotation

Longitudinal axis
of rotation

Fig. C2.2.   Moment of inertia (longitudinal axis of rotation)

and this places the mass further away from the axis of rotation. This increases the moment of inertia about this axis.

The angular analog of Newton's first law states that the **angular momentum** of a body will remain **constant** unless the body is acted upon by an **external torque**. At this point it is particularly important to express that angular momentum is related to a particular axis of rotation. If the body that is being considered is made up of several parts then the total angular momentum is the sum of all the individual momenta of each body part (that is acting about the same axis of rotation).

In angular terms the **angular momentum** can also be expressed by multiplying the **square of the distance** of the object from the axis of rotation (r^2) by the **mass** of the object (**m**) and its **angular velocity** (ω).

$$L = mr^2 \times \omega$$

Since mr^2 = the **moment of inertia** of an object then the **angular momentum** is the **moment of inertia** (**I**) multiplied by the **angular velocity** of the object (refer to the equation for angular momentum (**L = Iω**) shown previously). This angular momentum will also occur about a particular axis of rotation.

Now imagine a diver taking off from the springboard in an attempt to perform a double somersault before he/she enters the water. In order to perform this effectively the diver will create angular momentum while still in contact with the springboard. For example, he/she will have applied a force to the board which will create a torque or twisting moment on the body (because of its position in relation to the center of gravity or in this case the axis of rotation). This will enable the diver to leave the board with angular momentum. *Fig. C2.3* shows this in more detail.

Once the diver is airborne and in the absence of any external torque they will have a constant angular momentum (Newton's first law). This angular momentum will remain the same throughout the dive. Hence, in order to perform the necessary somersaults the diver will need to adjust his/her moment of inertia. The basic somersault takes place about the transverse axis (and in the sagittal plane) and the angular momentum at board contact will be created about this axis of rotation (although it could also be present about other axes of rotation). Since the amount of somersaulting angular momentum is constant and the diver is able to change his/her moment of inertia (by tucking up like a ball or by extending the limbs), the angular velocity must change in order to maintain this constant

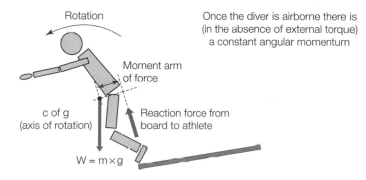

Note: in this example the c of g is located outside the body

Fig. C2.3.   Angular momentum and torque in diving (external forces acting)

angular momentum principle described by Newton's first law. If the diver reduces his/her moment of inertia (tucks up like a ball) the angular velocity will increase and he/she will be able to perform more somersaults in a short space of time (i.e., higher angular velocity). Conversely, if the diver wanted to enter the water in a straight and controlled position (with limited rotation) he/she would straighten the body and increase the moment of inertia. This would slow down their rotation (i.e., reduce the angular velocity). During all this activity the angular momentum of the diver will remain constant (*Fig. C2.4* helps to show this diagrammatically).

$L = I \omega$ (constant angular momentum in the absence of any external torques)

$L = \uparrow I$ and $\downarrow \omega$ (increased moment of inertia and decreased angular velocity)

$L = \downarrow I$ and $\uparrow \omega$ (decreased moment of inertia and increased angular velocity)

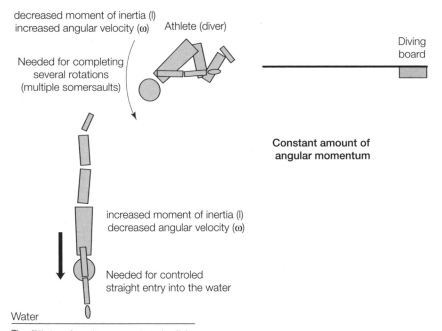

Fig. C2.4. *Angular momentum in diving*

Angular momentum at take-off (equals zero)

Newton's first law of motion states that in the absence of angular momentum and in the absence of any external torque the amount of angular momentum will remain constant (i.e., zero before take-off = zero after take-off). **If this is true, how is it that a diver without any angular momentum at take-off can still initiate twists and turns in the air during the dive?** Consider the case of the diver in *Fig. C2.3* and in particular what would happen if the athlete had zero angular momentum at take-off and yet the athlete was still required to twist and turn during the dive. This zero angular momentum would be where the reaction force (from the board) was applied directly through the center of gravity of the athlete (i.e., not at a distance (moment arm) from the axis of rotation). In this example the diver would leave the board with no angular momentum. The athlete could initiate twists and turns by using asymmetrical or symmetrical movements of the arms or legs. For example, the diver could pull one of his/her arms across the chest in a rotational movement about the longitudinal axis. This would create a certain angular momentum (for part of the body and an angular momentum in the opposite direction for the remainder of the body (i.e., summation of zero)) about a certain axes of rotation (i.e., the longitudinal axis) and in a certain direction (i.e., clockwise or anti-clockwise). The arm would have a small moment of inertia and would be moving with a large angular velocity. Hence, the body would have some angular momentum about this axis (because as we have seen before the equation for angular momentum is $L = I\omega$). However, since the diver had no initial angular momentum (at take-off) about the longitudinal axis, Newton's first law says that this directional angular momentum must be counteracted by angular momentum in an opposite direction (i.e., making the total angular momentum about this axis equal zero (the net angular momentum)). In diving the twist that is initiated by the asymmetrical or symmetrical arm movements is often counteracted by the legs in changing from a rotational twist to a pike and then back to a straight controlled extended position for entry into the water. In this application, there would be potentially more control of the rotation in the air and at entry to the water because the athlete controlled the amount of angular momentum he/she created by moving the arm. Hence, since this must be balanced by angular momentum in an oppo site direction the athlete will experience opposite controlled angular momentum to balance out the equation (because they had zero angular momentum to begin with). Therefore, just before contact with the water (which would be an external force) the net angular momentum would be equal to zero which is consistent with the angular analog of Newton's first law of motion.

Perhaps an easy way to remember and apply this is to imagine a cat falling out of a tree backwards. As the cat falls it will have zero angular momentum yet somehow it is still able to land on its feet (by twisting and turning in the air). As the cat falls backwards it brings its front legs close to the head. This reduces the moment of inertia about the longitudinal axis. This causes a large rotation (angular velocity) of the upper body clockwise (approximately 180°). At the same time the cat also extends the lower legs out away from the body creating a large moment of inertia (about the longitudinal axis). This results in only a small amount of angular rotation of the lower body in an opposite anti-clockwise direction (approximately 5°). This balances the angular momentum equation for both clockwise and anti-clockwise rotation about this specific axis. The upper body is now facing the ground and is ready for landing. However, in order to get the lower part of the body facing the ground the cat does exactly the same but opposite (it reduces the moment of inertia of the lower body and increases moment of inertia of upper body). This gives the lower body large clockwise rotation and the upper body only a small amount of anti-clockwise rotation. The final result is that the cat lands on both its front and back feet, and it does so with a net angular momentum of zero. However, like the diver, it was clearly able to initiate a twist and turn in mid-air without any initial angular momentum. Although this is not directly applicable to human movement, it demonstrates that Newton's first law applies to angular motion.

Newton's second law of motion

As a reminder, Newton's second law for objects in linear motion was concerned with bodies that are subjected to unbalanced forces and it is stated as follows:

When a force acts on an object the change in motion (momentum) experienced by the object takes place in the direction of the force and this is proportional to the size of the force and inversely proportional to the mass of the object.

In the angular analog of this law we replace the term **force** with **torque**, the term **mass** with **moment of inertia** and the term **momentum** with **angular momentum**. This law can now be reworded as follows:

When a torque acts on an object the change in angular motion (angular momentum) experienced by the object takes place in the direction of the torque, and this is proportional to the size of the torque and inversely proportional to the moment of inertia of the object.

This can be expressed algebraically by the equation $T = I \times \alpha$

where
T = net torque
I = moment of inertia
α = angular acceleration

Remembering that Newton's second law is concerned with unbalanced forces or torques we can see that if we apply a torque to an object, and the result of the **net torque** acting on the object is **not zero**, the object will have **angular acceleration** (Newton's first law: where we apply an external torque and cause a rate of change of angular velocity, i.e., an angular acceleration). The torque created a twisting or turning moment and the object moved with an angular velocity (i.e., it rotated). In the application of a net torque that is not zero the object would accelerate with angular acceleration (because it is rotating). The angular acceleration of the object will take place in the same direction as the applied torque. The amount of **angular acceleration** will be dependent on the amount of **applied torque** and the **moment of inertia** of the object. As we have seen, the human body is not a rigid body and as such it does not have a constant moment of inertia. The larger the moment of inertia of the object the less angular acceleration it will have for a given applied torque. Conversely, for a given applied torque (or net torque of greater than zero), the smaller the moment of inertia of the object the greater the angular acceleration it will have. *Fig. C2.5* helps to illustrate this in more detail with a diagram.

Considering *Fig. C2.5* we can see that the biceps brachii muscle in the upper arm exerts a force at a perpendicular distance to the axis of rotation (the elbow joint). This creates a torque (a twisting and turning moment) in an anti-clockwise direction. This torque causes the elbow joint to flex and the lower arm to rotate in an anti-clockwise direction (the same direction as the applied torque). As the arm rotates it will have a certain angular velocity. The arm was initially held stationary and the net torque applied to the arm in this case is not zero (i.e., the muscle will exert a torque on the arm). Hence the arm will accelerate with an angular acceleration anti-clockwise. The rate of change of angular velocity of the arm (or the angular acceleration) will be dependent upon how much torque is applied to the arm. The amount of angular acceleration will also be dependent on

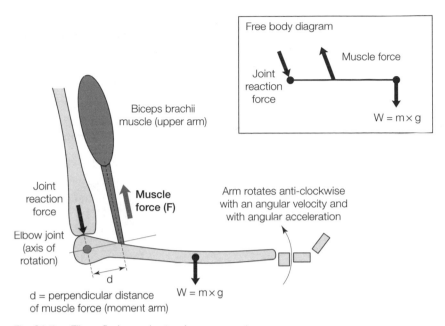

Fig. C2.5.   Elbow flexion and extension movement

the moment of inertia of the arm. As the arm is rotated in an anti-clockwise direction the moment of inertia of the arm will resist the rotation which is created by the torque. The greater the moment of inertia of the arm the smaller will be the angular acceleration of the arm. Conversely, the smaller the moment of inertia the larger will be its angular acceleration. Hence for a given applied torque there would be different angular accelerations for different moment of inertia values for the arm (like in the case of different sized and different shaped arms). This has important implications when exercising using weights and will be covered in more detail in Section C3.

In this context it is important to point out that, as we have seen, **angular momentum** (which is the object's **moment of inertia** multiplied by its **angular velocity**) is a **vector** quantity (i.e., it has **magnitude** and **direction**). Usually, we are normally concerned with rotation about one axis of rotation and it is applicable therefore to refer to angular momentum about a single origin or a single axis of rotation. In this manner we can consider it as a scalar quantity where we refer to its direction as either positive (anti-clockwise rotation) or negative (clockwise rotation). In addition, it is worth repeating that the total angular momentum of a body about any axis of rotation is made up by adding all the angular momenta of the various parts or segments of the body which are rotating about that axis. Within biomechanics this has important implications for understanding human movement and in more complex analyses the study of angular momenta about multiple axes of rotation is required.

As we have seen, a net torque (that is not zero) that acts on an object will cause an angular acceleration of the object in the direction of the net torque. The amount of net external torque will equal the rate of change of angular momentum (i.e., from the angular analog of Newton's second law).

$$\text{Torque}_{(net)} = \frac{\text{Change in angular momentum}}{\text{Change in time}}$$

$$\text{Torque}_{(net)} = \text{Rate of change of angular momentum}$$

The change in angular momentum of an object can be determined by examining the initial and final angular momentum possessed by the object:

$$\text{Change in angular momentum} = \text{Angular momentum}_{(final)} - \text{Angular momentum}_{(initial)}$$

Thus we can now include this in the equation for torque:

$$\text{Torque}_{(net)} = \frac{\text{Angular momentum}_{(final)} - \text{Angular momentum}_{(initial)}}{\text{Change in time}}$$

Mathematically this can be expressed as:

$$T_{(net)} = \frac{L_{(f)} - L_{(i)}}{t_2 - t_1}$$

Rearranging this equation produces the following:

$$T_{(net)} \times (t_2 - t_1) = L_{(f)} - L_{(i)}$$

This can now be expressed as the equation for angular impulse:

$T_{(net)} \times t$ = Angular impulse (torque multiplied by time)
$L_{(f)} - L_{(i)}$ = Change in angular momentum (final – initial)
Angular impulse = Change in angular momentum

This equation has important implications for the effective execution of rotational movements with human motion.

Application

Considering the diver in *Fig. C2.3* it is possible to see that the diver creates angular momentum by applying a force at a distance from an axis of rotation. Angular momentum is moment of inertia multiplied by angular velocity. As the diver creates torque he/she will create rotation. The amount of this rotation will depend on the torque created and the moment of inertia of the body. As we have seen, a net external torque that is not zero causes angular acceleration. Angular acceleration is the rate of change of angular velocity. The greater the angular acceleration of the body the greater the rate of change of angular velocity of the body (i.e., the more rotations we can create in a shorter space of time). As the diver creates torque he/she applies a force on the board for a certain length of time. In angular terms, since the reaction force is applied at a distance from the axis of rotation (the moment arm), a torque will be created. Now, as we have seen, the angular acceleration possessed by the object (the rate of change of angular velocity) is related to the torque and the moment of inertia. If the athlete has a large moment of inertia as he/she creates the torque, the rotational component will be small. As the rotational component of the action is small (i.e., reduced angular velocity) the athlete will be able to create a torque for a longer period of time. This is achieved because they will have less rotation effect and would be able to stay in contact with the board for longer before the rotation would cause them to have to leave it into the dive. This application of torque for a longer period of time will create a greater change in angular momentum (i.e., they will have more angular momentum). Consequently, the more angular momentum they have when they leave the take-off board into the dive, the more they are able to rotate in the air. For example, because they have a large angular momentum they can reduce their moment of inertia and perform more rotational somersaults in a given time (i.e., they will be rotating faster because once they leave the board and in the absence of external torque the amount of angular momentum they have will be constant).

This ability to create a torque in a controlled manner and to be able to apply this torque for a long period of time results in a greater change in angular momentum. If they had zero angular momentum before they started to apply the torque it follows that they will have more angular momentum the longer time that they can apply the torque (**Angular impulse = Change in angular momentum**). This has very important implications within human movement and is applicable in many examples within sport where the athletes use angular momentum and rotational movements to generate both linear and angular velocities. For example, the rotational component used by modern javelin throwers; the rotational running across the circle technique used in discus throwers; the golfer using rotational movements of hips and shoulders to generate torque that is transferred to the club to accelerate it quicker; and the tennis player serving with rotational movements about the longitudinal axis in order to impart large amounts of spin and velocity to the ball. All these examples and more utilize the angular analog of Newton's second law of motion.

Newton's third law of motion

In linear terms this law is stated as follows:

Whenever an object exerts a force on another there will be an equal and opposite force exerted by the second object on the first.

In angular terms this can be re-written as follows:

Whenever an object exerts a torque on another there will be an equal and opposite torque exerted by the second object on the first.

In the angular analog of this law the term **force** has been replaced by the term **torque**. Torque as we know is a turning or twisting moment which causes an angular acceleration of an object. In the context of this third law of motion it is important to remember (as with the linear version) it is the forces or torques that are equal and opposite and not the net effect of the forces or torques. The equal and opposite torques will act on each body differently (because the two bodies are different) and they will both act about the same axis of rotation. As with the linear analog of this law of motion it is important to remember the consideration of **external** and **internal** force or torque. If the body is in equilibrium (when the algebraic sum of all the torques acting is zero) under the action of external torque both the external and internal torque systems are **separately** in equilibrium. **In considering the net effect of external torques or forces acting on a body we would summate only the external torques that are acting on that body.** Hence, if body A exerts an external force or torque on body B, body B will exert an external reaction force or torque on body A. However, in considering the net effect on body B we would only consider the external forces or torques acting on body B (which is the external force or torque provided from body A – this is of course ignoring any other external forces such as gravity).

In the case of the arm in *Fig. C2.5* which is undergoing a flexion movement (an anti-clockwise rotation of the lower arm (the forearm)) the torque created by the biceps brachii muscle exerts a torque on the lower arm. This torque causes the lower arm to rotate in an anti-clockwise direction. The reaction torque to this will be a torque in the opposite direction created on the upper arm. As the torque is applied to the arm (to accelerate it anti-clockwise) there will be an equal and oppo-site torque acting on the upper arm. This is why when you conduct an arm curl during weightlifting you can feel the stress/strain in the upper arm. Because the torques are equal and opposite and act independently on two different bodies (the torque of the biceps acts on the lower arm (anti-clockwise) and the reaction torque acts on the upper arm and the net result is zero (equal and opposite torques)) but movement takes place because the torque acting on the lower arm is considered as an external torque acting on that body (Newton's second law). As we have seen this movement is dependent upon the moment of inertia of the object. The torque created is dependent upon the force applied and the moment arm at which the force is applied (the perpendicular distance from the axis of rotation). The angular analog of Newton's third law has important implications with human movement and in particular for consideration of injury. *Fig. C2.6* shows some further example of the action–reaction torque within sport and exercise.

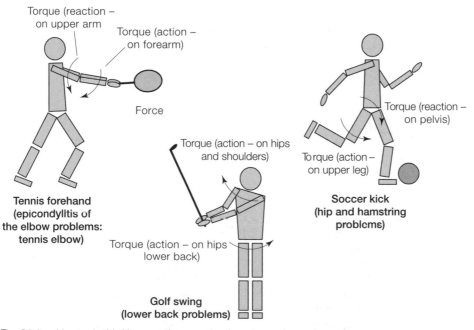

Fig. C2.6.   *Newton's third law – action–reaction (angular analogs – torque)*

Considering *Fig. C2.6*, it is possible to see three examples of the action–reaction torque situation: the tennis forehand; the soccer kick, and the golf swing. In the tennis forehand example, the torque provided by the muscles to the forearm will create an equal and opposite torque acting on the upper arm. This has impor-tant implications for the development of medial or lateral epicondylitis ("tennis elbow" injury). In the soccer kick example, the torque is created on the upper leg as it swings through in an anti-clockwise direction (created on the upper leg by the hip flexors). This causes a reaction torque at the pelvis. This has important implications for common hamstring injury problems experienced in soccer. Finally, in the golf swing the torque that is created by rotating the hips and shoulders through the backswing and downswing (both clockwise and anti-clockwise directions) causes a reaction torque on the lower back which has potential for the development of low back pain.

All the angular analogs of Newton's laws of motion are critical in the understanding and effective execution of human movement and performance. In addition they serve as a strong basis for more complex analyses of multi-axial and rotational movements that are common to many sporting actions.

C3 MOMENT OF INERTIA AND THE CONSERVATION OF ANGUALR MOMENTUM

Key Notes

Moment of inertia

Is the reluctance of an object or body to start rotating or change its state of rotation (i.e., speed up or slow down). It is measured as the product of the mass of the body and the distance of the mass from the axis of rotation squared ($I = mr^2$). Double the axis of rotation of the mass and you will quadruple the moment of inertia value. The moment of inertia value can be changed by redistributing the body mass about the axis of rotation (such as in the case of "tucking" or extending the body during a somersault).

Parallel axis theory

This is a method of calculation that is used to determine whole body or body part (such as the legs) moment of inertia. The parallel axis theory states that the moment of inertia about a parallel axis is equal to the moment of inertia of the segment in its original position plus the product of the mass and the square of the distance between the parallel axis.

Angular momentum

Is the product of the moment of inertia and angular velocity. It remains constant in the absence of any external force and the principle of conservation of angular momentum is derived from the angular analogue of Newton's first law of motion. In flight the angular momentum possessed by a body is constant and gravity is not considered to be an external force affecting the amount of angular momentum present in a system. In flight gravity will act through the center of gravity point and because the moment arm created will be zero – no external torque or moment is present. Athlete's can transfer angular momentum to different body parts and different axis and planes of movement/rotation. Athlete's can also initiate angular momentum of some body part from a state of having no whole body angular momentum (such as dropping an arm during the flight phase in diving). However, this angular momentum must be counter balanced by angular momentum of some other body part in the opposite direction (principle of conservation where angular momentum was zero to begin with and must be zero at the end). Individual segment angular momentum is determined by calculating the angular momentum of the segment about its own center of gravity plus the angular momentum of the segment about the whole body's center of gravity. Whole body angular momentum is the summation of all the individual segment momenta. Angular momentum is affected by the mass of the body, the distribution of this mass and its angular velocity and it is plane and axis specific. In addition, it has significant implications for performance in sport.

This section is concerned with the calculation of the moment of inertia and the interpretation of angular momentum and it is related to the topic of angular kinetics (concerned with forces and the effect of these forces on angular movement). As we have observed from Section B3 the linear momentum possessed by the body is defined as the product of its mass multiplied by its linear velocity and it is measured in the SI units of kg.m/s (kilogramme meter per second).

Linear momentum = mass × velocity (kg.m/s)

In angular terms, **angular momentum** is defined as the product of the moment of inertia of the body multiplied by its angular velocity, which is caused by the body's (or part of the body) mass and its distribution of mass in a circular motion about an axis of rotation. It is measured in the SI units of kg.m^2/s.

Angular momentum = moment of inertia × angular velocity (kg.m^2/s)

The angular momentum of an object about a particular axis will remain constant unless the object is acted on by an unbalanced eccentric force (such as another athlete, a ball, or an implement) or a couple (a pair of equal and opposite parallel forces).

The value in understanding angular momentum and its concepts within sport and exercise can be seen by considering how a soccer player learns to kick a ball effectively; how a golfer transfers angular movement of a club to the golf ball or indeed how a sprinter manages to move the limb quickly through the air in order to make the next contact with the ground that is needed to push off and move forward with speed.

Moment of inertia   The **inertia** of an object is referred to as the resistance offered by the stationary object to move linearly and it is directly proportional to its mass. The **moment of inertia**, however, is defined as the reluctance of an object to begin rotating or to change its state of rotation about an axis. Moment of inertia is related to the mass of the object (body or body part) and the location (distribution) of this mass from the axis of rotation. Without specific reference to a particular axis of rotation the moment of inertia value has little meaning.

Fig. C3.1 shows the moment of inertia values in some selected athletic situations during sport. It is important to reiterate that the moment of inertia values are specific to the axis of rotation about which the body is moving (e.g., either the center of gravity (transverse) axis of the body as in diving or the high bar (transverse axis) in gymnastics as portrayed in *Fig. C3.1*). Basically, as can be seen from *Fig. C3.1*, the greater the spread of mass from the rotation center (axis) the greater will be the moment of inertia. Note that the largest moment of inertia value is determined when the body is in the position when it is rotating about the wrist (hands) and the whole body is extended (i.e., the mass is distributed as far as possible away from the axis of rotation which in this case is about the hands (an axis of rotation through the hands)). Therefore, the moment of inertia of an object or body about a particular axis depends upon the mass of the object or body and the distribution of this mass about the axis of rotation. Specifically, an equation for moment of inertia about an arbitrary axis A can be give as:

Moment of inertia = mass × radius^2 (kg.m^2)
(About an axis A)
$$I_A = m \times r^2$$

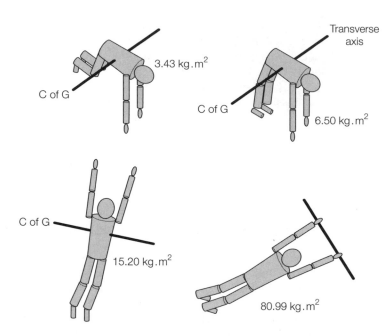

Fig. C3.1. Moment of inertia values in diving and gymnastic positions (recalculated from originals in Hay 1978, p. 147). (Figures redrawn from Hay, J. G. (1978) p 147, which unfortunately is now out of print [Hay, J. G. (1978) The Biomechanics of Sports Techniques. Prentice Hall, Inc. Englewood Cliffs, NJ].)

Fig. C3.2 shows the moment of inertia calculation for a 15 kg mass (concentrated point mass) when it is rotating about two different axes of rotation (6 m and 4 m from the same axis of rotation). This clearly indicates that as the rotation axis changes the mass is located farther away from the axis and as a result the moment of inertia changes. In *Fig.* C3.2 it is possible to see that when the 15 kg mass is moved closer to the axis of rotation (4 m away instead of 6 m away) the moment of inertia value decreases. This has important implications in sport and again looking at *Fig.* C3.1, it can be seen that the smallest moment of inertia value is achieved when the body forms a tight "tuck" about the center of gravity axis of rotation. In this case the mass distribution is close to the axis of rotation (the center of gravity) and the moment of inertia value is the least (3.43 kg.m² as opposed to 80.99 kg.m² in the extended position). Note that the calculation shown in *Fig.* C3.2 only works for a concentrated point mass such as the 15 kg mass used

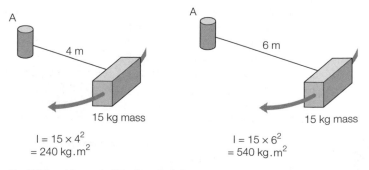

Fig. C3.2. Moment of inertia calculation

in this example. The distribution of mass in the human body is not considered to act as a concentrated point mass.

It is also important to note that because the equation for moment of inertia contains the r^2 (distance of the mass from the axis of rotation squared) component it is clear that changing the position of the mass has a much greater effect than changing the actual mass. For example, **doubling the distance** from the **axis of rotation** for a **given mass** will **quadruple the moment of inertia value**. This has important implications for human movement within sport and exercise.

Moment of inertia calculation (from *Fig. C3.2*)

$$I_A = mr^2$$

where

I_A = moment of inertia (kg.m²) about a particular axis A
m = mass (kg)
r = radius or distance of mass from axis of rotation (m)

For the 6m distance

I = mr^2
= 15×6^2
= **540 kg.m²**

For the 4 m distance

I = mr^2
= 15×4^2
= **240 kg.m²**

The moment of inertia value of regular shaped bodies about any arbitrary axis A, is determined by taking a number of measurements of the mass distribution about the axis of rotation and then by summating the result the moment of inertia of the whole body is determined. *Fig. C3.3* illustrates this is in a mathematically regular shaped body.

Moment of inertia $_{(A)}$ = $m_1r_1^2 + m_2r_2^2 + m_3r_3^2 + ... + m_nr_n^2$
where n = the number of samples taken
$$I_A = \sum m_n r_n^2$$

This process is difficult to do mathematically and for complex shapes, like the limbs of the human body, these values have been computed by researchers so that they are available for use by others. The data presented in *Table C3.1* represents values of moments of inertia for human limbs about their own center of gravity. These values can be used in further calculations.

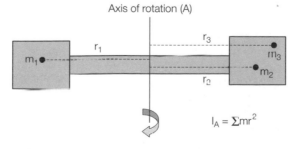

Fig. C3.3. *Moment of inertia calculation*

Table C3.1. Moment of inertia values of selected body segments about the transverse axis through the center of gravity of the segment (adapted and modified from Hay 1978, p. 145).

Segment	Moment of inertia
Head	0.024 kg.m²
Trunk	1.261
Upper arm	0.021
Forearm	0.007
Hand	0.0005
Upper leg (thigh)	0.105
Lower leg	0.050
Foot	0.003

(Adapted and modified from Hay, J. G. (1978) page 145, which unfortunately is now out of print [Hay, J. G. (1978) The Biomechanics of Sports Techniques. Prentice Hall, Inc., Englewood Cliffs, NJ]).

Considering the moment of inertia of the whole body, it is clear (from *Fig. C3.1*) that it is different depending on the axis about which the body is rotating. The moment of inertia of the whole body would be different if it was rotating about a longitudinal axis (such as in the case of a pirouette in ice skating) than it would be if it were rotating about the transverse axis (as in the case of a somersault). Similarly the moment of inertia of sports equipment can also vary for different axes and different planes of rotation. Imagine a young child trying to swing an adult golf club. In this case it is easy to see that the child has to move the hands down the club and shorten the lever in order to have any chance of swinging the club. In effect, the child is reducing the radius of rotation (i.e., the distance of the axis of rotation to the center of gravity of the club) and thus decreasing the moment of inertia (the resistance to change). In essence this is why it is easier for an adult to swing a 9-iron club than it is to swing a longer club such as a driver. The moment of inertia of the 9-iron when it is rotating about the frontal (anterior–posterior) plane axis (although strictly speaking it is not a single planar or single axis movement) is less than that of a driver and hence it is easier to swing. The same principle applies to a soccer kick. The athlete flexes the leg before the kick, thus shortening the radius of rotation and decreasing the moment of inertia so it is easier to accelerate the limb quickly in order to impart greater velocity to the stationary ball. As the leg is brought to the ball it is slightly extended (more so after contact), which will slow down its rotation and allow more control for an effective and accurate contact.

Determination of whole body or combined segment moment of inertia (parallel axis theorem)

The **parallel axis theorem** is one method that allows us to be able to calculate the moment of inertia of the whole body or the sum of several body parts (such as the leg of a soccer player before contact with the ball) about a particular axis and plane of rotation (e.g., rotation at the hip in the sagittal plane (transverse axis) as in the soccer kick). Although it is important to add that these movements are never truly single axes or single planar activities, and even the soccer kick would involve rotation about two or more planes and axes of movement (i.e., it is a three-dimensional movement). However, the theorem identifies that the moment of inertia about an axis that is parallel to the axis for which the moment of inertia was derived equals the moment of inertia of the body segment in its original position (i.e., about its own c of g) plus the product of the mass and the square of the perpendicular distance between the parallel axes. Thus the equation for the moment of inertia of the body or body segments is re-written as follows:

Parallel axis theory of calculating moment of inertia

$$I_A = I_{C of G} + md^2.$$

where

I_A = Moment of inertia of a body about an axis through a point, A
$I_{C of G}$ = Moment of inertia about a parallel axis through the center of gravity of the body or segment
m = the mass of the body or segment
d = the distance between the parallel axes

Fig. C3.4 illustrates this theory in more detail and helps to show how the parallel axes are determined in the example of the leg in the position of knee flexion before contact with the ball during a soccer kick. Furthermore, *Table C3.1* identifies the moment of inertia values for the selected body segments which represents the moment of inertia through an axis that is parallel to the axis of consideration and that is through the segment's center of gravity ($I_{C of G}$).

Calculation of moment of inertia of a leg segment (using the parallel axis theory)

Considering *Fig. C3.5*, it is possible to see that in the case of the leg in this position (which would be before the contact phase in a soccer kick) the center of gravity of each segment is given as a distance from the center of rotation (i.e., the hip joint). When the position of each segment's center of gravity is given it is possible to use the parallel axis theory to determine the moment of inertia about any axis if the moment of inertia of each segment about their respective parallel axis is known. This is an alternative method to using the radii of gyration measure (which is essentially used for single segment moment of inertia calculations) and allows the calculation of the whole body or whole limb moment of inertia (such as in the case of the leg).

Parallel axis theory of calculating moment of inertia

$$I_A = I_{C of G} + md^2$$

The above equation is applied individually to calculate the moment of inertia of the upper leg, lower leg, and foot (*Fig. C3.5*) separately and the three values are

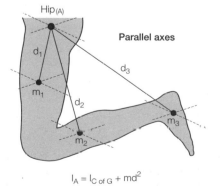

$$I_A = I_{C of G} + md^2$$

Fig. C3.4. Parallel axis theorem to determine whole leg moment of inertia (transverse axis)

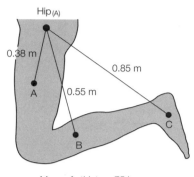

Mass of athlete = 75 kg

Fig. C3.5.   Moment of inertia of leg during a soccer kick (transverse axis – sagittal plane)

then summated to represent the moment of inertia of the whole leg in this position. However, in order to determine the mass of each segment in this 75 kg athlete it is necessary to use the anthropometric data presented in the work of Winter (1990). *Table C3.2* illustrates these values in more detail.

Table C3.2.   Anthropometric data where m is body mass in kg (derived from data presented in Winter 1990, p. 56)

Hand	$0.006 \times m$
Forearm	0.016
Upper arm	0.028
Forearm and hand	0.022
TOTAL ARM	0.050
Foot	0.0145
Lower leg	0.0465
Upper leg	0.100
Foot and lower leg	0.061
TOTAL LEG	0.161

(The values in the table are derived (and reproduced with permission) from data presented in Winter, D. A. (1990) Biomechanics and Motor Control of Human Movement (2nd edition). Wiley-Interscience Publishers, New York. (3rd edition published 2004))

Determine the moment of inertia of the leg in this position (*Fig. C3.5*)

A.   Upper leg

$I_A = I_{C\,of\,G} + md^2$
$I_A = 0.105 + ((0.100 \times 75) \times 0.38^2)$
$I_A = 0.105 + 1.083$
$I_A = \underline{\textbf{1.188 kg.m}^2}$ (upper leg moment of inertia)

B. Lower leg

$I_A = I_{C\,of\,G} + md^2$
$I_A = 0.050 + ((0.0465 \times 75) \times 0.55^2)$
$I_A = 0.050 + 1.055$
$I_A = \underline{\textbf{1.105 kg.m}^2}$ (lower leg moment of inertia)

C. Foot

$$I_A = I_{C\,of\,G} + md^2$$
$$I_A = 0.003 + ((0.0145 \times 75) \times 0.85^2)$$
$$I_A = 0.003 + 0.786$$
$$I_A = \underline{0.789\ kg.m^2}\ \text{(foot moment of inertia)}$$

Total moment of inertia of leg in this position

$$I_A = I_{A\,(upper\,leg)} + I_{A\,(lower\,leg)} + I_{A\,(foot)}$$
$$I_A = 1.188 + 1.105 + 0.789$$
$$I_A = \underline{3.08\ kg.m^2}$$

The moment of inertia of the leg in this position and about the transverse axis just before kicking a soccer ball would be 3.08 kg.m². It is important to understand that if the athlete could reduce the rotation axis (distribute the mass differently), by flexing the leg more, the moment of inertia would be reduced and the limb would be able to be moved (accelerated) much faster (rotationally) to generate potentially more velocity that could be imparted to the ball. The same principle will apply to a sprinter who wishes to bring the leg through quickly in order to make contact with the ground again, or indeed the golfer while swinging the golf club could bend the elbows to reduce the moment of inertia and hence increase the angular velocity of the swing (reducing the resistance to change). This understanding will now be developed further with specific reference to angular momentum.

Angular momentum

Angular momentum is represented by the letter L or H and is determined by the product of the moment of inertia and the angular velocity (measured in radians/s) of a body or segment. It is expressed in the units of **kg.m²/s** (kilogram meter squared per second).

Angular momentum = moment of inertia (kg.m²) × angular velocity (rads/s)

L = I ω (kg.m²/s)

Consider *Fig. C3.6*, which represents the 15 kg mass that was used previously as an example to calculate the moment of inertia values. In *Fig. C3.6*, the mass is now given an angular velocity of 3.5 rads/s. The angular momentum is the product of moment of inertia (15 × 6²) and angular velocity (3.5) and is expressed as **1890 kg.m²/s.** If there was no angular velocity then there would also be zero

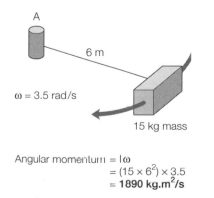

A

6 m

ω = 3.5 rad/s

15 kg mass

Angular momentum = $I\omega$
$$= (15 \times 6^2) \times 3.5$$
$$= 1890\ kg.m^2/s$$

Fig. C3.6.   Calculation of angular momentum

angular momentum. The radius of rotation also determines the angular momentum and is a very important factor because of the mathematical squared component. As we have seen before with moment of inertia, if we double the distance (radius of rotation = 12 m) we quadruple the angular momentum to **7560 kg.m²/s** (which is $4 \times$ the previous 1890 kg.m²/s value).

Angular momentum (as with the moment of inertia) must be expressed with reference to an axis of rotation and the calculation of the whole body angular momentum is the sum of all its individual body segment momenta. For human body segments that rotate about an axis other than their center of gravity the parallel axis theorem is used and this is also applied to angular momentum calculations. We have seen previously that for rotation about an arbitrary axis A, the moment of inertia is:

$$\mathbf{I_A = I_{CofG} + md^2}$$

Hence, angular momentum can also be expressed as:

$$\mathbf{L_A = (I_{CofG} + md^2)\,\omega}$$
$$\mathbf{L_A = I_{CofG}\cdot\omega + md^2\,\omega}$$

for the segment about its own C of G and for the segment about a parallel axis

where

L_A = Angular momentum (kg.m²/s) about axis of rotation A
I_{CofG} = Moment of inertia about an axis through the center of gravity of the segment
m   = the mass of the body or segment
d   = the distance between an axis through point A and a parallel axis through the center of gravity of the segment
ω = angular velocity (rads/s)

The calculation of whole limb (e.g., the leg complex) angular momentum about the hip axis of rotation is determined by summating the angular momentum of individual segments about an axis through their own center of gravity and the angular momentum of the segment about the axis of rotation (i.e., the hip axis). *Fig. C3.7* helps to illustrate this concept in a little more detail. It is important to reiterate that this calculation is axes and plane specific.

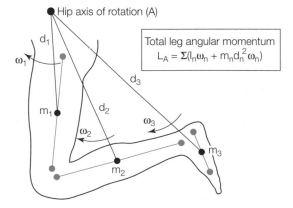

$L_{1\ (about\ hip\ axis)} = I_1\omega_1$ (for thigh segment about its own axis c of g) +
$\quad\quad m_1 d_1^2 \omega_1$ (for segment about hip axis (A))

Fig. C3.7.   Angular momentum of the limb (transverse plane)

Conservation and transfer of angular momentum

During movement of the body the total angular momentum possessed by the body (in the absence of any external force) remains constant. When the body is in flight, gravity is not considered to be an external force that affects angular momentum because it does not cause any rotational component (i.e., it will act through the center of gravity and have a moment arm of zero). This conservation of angular momentum principle is derived essentially from Newton's first law of motion, which in angular terms can be expressed as follows:

> **Every object will remain in a non-rotating state or will continue to rotate about a particular axis with uniform angular velocity unless acted upon by an unbalanced eccentric force or couple.**

As an athlete prepares to jump into the air (such as in the case of a diver on a springboard) the athlete, while in contact with the ground (or board), will generate angular momentum (creating rotation – by applying forces at perpendicular distances to axes of rotation – thus creating reaction torques or moments). Once he/she is in the air (during flight as in the case of the diver) and in the absence of external forces (neglecting gravity effects) the athlete's angular momentum will remain constant. In this respect it is possible for the athlete to change his/her moment of inertia in order to increase or decrease his/her angular velocity. The diver will form a tight tuck around the center of gravity rotation axis (transverse axis – sagittal plane) and the angular rotation (causing somersaults) will increase because the athlete has decreased the moment of inertia of the body. Similarly, if the athlete extends the body the moment of inertia will increase and the angular rotation (angular velocity) will be reduced. This is one of the reasons why the diver will extend suddenly at the end of the dive (i.e., to reduce the angular rotation and enter the water with minimal rotation in a straighter, more controlled, aesthetic position).

Similarly, it is possible to transfer the angular momentum possessed by the body in one axis to another different axis of rotation within the body. This is how divers initiate twist and tilt maneuvers in the air when they only seem initially to have rotation about one axis and in one plane of movement. In addition, it is also possible to see that mechanically divers can initiate angular momentum for one part of the body while in the air (remembering that the overall momentum of the body must remain constant). For example, a diver who leaves the board with zero angular momentum can vigorously move or rotate an arm in a particular axis of rotation. The corresponding effect will be that the athlete will generate angular momentum (for this segment) about this axis and plane of motion. This angular momentum must therefore (because of the conservation of angular momentum principle) be balanced by an angular momentum in an opposite direction by another part or segment (i.e., zero or fixed whole body angular momentum). Hence, it may be the case that the athlete is seen to have to counteract this new angular momentum created in the arms by a simultaneous rotation and movement in the legs. In this context it is important to reiterate the effects of moment of inertia. Segments or body parts (such as the legs) that have large moments of inertia will thus have smaller angular rotations. Therefore, the high angular rotation created by a diver vigorously dropping an arm in mid-flight may be balanced by what appears only a small rotation of the legs (because the legs can have a much larger moment of inertia than the arms and the amount of angular momentum must remain constant).

Fig. C3.8 shows a diagram that helps to explain this principle of conservation of angular momentum in more detail. It is important to remember when considering this diagram that the angular momentum is determined from the product of moment of inertia and angular velocity **(L = Iω)** and it is constant in the absence of any external force or torque. Thus, it can be seen that if **I increases**, then **ω must decrease**. Similarly, if **I is decreased**, then **ω must increase**.

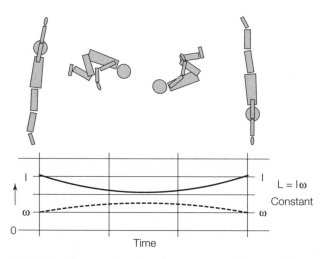

Fig. C3.8.   Conservation of angular momentum (a forward 1½ somersault during diving)

There are many examples in sport and exercise where moment of inertia and angular momentum are important and throughout this section we have seen examples such as soccer kicking, the leg drive in sprinting, the golf swing, and diving. However, the following provides some more examples of these principles and concepts in application.

- The ice skater pirouetting on the ice with the arms out would have a high moment of inertia and small angular velocity. Bringing the arms in will reduce the moment of inertia and increase angular velocity (rotation speed).
- The gymnast performing several somersaults and twists in the air would need to reduce the moment of inertia about different axes in order to increase rotation velocity and have any chance of performing the number of twists and somersaults before landing.
- The downhill racing skier using long skis (having a large moment of inertia) will find it difficult to initiate turns easily (neglecting friction effects). Whereas the slalom skier with shorter skis (smaller moment of inertia) will find turning much easier.
- The young child using a set of shorter shafted golf clubs or the tennis racket with a shorter handle and lighter head will find it easier to play golf or tennis.

- The basketball player performing a "slam dunk" shot is forced to counteract the rotation initiated in the arms by movement in the legs.
- The elite tennis player serving with a bent arm action in order to reduce the moment of inertia of the arm and racket, and increase rotational velocity which results in a faster serve.

C4 CENTER OF GRAVITY AND CENTER OF MASS

Key Notes

The center of gravity

This is the point in an object at which the entire weight of an object is assumed to act. More precisely it is the point at which the force of gravity for the entire object is considered to act such that it would behave in the same way as when the force of gravity is distributed across the entire object.

The center of mass

This is defined as the point which represents the mean position for the concentration of the entire mass of the body.

Imaginary points

The center of gravity and center of mass are imaginary points, and they do not physically exist and hence they cannot be seen.

Human body center of gravity position

The center of gravity position in the human body may continually change (or it may also be stationary). It is a point that can be located within the body or it is a point that can be located outside of the body. The high jumper will allow the center of gravity to pass under the bar during the high jump clearance phase. Stability is the ability of an object to maintain its static equilibrium. The stability of an object is affected by the position of the center of gravity. Stability is achieved when the center of gravity (c of g) of the object lies nearer the lower part of the object. If the vertical line through the c of g falls within the base of support the object is considered to be stable. If the c of g falls outside the base of support then the object is considered to be unstable.

Calculation of human body c of g

The center of gravity of the human body can be calculated by various methods. The center of gravity board is one method that can easily determine the position of the center of gravity in static postures. Alternative methods involving 3D computation using video digitization of body landmarks can determine c of g position used in the analysis of human movement. The center of gravity position in either two or three dimensions or in static or dynamic conditions is determined from the principle of moments. Moments about different axes are taken in order to calculate the whole body center of gravity position/location. Individual segment center of mass data is used in the determination of whole body center of gravity calculations. Different techniques exist that are used to determine individual segment center of mass positions.

Application

An understanding of the center of gravity position within the human body is critical for the application of biomechanics to the study of human movement. The center of gravity movement patterns over the hurdle are an important consideration for the athletics coach. Similarly, the center of gravity movement pattern in the child's pathological walking gait has important implications for clinical assessment.

The center of gravity

The **center of gravity (c of g)** of an object is defined as the **point** at which the entire **weight** of an object is assumed to be **concentrated**. This can be further clarified to mean the following: the center of gravity of an object is defined as **the point** at which the force of gravity for the entire object can be placed so that the object will behave the same as in the actual case when the force of gravity is distributed across the entire object. The term **center of mass (c of m)** is defined as **the point** which corresponds to the mean (average) position for the **concentration** of the entire **matter** in the body. Within biomechanics the two terms are often used synonymously (i.e., having the same meaning). The terms center of gravity and center of mass are used for imaginary points (**i.e., they do not physically exist as a point that can be seen**) that describe concentrations of weight or matter.

Every object has a center of gravity and for bodies of uniform density (where density is defined as the mass per unit volume) the center of gravity is at the geometric center of the object. However, for bodies with non-uniform density the center of gravity has to be calculated. As we know the human body has non-uniform density and it is an irregular shape and it assumes many different positions during sport and movement. Hence, the center of gravity may be constantly moving (although it can also be stationary). This center of gravity position can be either within or outside of the body. For example, in the high jump athletes often allow the center of gravity to move outside of the body so it can pass under the bar while the athlete travels over the bar. *Fig. C4.1* shows some of the considerations for the center of gravity in more detail.

Stability is the ability of an object **to maintain** its beginning **static equilibrium posture**. The stability of an object is affected by the position of the center of gravity and generally if a vertical line through the **c of g passes within** the base of support the object is stable and it is unstable if the **c of g falls outside** the base of support. Stability is achieved when the c of g lies near the lower part of the object. An increase in the area of the base of support will provide greater stability. Additionally, a heavier object is generally more stable than a lighter object because the torque needed to topple the heavier object would be greater. For

Regular shapes (objects of uniform density assumed)   Irregular shape

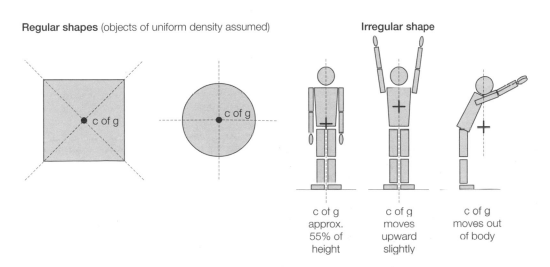

c of g

c of g

c of g
approx.
55% of
height

c of g
moves
upward
slightly

c of g
moves out
of body

Fig. C4.1.   The center of gravity (c of g)

example, imagine the position of the c of g while you are standing upright (i.e., it will be at approximately 55–57% of your height (vertically) if you are standing symmetrically) and then imagine the position of the c of g if you lie flat on the floor. The example where you are lying flat will cause the c of g to be closer to the ground and this will offer a much greater degree of stability than when you are standing upright. For example, it will be more difficult to topple you over when you are lying flat (i.e., an increased base of support and a lower c of g position (closer to the base of the support)). Within human movement many situations require the body to be able to move from a stable to an unstable situation. For example, the 100 m sprinter is required to be in an unstable a position at the start of the race (without causing a false start) so that they can quickly move off into the sprint race. Similarly, once the athlete is moving ideally they need to also be in an unstable position so that it is easier to move quickly into each different stride. *Figs C4.2* and *C4.3* illustrate some examples of static and dynamic (moving) stability in relation to the c of g.

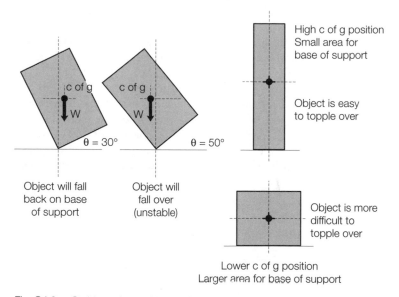

Fig. C4.2.   Stable and unstable positions

Calculation of the center of gravity

The vertical position of the center of gravity of the body can be calculated by using the principle of moments. Consider *Fig. C4.4* where the human body is lying flat on what is termed a **center of gravity board**.

Using the principle of moments ($\sum$CWM = $\sum$ACWM from section C1) the following equation can be developed (in this equation the full stop is used to represent a multiplication process). Remember that a moment is defined as a force (weight) multiplied by a perpendicular distance from an axis of rotation (the fulcrum in this case).

$$W_1 . x_1 + W_2 . x_2 = W_3 . d$$

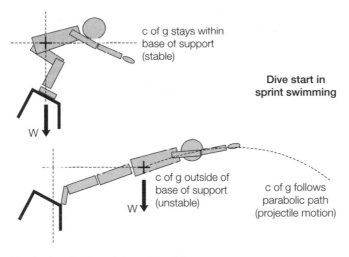

c of g stays within
base of support
(stable)

**Dive start in
sprint swimming**

W

c of g outside of
base of support
(unstable)

W

c of g follows
parabolic path
(projectile motion)

Fig. C4.3.   Static and dynamic stability

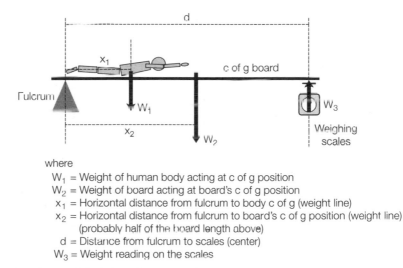

Fig. C4.4.   Calculation of the center of gravity

When W_3 (the recorded weight (mass × gravity) on the scales) equals zero and the subject is **not lying** on the c of g board, then the equation can be rearranged as follows. Note: it is important to clarify that the scales are actually set to zero (when the board alone is in place) in this instance (and if not set to zero) they would actually record in part the weight of the board (i.e., the effect of gravity on the board acting at a position from the fulcrum because one end of the board is already supported).

$$W_2 \cdot x_2 = 0$$

Hence, we can now eliminate ($W_2 \cdot x_2$) from the previous equation (because it is zero).

$$W_1 \cdot x_1 = W_3 \cdot d$$

Rearrange this equation to produce the formula for x_1 (the position of the body's center of gravity from the fulcrum) and we have the following.

$$x_1 = \frac{W_3 \cdot d}{W_1}$$

x_1 represents the vertical height (actually expressed as horizontal position in our example because the athlete is lying flat) of the subject's center of gravity from the feet. Note: in this example we are only calculating the vertical position of the c of g from the feet. The effect (i.e., the difference between one position and another) on the c of g position by movement of the legs and arms can then be calculated by the same method but by taking the difference between the scale readings for the two relative positions. For example, if the arms are moved down to the side of the body we would expect the c of g position to move lower down the body. Similarly the athlete could stand on the board or lie in a different direction on the board to determine the c of g position in other planes and axes of the body (but this method would still only give one position in one direction (i.e., in one plane and one axis) at a time).

Within human movement the c of g has a location that will depend on the position of the body. The c of g is constantly moving (although remember it can also be stationary) depending upon the body position and as we have seen it can often lie outside the body. In this case we would need to modify the c of g board experiment in order to take account of this lateral or medial deviation of the c of g (i.e., we need to use a c of g board that has both vertical and horizontal axes (i.e., x and y axes)).

In this case the same method of calculation applies but the athlete would assume a position on a board that had weighing scales at each of two positions on the board (i.e., a three-point reaction board – two scales and a fulcrum). Using the same principle of moments described previously the two-dimensional (2D) position (x and y coordinates) of the c of g for various body orientations on the board can now be calculated. *Fig. C4.5* illustrates this application in more detail.

In *Fig. C4.5* moments are taken about both OX and OY axes in order to calculate the position of the c of g during various body orientations that are assumed on the c of g board. The same method of calculation used for *Fig. C4.4* is used but with the addition of taking the moments about each different axis (i.e., length and breadth of the c of g board or the OX and OY axes).

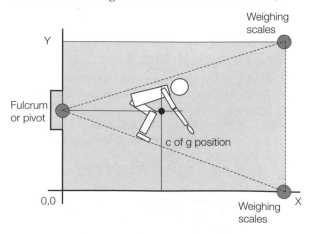

Fig. C4.5.   2D calculation of the center of gravity

Throughout human movement the c of g location may be constantly changing and in order to determine the c of g in various body positions throughout human movement we must use an alternative method to the c of g board. Usually this is achieved by taking a digital video sequence of the athlete performing an activity and then digitizing (the location of body landmarks – shoulder, elbow, hip, knee, ankle, etc.) each field of this sequence in order to produce a two-dimensional "stick figure" model of the human body. This can also be achieved in a 3D analysis of human movement (section F on measurement techniques explains this in more detail). In this respect the human body is drawn as a "stick figure" that is made up of various body segments represented by lines (or "sticks"). The c of g for each of these segments is plotted along each "stick" length that represents a limb or segment of the body. The location of the joint positions needed to create the "stick figures" are presented in *Table C4.1* (which are derived (and reproduced with permission) from data presented in the book *Biomechanics and Motor Control of Human Movement* (2nd edition) by David Winter published in 1990 by Wiley-Interscience publishers, New York, 3rd edition, 2004).

Table C4.1

Segment	Location
Head and neck	C7–T1 (vertebra) and 1st rib/ear canal
Trunk	Greater trochanter/glenohumeral joint
Upper arm	Glenohumeral axis/elbow axis
Forearm	Elbow axis/ulnar styloid
Hand	Wrist axis/knuckle II middle finger
Upper leg (thigh)	Greater trochanter/femoral condyles
Lower leg	Femoral condyles/medial malleolus
Foot	Lateral malleolus/head metatarsal II

The c of g of each segment (*Table C4.2*) can now be expressed at a point that is a percentage of the length of the segment (from both a proximal and a distal location). These percentage lengths are again derived from data presented by David Winter in the book described above. *Figs C4.6* and *C4.7* illustrate the

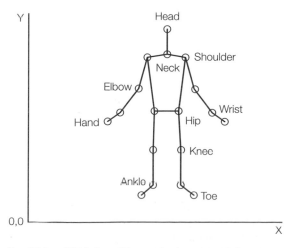

Fig. C4.6. "Stick figure" human body representation

Table C4.2

Segment	Location of center of mass (expressed as a % of the distance from each end)		Proportion of total body weight (expressed as a % for each individual segment)
	Proximal	Distal	
Head and neck	100	0	**8.1**
Trunk	50	50	**49.7**
Trunk, head, and neck	66	34	57.8
Upper arm	43.6	56.4	**2.8**
Forearm	43	57	**1.6**
Hand	50.6	49.4	**0.6**
Upper leg (thigh)	43.3	56.7	**10.0**
Lower leg	43.3	56.7	**4.65**
Foot	50	50	**1.45**
Total (all segments)			**100.00**

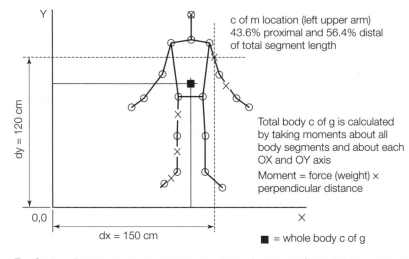

Fig. C4.7.   Center of mass (c of m) location for each segment (frontal plane – anterior view of body)

meaning of these two tables in more detail. Note: it may be important to indicate that there are also other methods of locating center of mass positions which are derived from techniques such as magnetic resonance imaging (MRI) and mathematical modeling.

In order to determine the center of gravity position of the whole body in the example shown in *Fig. C4.7* moments about each axis for each body segment are taken. For example, the **moment** about the OX axis for the **left upper arm** (because it is a frontal plane anterior view of the body) would be **the distance from this axis** to the **segment center of mass position** multiplied by the **mass of this particular segment**. For a 75 kg athlete and the left upper arm position shown in *Fig. C4.7* this would be as follows.

Taking moments
about OX

Moment = force × perpendicular distance from axis of rotation
Moment (upper arm) – ((0.028 × 75) × 9.81) × 1.20

where
0.028 = the percentage of mass of the whole body for this segment expressed as a
decimal (i.e., a proportion of 1.0)
75   = athlete's mass in kilograms (kg)
9.81 = acceleration due to gravity expressed in m/s^2
1.20 = perpendicular distance from OX which is 120 cm expressed in meters (m)

Moment about **OX** (**left upper arm**) = ((0.028 × 75) × 9.81) × 1.20
= **24.72 Nm**
Moment about **OY** (**left upper arm**) = ((0.028 × 75) × 9.81) × 1.5
= **30.90 Nm**

In order to calculate the center of gravity of the whole body in this position **all** the moments for each segment about each axis of rotation are calculated and summated (i.e., sum the moments about OX and OY separately). Next, since we know that the sum of the center of gravity of all the segments will equal the total center of gravity of the whole body (i.e., 100% of the weight) we can divide these respective summed moment values by the athlete's body weight in order to find the respective OX and OY coordinate positions of the whole body center of gravity. This can then be drawn on the diagram shown in *Fig. C4.7* as the intersection of two lines from the respective OX and OY axes (see *Fig. C4.7*).

This method of determining the center of gravity of the whole body is used extensively in many software digitization packages that present whole body c of g calculations. The same principle is applied in 3D analysis of human movement but in this case moments are also taken about a third OZ axis. Finally, it may be important to add that there are a number of different anthropometric tables that are used to determine the relative mass and relative center of mass positions for the various body segments. These are often evident in different software packages that are used to calculate the whole body center of gravity and hence as a result they present slight differences in each of their respective calculations of the location of the whole body center of gravity (c of g).

Application

An understanding of the c of g calculation and the movement patterns associated with the whole body c of g during activity is critical. Such applications range from the learning of how a child begins to walk to how an athlete can effectively clear a hurdle in athletics. The movement patterns of the c of g of the whole body are used extensively in coaching and sports science. Furthermore, they are frequently used in the study of pathological movement disorders by doctors, physiotherapists, and biomechanics specialists. Hence, it is important in a text of this nature that the meaning of the center of gravity is understood in detail. This indepth study will be incorporated throughout this text as many of the sections will use the center of gravity and center of mass of both whole body and individual body segment analysis.

C5 EQUILIBRIUM, BALANCE, AND STABILITY

Key Notes

Equilibrium

When a body is in a state of rest or is in a state of uniform motion (i.e., constant velocity) it is in a condition of equilibrium. Static equilibrium is concerned with when the object is not moving and dynamic equilibrium is a condition where the object is in a state of movement with constant velocity (i.e., not accelerating).

The first and second condition of equilibrium

These conditions can be used to solve for forces that are needed to maintain equilibrium. The first condition of equilibrium is concerned with when the sum of the external forces acting on an object is zero ($\sum F = 0$) whereas the second condition of equilibrium is concerned with when the sum of the moments (force × perpendicular distance from axis of rotation to the applied force) acting on a system is zero ($\sum CWM + \sum ACWM = 0$).

Balance and stability

Balance is described as the state of equilibrium and it can be used to describe how the condition of equilibrium can be controlled. Stability is the quantity of being free from any change (i.e., a resistance to a disruption in equilibrium). Balance and stability can be affected by the position of the center of gravity in relation to the base of support. If the vertical line through the center of gravity falls outside the base of support the object will be unstable. Alternatively, the larger the base of support and the more massive the object, generally the more stability the object will possess.

Application

The swimmer on the blocks at the start of the sprint race will assume an unstable position such that he/she will easily be able to move from this starting position into the dive (i.e., quickly). Similarly, the sprinter at the start of the 100 m sprint race will be in an unstable position ready to drive off into the sprint at the sound of the starting signal. However, the boxer will assume as stable a position as possible in order to prevent being knocked over. Often, however, it is necessary throughout sport to be able to change the stability possessed by the body. For example, in gymnastics it is necessary to have both stable and unstable positions during different movements and exercises.

Equilibrium

When a body is in a state of rest (not moving) or is in a state of constant velocity (moving but not accelerating) it is said to be in a condition of **equilibrium**. When the body is at rest it is in a state of **static equilibrium** and when it is moving with constant velocity it is in a state of **dynamic equilibrium**. The first condition of equilibrium is concerned with the sum of forces acting on a body ($\sum F = 0$). When the sum of all the **external forces** acting on a body is equal to zero and the body does not translate (linear motion) the object is in **static equilibrium**.

Considering *Fig. C5.1*, when an object is placed on a table the external forces that are acting consist of the weight of the object acting downward (caused by gravity) and the force from the table acting upwards on the object. The sum of the external forces acting on the object is equal to zero and the object does not move (static equilibrium).

The object in *Fig. C5.2* is subjected to several forces simultaneously and the sum of these forces acting on the object is zero. In this case the object will not move and it will stay in its state of equilibrium. The forces can be expressed as a **polygon** (a plane figure formed by three or more segments (lines) – or they can also be expressed by any shape (and in any order) where they are presented in the "tip to tail" closed method of representation of force vectors). If two or more forces act on an object and the object does not move, the first condition of equilibrium can be used to solve the situation for the resultant of these two forces.

Where (external forces acting)

W = weight of the object (caused by gravity)
20 kg × 9.81 m/s^2 is acting on the table.
The force from the table (F$_2$) is acting on the object

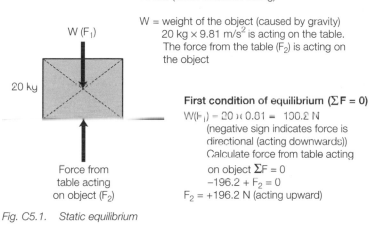

First condition of equilibrium ($\Sigma F = 0$)
$W(F_1) = 20 \times 0.81 = 100.2$ N
(negative sign indicates force is directional (acting downwards))
Calculate force from table acting on object $\Sigma F = 0$
$-196.2 + F_2 = 0$
$F_2 = +196.2$ N (acting upward)

Fig. C5.1. Static equilibrium

Concurrent application of four forces to an object. Object does not move and is in a state of static equilibrium

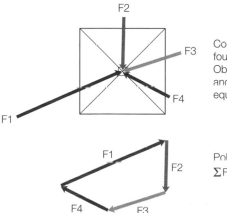

Polygon of force vectors
$\Sigma F = 0$ static equilibrium

Fig. C5.2. Static equilibrium (polygon of forces)

In *Fig. C5.3*, two co-planar (occupying the same plane), forces act on an object and the object does not move. Hence, there must be a third force that acts (unless the forces applied are equal and opposite) in order to maintain equilibrium. To calculate the third force, the **resultant** (one force that would have the equivalent effect of the two forces, i.e., the sum of two or more **vectors** (having both magnitude and direction)) of the two forces must be determined. This can either be solved graphically (by drawing a scaled diagram) or mathematically by resolving the forces and then using the first condition of equilibrium ($\sum F = 0$) to determine the third force (the force needed to maintain equilibrium).

In order to determine the resultant force (R) of these two forces (F1 and F2) in *Fig. C5.3*, the concurrent forces are expressed at a point using a **free body diagram** (which is a diagram where all external forces on the body/object are represented by vectors). The resultant of these two forces can now be determined graphically or mathematically (*Figs C5.4, C5.5* and *C5.6*).

First and second condition of equilibrium

In the previous examples the forces that were examined were represented within the linear force system with concurrent, co-planar forces. However, within human motion forces do not always act in the same plane and they often act on bodies as force couples (pairs of equal and opposite parallel forces). In this latter case we use the parallel force system to determine static equilibrium. The forces that act at parallel positions will often cause rotations of objects about specific

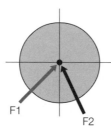

Object remains in static equilibrium

Two co-planar forces are acting on an object

What is the force acting against these two forces which is needed to maintain static equilibrium ($\sum F = 0$)?

Determine the resultant of the two forces F1 and F2

The force will be equal in magnitude and opposite in direction to the resultant force

Fig. C5.3.   Co-planar forces acting on an object (object does not move)

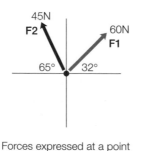

Forces expressed at a point (free body diagram)

Graphical representation where scaled drawing of the forces allows the resultant (R) to be determined.

Force is equal and opposite to resultant force

Fig. C5.4.   Free body diagram of concurrent forces

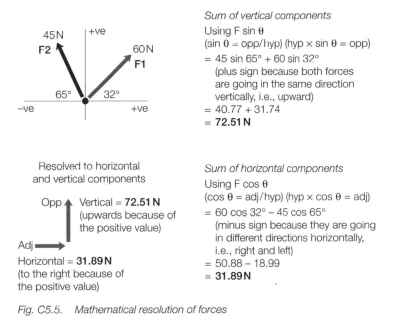

Sum of vertical components

Using F sin θ

(sin θ = opp/hyp) (hyp × sin θ = opp)

= 45 sin 65° + 60 sin 32°

(plus sign because both forces
are going in the same direction
vertically, i.e., upward)

= 40.77 + 31.74

= **72.51 N**

Resolved to horizontal
and vertical components

Opp ↑ Vertical = **72.51 N**
(upwards because of
the positive value)

Adj →
Horizontal = **31.89 N**
(to the right because of
the positive value)

Sum of horizontal components

Using F cos θ

(cos θ = adj/hyp) (hyp × cos θ = adj)

= 60 cos 32° – 45 cos 65°

(minus sign because they are going
in different directions horizontally,
i.e., right and left)

= 50.88 – 18.99

= **31.89 N**

Fig. C5.5.   Mathematical resolution of forces

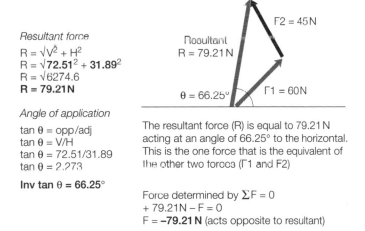

Resultant force

R = $\sqrt{V^2 + H^2}$

R = $\sqrt{72.51^2 + 31.89^2}$

R = $\sqrt{6274.6}$

R = 79.21 N

Resultant
R = 79.21 N

θ = 66.25°   F1 = 60 N

Angle of application

tan θ = opp/adj

tan θ = V/H

tan θ = 72.51/31.89

tan θ = 2.273

Inv tan θ = 66.25°

The resultant force (R) is equal to 79.21 N
acting at an angle of 66.25° to the horizontal.
This is the one force that is the equivalent of
the other two forces (F1 and F2)

Force determined by ΣF = 0
+ 79.21 N – F = 0
F = **–79.21 N** (acts opposite to resultant)

Fig. C5.6.   Mathematical resolution of forces

axes. Hence, in this situation we use **moments (force × perpendicular distance)**
to solve the equilibrium condition. The **second condition of equilibrium** states
that the sum of the moments (torques) acting on an object is equal to zero
(**∑Clockwise moments (CWM) + ∑Anti-clockwise moments (ACWM) = 0)**. *Fig.
C5.7* illustrates this in more detail in an example using the flexion and extension
movement of the arm. This figure shows a position (approximately 90°of elbow
flexion) of the arm during the flexion/extension movement. The biceps muscle is
exerting a force (F$_1$) which holds the arm in this position. The weight of the arm
and hand create a force (F$_2$) acting at a distance from the axis of rotation which
opposes the moment created by the muscle force. The weight of the arm creates a
clockwise moment (i.e., it would have a tendency to rotate the arm in a clockwise

direction) whereas the muscle force creates an anti-clockwise moment or torque. *Fig. C5.8* shows the second condition of equilibrium (ΣMoments = 0) which is used to solve the condition of equilibrium (no movement) and calculate the muscle force needed to hold the arm stationary in this position.

The muscle force needed to hold the arm in this static equilibrium is 382.59 N. This force creates an anti-clockwise moment that balances the clockwise moment created by the weight of the arm and hand. The result is that the arm remains stationary and the limb is in a position of static equilibrium.

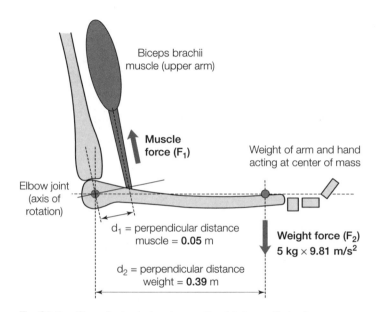

Fig. C5.7.   Elbow flexion/extension position (static equilibrium)

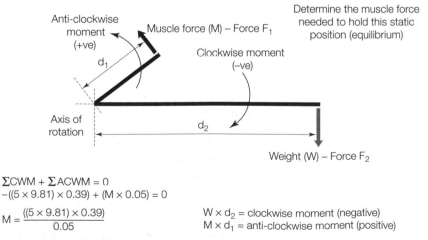

ΣCWM + ΣACWM = 0
$-((5 \times 9.81) \times 0.39) + (M \times 0.05) = 0$

$M = \dfrac{((5 \times 9.81) \times 0.39)}{0.05}$ $W \times d_2$ = clockwise moment (negative)
          $M \times d_1$ = anti-clockwise moment (positive)

Muscle force = 382.59 N

Fig. C5.8.   The second condition of equilibrium ($\Sigma M = 0$)

Dynamic equilibrium

When a body is in motion and it is not accelerating it is said to be in **dynamic equilibrium**. The same conditions as for static equilibrium apply and can now be re-written to include this motion component. The first and second conditions of equilibrium are presented as follows for dynamic equilibrium. This approach can also be used to determine the **force necessary to accelerate an object** (hence the inclusion of the acceleration component shown for dynamic equilibrium equations) and will be developed in more detail in section C9.

First condition of equilibrium (dynamics)

$$\sum F - ma = 0$$

(however it is relevant to note that if the object is in equilibrium the acceleration of the object will be zero and this equation will become $\sum F = 0$).

Second condition of equilibrium (dynamics)

$$\sum M - I\alpha = 0$$

where
$\sum$ = the sum of
F = forces
M = moments
m = mass
a = linear acceleration
I = moment of inertia
α = angular acceleration

Similarly, in equilibrium this angular acceleration component will be zero and the second condition is written as $\sum M = 0$.

Balance and stability

Balance is defined as a state of equilibrium and it can often be used to describe how the condition of equilibrium is controlled. **Stability** is the quality of being free from any change. More specifically this can be classed as the resistance to the disruption (disturbance) of equilibrium. The more stable an object the more resistance it will offer to being disturbed (i.e., moving out of the state of equilibrium). Balance and stability within sport are important concepts and an athlete will often use these components to achieve specific movement patterns. The swimmer on the blocks during a sprint start will have a small degree of stability such that they can easily be disturbed from their state of equilibrium (i.e., the ability to be able to move quickly into the dive from reacting to the starting signal). Similarly, the 100 m sprinter will be in the same situation at the start of the race (balanced but with a small amount of stability) such that they can easily move into the race by driving off from the blocks (this point of limited stability is often very close to the point of making a false start). The boxer will create a high level of stability such that they are unable to be knocked over during the fight, and the gymnast will often need to be in both stable and unstable positions depending on the particular task that they are required to perform. Quick movements from one activity to another in gymnastics would require less stability whereas slow controlled movements (i.e., like during a landing) require greater stability. *Fig. C5.9* illustrates some examples of balance and stability within human movement.

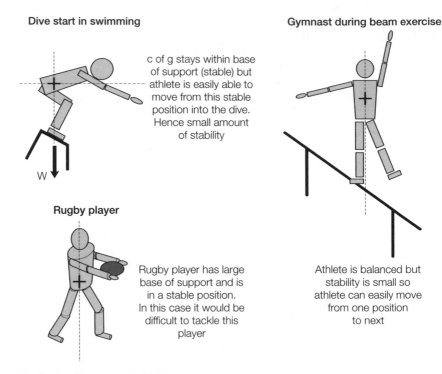

Dive start in swimming

c of g stays within base of support (stable) but athlete is easily able to move from this stable position into the dive. Hence small amount of stability

W

Gymnast during beam exercise

Athlete is balanced but stability is small so athlete can easily move from one position to next

Rugby player

Rugby player has large base of support and is in a stable position. In this case it would be difficult to tackle this player

Fig. C5.9.   Balance and stability

Application

The mass of an object affects its stability and generally the more mass an object has the greater will be its stability. The more mass possessed by an object the more force will be required to move it (i.e., disturb its equilibrium). The base of support of the object is also related to the amount of stability offered by an object. The larger the area for the base of support of an object and generally the more stable the object becomes. Try standing on one leg and then see the difference in your balance and stability when you stand on two legs. In the example, where you stood on two legs you increased the base of support and are in a more stable position. In terms of balance when the center of gravity of the object moves out from being over the base of support the more unbalanced and unstable the object becomes. The closer the center of gravity is to the extremities (outer edges) of the base of support the less stable the object. Finally, the lower the center of gravity (i.e., the nearer to the base of support) the more balanced and stable the object. *Fig. C5.10* identifies some of these concepts in more detail.

Equilibrium, balance and stability are critical within the study of sport and human movement and these concepts will be discussed in more detail throughout many of the sections within this text. Hence, it is important that you have a good understanding of their application.

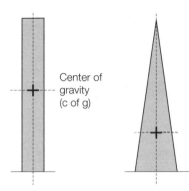
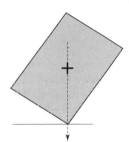

Center of
gravity
(c of g)

| Small base of support. C of g is high and object is unstable (easily toppled over) | Larger base of support. Lower c of g and object is more stable (more difficult to topple over) | C of g is just on the inside edge of the base of support and object is unstable (i.e., it will fall back down but it can also be toppled over easily) | Similarly swimmer's c of g is just outside base of support. Swimmer is unstable and can easily move into the dive start |

Fig. C5.10.　Balance and stability

C6 LEVERS

Key Notes

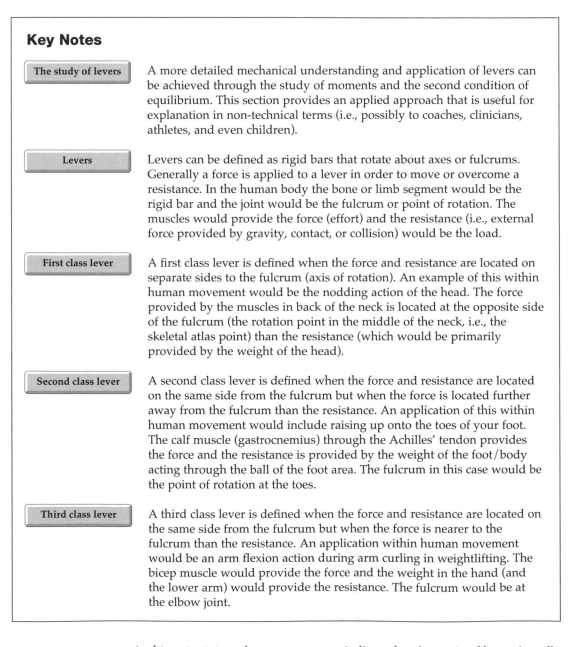

The study of levers

A more detailed mechanical understanding and application of levers can be achieved through the study of moments and the second condition of equilibrium. This section provides an applied approach that is useful for explanation in non-technical terms (i.e., possibly to coaches, clinicians, athletes, and even children).

Levers

Levers can be defined as rigid bars that rotate about axes or fulcrums. Generally a force is applied to a lever in order to move or overcome a resistance. In the human body the bone or limb segment would be the rigid bar and the joint would be the fulcrum or point of rotation. The muscles would provide the force (effort) and the resistance (i.e., external force provided by gravity, contact, or collision) would be the load.

First class lever

A first class lever is defined when the force and resistance are located on separate sides to the fulcrum (axis of rotation). An example of this within human movement would be the nodding action of the head. The force provided by the muscles in back of the neck is located at the opposite side of the fulcrum (the rotation point in the middle of the neck, i.e., the skeletal atlas point) than the resistance (which would be primarily provided by the weight of the head).

Second class lever

A second class lever is defined when the force and resistance are located on the same side from the fulcrum but when the force is located further away from the fulcrum than the resistance. An application of this within human movement would include raising up onto the toes of your foot. The calf muscle (gastrocnemius) through the Achilles' tendon provides the force and the resistance is provided by the weight of the foot/body acting through the ball of the foot area. The fulcrum in this case would be the point of rotation at the toes.

Third class lever

A third class lever is defined when the force and resistance are located on the same side from the fulcrum but when the force is nearer to the fulcrum than the resistance. An application within human movement would be an arm flexion action during arm curling in weightlifting. The bicep muscle would provide the force and the weight in the hand (and the lower arm) would provide the resistance. The fulcrum would be at the elbow joint.

At this point it is perhaps important to indicate that the study of levers is really only an applied practical example of the study of moments. All the problems and considerations concerned with levers can actually be solved by using clockwise and anti-clockwise moments and the second condition of equilibrium (as we have seen in previous sections within this text). The student of biomechanics should become more familiar with the application and understanding of moments than with the classification of simple lever systems. Nevertheless, often as bio-mechanists' and exercise scientists we are regularly required to explain principles (i.e., to coaches, to clinical practitioners, to athletes, and even to children) in a language that is not technical. This application of levers would be a non-technical example of expressing the principle of moments.

Levers

Levers can be defined as **rigid bars** that rotate about **axes** or **fulcrums**. In the human body the **bone** or the limb/segment would act as the **rigid bar**, and the **joint** would act as the **axis of rotation** or **fulcrum**. A **fulcrum** can be defined as the pivot about which a lever turns. Generally, a force is applied to a lever in order to move or overcome a resistance (i.e., another force). Within the human body the muscles often provide the force and the resistance (i.e., the other force) is provided by other external forces acting on the system (i.e., gravity, collision, contact, and load). *Fig. C6.1* illustrates this in diagrammatic form.

Levers are classified into one of three types. These are termed **first**, **second**, or **third** class levers. A **first class lever** is when the force and the resistance are located at separate sides of the fulcrum. A **second class lever** is when the force and the resistance are located on the same side from the fulcrum position. However, in this case the force is further away (at a greater distance) from the fulcrum than the resistance. A **third class lever** is similar to the second class lever (with the force and resistance on the same side from the fulcrum position) but this time the force is nearer to the fulcrum than the resistance. In the first class lever system the distances of the force and the resistance that act either side of the fulcrum do not need to be equal. *Fig. C6.2* shows the three lever systems in more detail.

Within the human body, levers play an important role in the application of force and overcoming resistance (another force) in order to initiate or continue movement. Each lever classification (first, second, and third) can have different applications. The following identifies some of the applications both in general and within human movement.

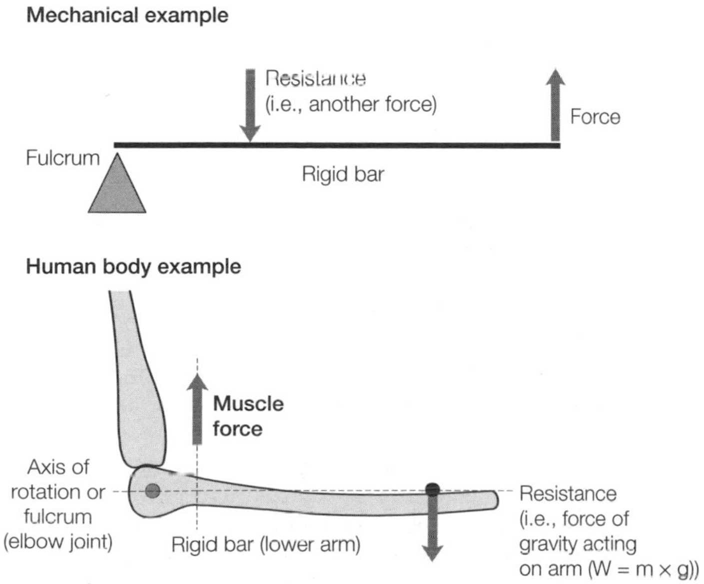

Mechanical example

Human body example

Fig. C6.1.   Levers (mechanical and within the human body)

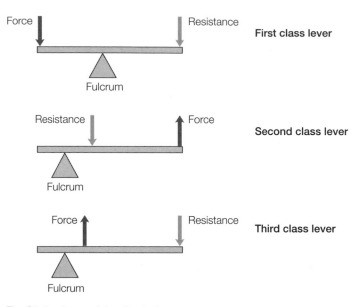

Fig. C6.2.   Levers (classification)

First class lever   This lever classification is similar to the seesaw that children would play on in the park or playground. In this situation you would observe that a child is sitting at each side from the point of rotation of the seesaw (i.e., the fulcrum). The children (depending on their respective mass) would move further in or further away from the fulcrum in order to balance the see-saw. If they both sat at equal distances from the fulcrum the seesaw would move downward at the end with the heavier child. In more general terms a pair of scissors would represent a situation in which a first class lever is used. The fulcrum would be the axis of rotation of the scissors (i.e., approximately in the middle), and the force would be provided at one end by the hands in order to overcome a resistance at the other end (i.e., the cutting of the paper or object). Within the human body there are many applications of the first class lever. For example, the rotating of the head (or "nodding") forward and backward would be an example of a first class lever within the body. Similarly the overarm throwing action of a ball would also be an example of a first class lever within human movement. These examples can be seen in more detail in *Fig. C6.3.*

Second class lever   In the second class lever the force and the resistance are located on the same side (from the fulcrum position) but the force moment arm (distance) is greater than the resistance moment arm. For an equal and opposite resistance this would mean that the force required to move the resistance would be less than the resistance. In this case there would be a mechanical advantage in the favor of the force being applied. Within human movement examples of second class levers include raising up onto your toes or the simple action of lifting the screen on your laptop computer. However, the more common general example of the second class lever shown in many biomechanics text books would be the use of a wheelbarrow to move a load. *Fig. C6.4* shows these examples in more detail.

Head movement

Throwing a ball

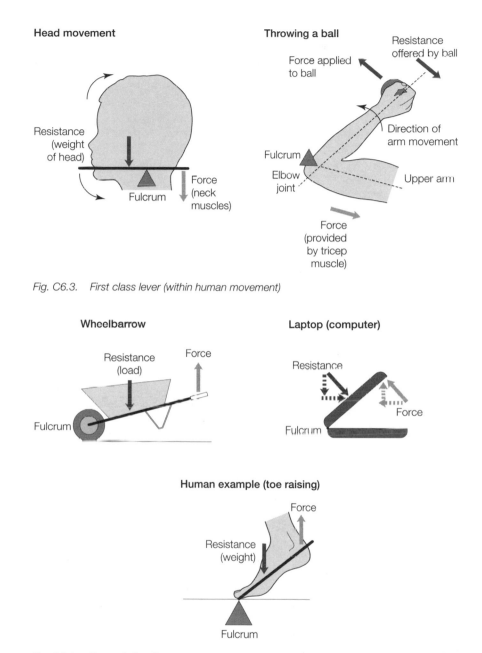

Fig. C6.3.   First class lever (within human movement)

Wheelbarrow

Laptop (computer)

Human example (toe raising)

Fig. C6.4.   Second class lever

Third class lever   In the third class lever system again we can see that the force and the resistance are located on the same side from the fulcrum. However, this time the force is nearer to the fulcrum than the resistance. This would mean that more force would be required to move a given resistance. One of the most common examples of this within human movement is the action of flexion of the elbow joint. The bicep muscle acts at a position that is close to the fulcrum (i.e., the elbow joint) whereas the resistance acts at a point further away from the fulcrum (usually a load held in the hand such as an object or weights when doing arm curl exercises). In more general terms the use of a shovel would be a good example of a third class lever and the use of a paddle while canoeing would be another example within human movement. *Fig. C6.5* shows these examples in more detail.

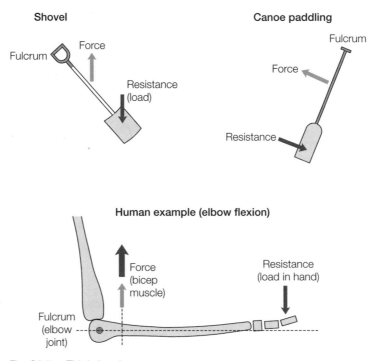

Fig. C6.5. Third class lever

Mechanical advantage

Mechanical advantage is defined as the ratio of the force moment arm (distance from the fulcrum) divided by the resistance moment arm (distance from the fulcrum). When the moment arm (perpendicular distance) of the **applied force** is **greater than** the moment arm of the **resistance** (i.e., for a given resistance/force) then the force needed to move the resistance is less than the force offered by the resistance (i.e., there is a mechanical advantage). This is the case in the second class lever example. Conversely, when the moment arm of the **applied force** is **less than** the moment arm of the **resistance** then more force is needed to move the given resistance (i.e., there is a mechanical disadvantage). This is the case in the third class lever example. A further explanation of this can be observed using the previously identified example of the seesaw in the children's playground. *Fig. C6.6* represents a first class lever as in the case of the seesaw on the children's playground.

Two children sit one at each end of a seesaw. One of the children has a mass of 28 kg and the other a mass of 35 kg. The child who weighs 35 kg is sitting at a point that is 1.2 m away from the fulcrum position. This child is sitting to the right-hand side of the fulcrum and would cause a clockwise moment (i.e., a tendency to cause a clockwise rotation of the seesaw). At what distance must the child who is 28 kg sit in order to balance the seesaw? This problem can be

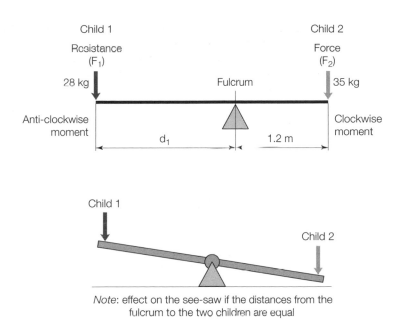

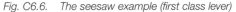

Fig. C6.6. The seesaw example (first class lever)

solved by using the second condition of equilibrium ($\sum M = 0$) and considering the mechanical advantage principle associated with levers.

$$\text{Mechanical advantage} = \frac{\text{Force distance (moment arm)}}{\text{Resistance distance (moment arm)}}$$

Considering *Fig. C6.7* we can see that when the force moment arm is greater than the resistance moment arm there is a mechanical advantage and less force is required to overcome a given resistance. However, it may be important to add that the use of the terms mechanical advantage and mechanical disadvantage in this context are very much specific to the function and purpose of the levers that

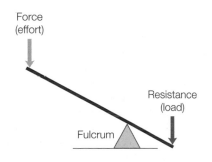

The **load** is **closer** to the **fulcrum** than the **effort**. Hence **less effort** would be required to move a **given load** (resistance). This allows **heavy objects** to be lifted with a **smaller effort**. There is a **mechanical advantage** in the favor of the **effort** (force). This would be analogous to the use of a **car jack** used to lift a car in order to change a wheel (i.e., **small effort** (human) to **lift a large load** (motor car))

Fig. C6.7. Mechanical advantage (first class lever)

are being used. For example, although there may be mechanical advantages to using a certain type of levers in particular applications it may be the case that such an application is not the most appropriate (i.e., lifting within industry is an example where it is not always possible to use the most mechanically advantageous lever system).

Considering this principle it is clear that the child who is sitting to the left must be sitting at a distance greater than 1.2 m in order to overcome the moment created by the child that is sitting to the right (because the child on the left has much less mass). The child who is sitting at a distance of 1.2 m to the right has a mass of 35 kg which would represent a weight of 343 N (weight = mass × acceleration due to gravity). This would create a clockwise moment of 412 Nm (343 N × 1.2 m). In order to balance the seesaw we can use the second condition of equilibrium to solve this problem. The child who is sitting to the left and who is 28 kg will create a weight force of 274 N.

$$\text{Clockwise moments} + \text{Anti-clockwise moments} = 0$$
$$F_1 \times d_1 + F_2 \times d_2 \qquad\qquad = 0$$

Re-arrange equation

$$F_1 \times d_1 = F_2 \times d_2$$
$$d_1 = \frac{F_2 \times d_2}{F_1}$$
$$d_1 = \frac{412 \text{ Nm}}{274 \text{ N}}$$
$$d_1 = \underline{\mathbf{1.50 \text{ m}}}$$

We can now see that the child on the left must sit at a distance of 1.5 m from the fulcrum in order to balance the seesaw. In terms of mechanical advantage this can now be expressed in the context of the formula that is related to levers.

$$\text{Mechanical advantage} = \frac{\text{Force distance (moment arm)}}{\text{Resistance distance (moment arm)}}$$

For the seesaw example in *Fig. C6.6* this is as follows (considering Child 1 as the force and Child 2 as the resistance).

$$\text{Mechanical advantage} = \frac{1.5 \text{ m}}{1.2 \text{ m}}$$

$$= \underline{\mathbf{1.25}}$$

When the ratio of the force moment to resistance moment arm is **greater than 1.0** then there is a **mechanical advantage** in favor of the **force**. In this case the ratio is 1.25 and the force required to move (or in this case balance) the resistance would be less than the resistance. As we can see by considering the actual figures this is correct as the force created by the child on the left (child 1) sitting at a distance of 1.5 m is only **274 N**, while the force created by the child on the right (child 2) who is sitting at a distance of 1.2 m is **343 N**. There is a mechanical advantage in favor of the child who is sitting on the left (child 1). This can be applied to any situation concerning levers and the movement or overcoming of a resistance (load).

In the second and third class lever systems the same principle for the calculation of **mechanical advantage** applies (see *Figs C6.8* and *C6.9*). However, in **second class levers** the ratio would always be **greater than 1.0** (a mechanical advantage in favor of the force (effort)). Whereas in **third class levers** the ratio

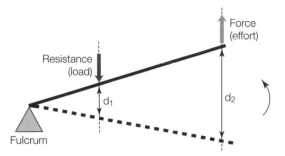

Mechanical advantage in favor of the force (effort). The moment arm of the force (distance from fulcrum) is **greater than** the **moment arm** of the **resistance**. The **force** required to move a given resistance would be **less than** the **resistance**. Note that the smaller **force** is required to move through a **larger *linear* range of motion** (ROM) than the **resistance**. Hence although the given force needed to move the resistance is less it would need to be applied through a larger ROM. Although the **angular motion** of all points on the rigid bar (lever) is **the same** the **linear motion** of the two points: load and effort (d_1 and d_2) is **different**

Fig. C6.8.   Mechanical advantage (second class lever)

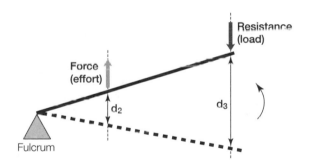

Mechanical advantage In favor of the **resistance** (load). The moment arm of the **resistance** is **greater than** the **moment arm** of the **force**. The **force** required to move a given resistance would be **more than** the **resistance**. Note that the **larger force** is only required to move through a **small linear ROM** (d_2) in order to move the **load a larger linear distance** (d_3). Hence although the given force needed to move the resistance is more it has the effect of moving the resistance through a large ROM

Fig. C6.9.   Mechanical advantage (third class lever)

would always be **less than 1.0** representing a mechanical advantage in favor of the resistance (load). In this context it may seem that the application of the third class lever would always be inefficient. However, *Figs C6.8* and *C6.9* show some further considerations for the use of these two lever systems in terms of mechanical advantage. In the second class lever system, although there is a mechanical advantage in favor of the force (effort), it is clear that the force must be applied through a large (linear) displacement in order to move the resistance only a small amount (linear displacement). In the case of the third class lever the effect is opposite and, although a larger force is needed to move a given resistance, this force is only applied over a small displacement (linear) in order to move the resistance a larger linear displacement. Hence, there are advantages and disadvantages to using each lever system within human movement.

Application

Within the body, muscles can often operate at different mechanical advantages and there are applications when the muscle is required to operate at an advantage and also at a disadvantage. For example, the bicep muscle in the upper arm which causes a flexion rotation about the elbow joint will function throughout the movement creating both maximum and minimum amounts of torque, i.e., the twisting moment that may cause rotation (flexion). This mechanical advantage and disadvantage can be seen in more detail by considering section C1 where *Fig. C1.8* shows the bicep muscle in two positions of different mechanical advantage. When the perpendicular distance of the muscle line of pull from the joint axis (elbow joint) of rotation is larger the more torque will be created for a given muscle force. When the perpendicular distance to the muscle line of pull is less the torque that is created (as in the case of an extended arm position shown in *Fig. C1.8*) is much smaller.

The moment arm of the bicep's force at different points in the movement of flexing the arm will change. This will create more or less torque for a given effort. This principle of torque generation and mechanical advantage through levers is applied in the design of many modern exercise machines. For example, many devices will be able to accommodate the different torque generating capacities of a muscle during exercise. The machines will often change the loading patterns at various points in the exercise in order to accommodate to the changing torques. Similarly, machines will often load muscles more at the points of their greatest torque generating capacity. An example of the application of levers and exercise devices will be identified in section F8 on Isokinetic Dynamometry.

Within the human body there are many applications of first, second, and third class levers. Often muscles will act in opposition to each other during the

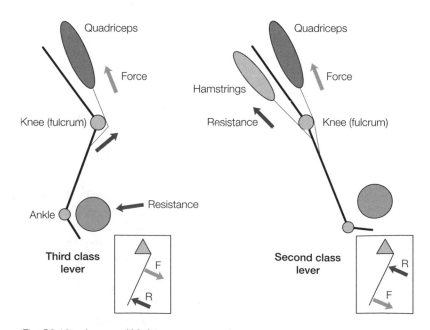

Fig. C6.10.   Levers within human movement

operation of a lever system (agonist and antagonist muscle actions). For example, the quadriceps muscle in anterior part of the upper leg will extend the lower leg at the knee joint, whereas the hamstring muscle which is in the posterior part of the upper leg will oppose this movement acting as an antagonist muscle to the quadriceps (the hamstring muscle causes flexion of the lower leg at the knee joint). *Fig. C6.10* shows the action of kicking a ball in more detail with respect to anatomical levers within the body.

In the action of kicking a ball the quadriceps muscle creates a third class lever system (see *Fig. C6.10*) however, after the kick is completed the antagonist action of the hamstrings creates a second class lever system by providing some resistance (control) to this motion (to slow the leg rotation (extension caused by quadriceps) and control the follow through phase of the kick). Within human movement there are many applications of levers and some of these will be used as further examples within this text.

C7 CENTRIPETAL FORCE AND CENTRIPETAL ACCELERATION

Key Notes

Centripetal force

The centripetal force ($F_{centripetal}$) is a force directed to the center of rotation when objects move in a circular path. This force is always required to make objects move in a circular path. Conversely, if an object moves in a circular path, the centripetal force must be acting.

Centripetal acceleration

Newton's second law (F = ma) tells us that whenever a force (F) acts this produces an acceleration (a). So when the centripetal force acts this must produce a centripetal acceleration.

What is the centripetal force dependent on?

There are two main formula to describe the centripetal force. These are:

$$F_{centripetal} = m. r. \omega^2$$

$$F_{centripetal} = m. v^2 / r$$

It can be seen that the centripetal force is dependent on the radius of rotation of the circle (r), and the angular (ω) or linear (v) velocity. The force is also dependent on mass (m). So the heavier, faster, and more distant the object is to the center of rotation, the greater the centripetal force needs to be in order to produce circular motion.

Centripal force

Newton's first law states that an object will continue in at rest or in uniform motion unless some external force acts on it to change its state of motion. This implies that if an object deviates from a straight line (i.e., moves in a curved path) then some force must act to cause this to happen. When an object moves in a circular path (e.g., when a hammer thrower rotates the hammer before release, *Fig. C7.1*), the force causing circular motion is said to be the **centripetal** force ($F_{centripetal}$). This term describes that the force is a **center-seeking** force.

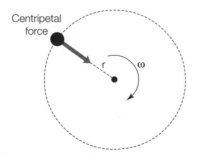

Fig. C7.1. Centripetal force during the rotation of a hammer

The centripetal force will cause an acceleration in the direction of the force according to **Newton's second law** and this is called **centripetal acceleration** ($a_{centripetal}$) which is defined by

$$F_{centripetal} = m. \, a_{centripetal} \qquad\qquad (C7.1)$$

This centripetal acceleration causes the object to move in a circle. If an object rotates about a circle of **radius of rotation** (r) and **constant angular velocity (ω)** then the centripetal acceleration is given by the equation

$$a_{centripetal} = r. \, \omega^2 \qquad\qquad (C7.2)$$

As $v = \omega . \, r$ (see Section A4) then equation C7.2 can be developed to give an alternative expression for the centripetal acceleration as

$$a_{centripetal} = r. \, (v^2 / r^2) = v^2 / r \qquad\qquad (C7.3)$$

Equations C7.2 and C7.3 can be substituted into equation C7.1 as appropriate to provide two expressions for the centripetal force.

$$F_{centripetal} = m. \, r. \, \omega^2 \qquad\qquad (C7.4)$$

$$F_{centripetal} = m. \, v^2 / r \qquad\qquad (C7.5)$$

How is the centripetal force applied? In sports and exercise this is normally provided by some linkage which can physically apply a force to the object of interest. Consider the case of a hammer thrower (*Fig. C7.2a*) at the start of the

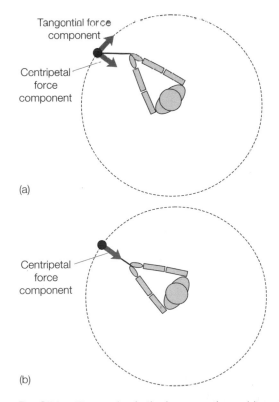

(a)

(b)

Fig. C7.2. The tension in the hammer wire and its components (a) during the increase in hammer head speed and (b) at maximum velocity before release

movement when the hammer is increasing in speed. The force that causes the hammer to rotate around the athlete is supplied by the wire. This force has two components, one acting in the direction toward the center of rotation (the centripetal force) and the other tangential to the circular path. The centripetal force causes motion of the hammer head in a circle while the tangential force component causes its acceleration (i.e., increase in speed) around the circle. Once hammer head speed has been developed, the athlete must maintain that speed in preparation for release. Constant speed of rotation is achieved when the centripetal force alone acts (*Fig. C7.2b*).

The term **centrifugal force** (center-fleeing force) often crops up in the literature and much unnecessary confusion exists over the correct use of the term. From **Newton's third law** every force has to have an equal and opposite reaction force. The tension in the hammer wire supplies the centripetal force to the hammer head (*Fig. C7.3*). The same tension also applies an equal and opposite reaction force to the thrower and, as this is directed away from the center, it is called the centrifugal force. Thus, the hammer thrower experiences a force pulling away from the hands and it is this sensation which sometimes makes people think the force acting on the hammer is also in the same center-fleeing direction.

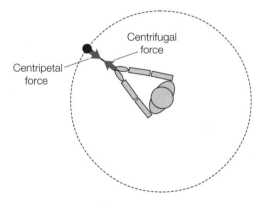

Fig. C7.3. The centrifugal force acting on the hammer thrower

Application 1 – Gymnastics – The grand (or giant) circle

In the performance of the grand circle in men's gymnastics, the body rotates about the high bar (*Fig. C7.4*). The force which keeps the body moving in a circle is supplied by the tension in the arms and is the source of the centripetal force. This tension will vary during the performance but will be a maximum when the angular velocity is a maximum (equation C7.4) which is when the gymnast swings directly beneath the bar.

The centripetal force acting at the gymnast's hands pulls the body towards the bar and is given by equation C7.4. The component of the gravitational force acting on the center of mass and along the direction of the body inclined at an angle θ to the vertical (m.g.cos θ) pulls the gymnast away from the bar. This increases the tension (T) in the arms which is given by the sum of these two opposing forces as

$$T = m.\ R.\ \omega^2 + m.g.\cos\theta$$

The tension in the arms is maximal when $\theta = 0$ and the equation can be written in terms of body weight (m.g)

$$T = m.g\ [\ R.\ \omega^2/g + 1]$$

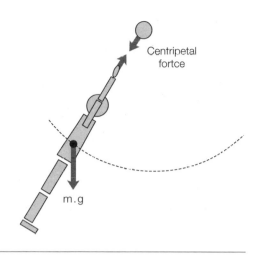

Fig. C7.4. Free body diagram of a gymnast

For typical values of R (1.3 m) and ω (5.5 rad/s) the maximum value of T is about five times body weight. Such a high force explains why the gymnast needs to be very strong, particularly in grip strength, and can sometimes fall off if the grip is not strong enough.

Application 2 – Running around a bend on a flat track

When running around the bend on a flat track, athletes are seen to lean into the bend. This leaning action causes a lateral frictional force which is the source of the centripetal force. The lean will continue until the athlete reaches an angle θ where there is sufficient friction force produced to enable the athlete to round the bend comfortably. This situation can be analyzed by considering the moments set up by the frictional force (F) and the normal reaction (N) force about the center of mass.

The free body diagram is drawn in *Fig. C7.5*. As the athlete is balanced the moments about the center of mass are equal':

Equating moments $\qquad$ $N.x = F.y$

Therefore $\qquad$ $F / N = x / y = \tan \theta$ $\qquad\qquad$ (1)

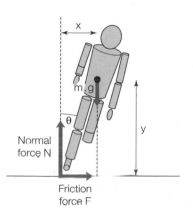

Fig. C7.5. Free body diagram of an athlete rounding a bend

Resolving vertically $\qquad$ $N - m.g = 0$ $\qquad$ (2)

The friction force is the source of the centripetal force

so $\qquad$ $F = m . v^2 / r$ $\qquad$ (3)

Combining (1), (2) and (3) gives

$$F / N = v^2 / r.g = \tan\theta$$

and so $\qquad$ $\theta = \tan^{-1} (v^2 / r.g)$

This equation allows the prediction of the angle of lean of a runner for any velocity, v, and cornering radius r. Typically for an athlete running at 10 m/s (a 200 m race) around an athletics track of radius of rotation 40 m, the angle of lean would be 14°. If the athlete performed on an indoor track of radius of rotation 20 m then the angle of lean would increase to 27°.

Application 3 – A runner cornering on a banked track   In indoor athletics, the track is often banked on the bends so that the athlete does not have to lean at large angles and run the risk of slipping and falling, possibly into other competitors. The ideal angle of banking is such that there is no lateral friction force at the athlete's foot. The analysis is performed in the same way as in the previous example using the following relationships from *Fig. C7.6*.

Resolving vertically $\qquad$ $N \cos\theta - m\,g = 0$ $\qquad$ (1)

The component reaction force is the source of the centripetal force

so $\qquad$ $N \sin\theta = m\,v^2 / r$ $\qquad$ (2)

Dividing (2) by (1) gives $\qquad$ $\tan\theta = v^2 / r.g$

Which is the same as found in the previous example.

Note that the angle of banking indoors will be greater than the angle of lean outdoors, due to the reduced radius of the bend for indoor athletics. For indoor cycling the angle of banking must increase as the speed of the cyclist increases, leading to a curved banking profile, where slow speeds are completed lower down the bank which is more shallow, while sprints are completed higher up the bank where the angle is greater.

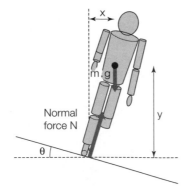

Fig. C7.6.   Free body diagram of an athlete rounding a bend on a banked track

Application 4 – Golf

In the examples considered previously, the curved motion was achieved by the application of a force which was the source of the centripetal force (tension in the wire for the hammer thrower; friction at the foot for the runner). There is an important class of movements in sports that are used to attain high end speed velocity by controlling the point at which the centripetal force is released. This occurs in ball kicking, the golf drive, the tennis serve, throwing actions, and many more. This type of action is also characterized by a proximal-to-distal sequence.

In all of these movements the active limb is made up of two or three segments (e.g., thigh and shank in kicking; upper arm, lower arm, and racket in tennis). The segments are flexed and constrained to rotate about the body during the initial stages of the movement. At some critical point in the movement the centripetal force responsible for holding the segments in place is released and the end segment moves outwards from the circular path and in doing so increases the radius of rotation and hence the linear velocity of the end point.

This is seen in golf, where the downswing is considered to occur in two parts: the first part is where the arms and club retain the same orientation to each other with a constant wrist angle; the second is where the wrist angle is allowed to increase enabling the club to extend prior to impact (*Fig. C7.7*).

(A) (B) (C)

Fig. C7.7. A–B The first part of the downswing in which the arms and club remain fixed with respect to each other. B–C The wrist is relaxed removing the centripetal force and so the club head moves outwards to increase its end point velocity

The movement of the club head outward is sometimes wrongly thought to be due to a force acting to pull it outward, and this "center-fleeing" force is mistakenly identified as the centrifugal force. It should be noted that it is not the centrifugal force causing outward motion but the **absence of the centripetal force**.

C8 THE ESTIMATION OF MUSCLE AND JOINT FORCES – STATIC APPLICATION

Key Notes

Static calculations

Static calculations are related to the dynamic calculations that form the basis of complex inverse dynamics approaches that are used in biomechanics for the modeling of joint and muscle forces. They are valuable in providing an understanding of injury potential and performance characteristics.

Free body diagrams

A free body diagram is defined as a picture (a diagram) of all the forces that act on an object or mass. This is probably the most important method for the representation of forces acting on any system. If all the external forces acting on the system (i.e., an object or selected mass) are drawn or represented by vectors then the first ($\sum F = 0$) and second ($\sum M = 0$) conditions of equilibrium can be used to solve force and moment problems.

Muscle forces

Determined using the second condition of equilibrium where the sum of the clockwise and anti-clockwise moments is zero. Muscles are the active stabilizers of joints.

Joint forces

Determined using the first condition of equilibrium where the sum of the forces is zero. Ligaments offer passive restraint to motion in a joint (i.e., they provide passive stability). Passive is defined as receiving or subjected to an action without responding or initiating an action. Muscles provide dynamic (active) stability.

Summary
 • **Injury**

Such muscle and joint forces in a single leg standing posture can exceed body weight and can result in the potential for injury.

 • **Performance**

It is important to understand the muscle forces in static calculations in order to be able to assess the training effect in activities such as weightlifting.

Static calculations

Static calculations are useful in that they form the basis of inverse dynamics calculations that are used in many modeling programs in today's biomechanics world of computer simulation. Understanding the basic calculation of the forces that act on a joint during various forms of human movement is essential if we are to have any understanding of injury mechanisms and/or performance characteristics. Although, these calculations are primarily based on a two-dimensional approach and are presented in the sagittal plane they are nevertheless valuable for the student of biomechanics in providing an understanding for more complex three-dimensional problems. In some applications, they are valid, as in the case of estimating the loads on the joint during squatting in weightlifting or alternatively estimating the forces on the elbow during the holding of the arm in the 90° position (as when doing an arm curl in weightlifting). The methods used are derived from standard mechanics, the application of trigonometry and the use of free body diagrams and they provide the reader with skills that can be used to model loads and forces that could be responsible for injury.

Application of statics (knee joint and quadriceps muscle forces)

In the upright standing posture as viewed in the sagittal plane the line of gravity falls approximately through the knee joint axis of rotation as shown in *Fig. C8.1*. The moment arm at the knee joint (force × perpendicular distance) is considered to be zero (0) and at this point there is no or minimal muscular force required to maintain this position. The limbs are said to be in a position of equilibrium (balanced). Electromyographic activity of the quadriceps and hamstring muscles during standing in this position has been shown to be negligible. In order to calculate the static muscle and joint forces around the knee during standing or balancing on one leg the following sequence of calculation is required.

F = force acting directly over knee joint through the center of rotation of the knee joint (i.e., does not cause any rotational component). Force is derived from the proportion of the body (mass) above knee joint that is acting through the thigh (upper leg)

Fig. C8.1.   Force acting over knee in static standing posture

Q. What is the single joint compressive force that is acting on the knee in this standing position in a 75 kg athlete?

Weight (W) = mass (kg) × acceleration due to gravity (m/s^2)

Acceleration due to the gravitational attractive force from the Earth varies by a small amount (negligible) according to your position on the surface of the Earth however, in this example we will consider it to be 9.81 m/s^2. Therefore we can now calculate the weight of the athlete. It is important to note that weight will have the units of Newtons (N) because it represents the force acting on the athlete by virtue of gravitational effect caused by the mass of the Earth.

Free body diagrams

The free body diagram represents a picture (diagram) of **all** the forces (external forces) which are acting on a system. This method of force representation (discussed elsewhere) is probably the first and most important process for solving force problems. In a free body diagram **all** the forces acting on the system (i.e., an object or a mass) are represented graphically by drawing vectors (i.e., lines representing force with both magnitude and directional components shown). Then by using the first ($\sum F = 0$) and second ($\sum M = 0$) conditions of equilibrium (derived from Newton's laws) it is possible to analyze and describe the resulting force actions and motions. Some of the **types of forces** that can be expressed using free body diagrams include **weight, applied, contact, normal, tensional, compressive, joint, frictional, ground reaction**, and **muscle**. In drawing or developing free body diagrams it is important **first to isolate the body**, then **draw and label all the external forces acting on the body**, then **mark all the angles and magnitudes of force**, and finally choose (or use a conventional) a **coordinate system** for positive and negative forces or moments. By adopting this approach to the solution of force and torque it is possible to analyze any system in either two or three dimensions.

Determine the weight of the athlete

Weight = mass × acceleration
W = m × g

In the case of our 75 kg athlete on the Earth with an acting acceleration due to gravity of 9.81 m/s^2 this is determined as follows:

W = m × g
W = 75 × 9.81
W = 735.75 Newtons (N)
W = 736 N

This is the weight that is experienced by the athlete. If the athlete stood on a force platform (see section F5) the weight recorded by the platform would be 736 N. The weight is, however, acting over both feet and if the subject were to stand on two force platforms (one under each foot) separate forces would be recorded of **368 N**. *Fig. C8.2* illustrates this in more detail.

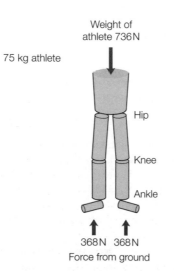

Weight of
athlete 736N

75 kg athlete

Hip

Knee

Ankle

368N 368N
Force from ground

Fig. C8.2.   Lower body frontal plane view

In order to determine what is the single compressive force acting over each knee joint it is necessary that we take off the weight (or effect) of one lower leg and foot. The anthropometric details for the weight of one lower leg and foot can be determined from the data presented in *Table C8.1*.

Table C8.1.   Anthropometric data where m is body mass in kg (derived from data presented in Winter 1990, p. 56)

Hand	$0.006 \times$ **m**
Forearm	0.016 m
Upper arm	0.028 m
Forearm and hand	0.022 m
TOTAL ARM	0.050 m
Foot	0.0145 m
Lower leg	0.0465 m
Upper leg	0.100 m
Foot and lower leg	0.061 m
TOTAL LEG	0.161 m

(The values in the table are derived (and reproduced with permission) from data presented in Winter, D. A. (1990) Biomechanics and Motor Control of Human Movement (2nd edition). Wiley-Interscience Publishers, New York (3rd edition published 2004))

From the table we can see that the anthropometric mass for one lower leg and foot segment is presented as $0.061 \times m$ (where m is the body mass of the athlete which is 75 kg in this case). Hence the weight of one lower leg and foot is determined as follows:

$$\text{Foot and lower leg} \quad = 0.061 \times m$$
$$= 0.061 \times 75$$
$$= \underline{\textbf{4.58 kg}}$$

To determine the weight of this foot and lower leg we multiply by the acceleration due to gravity.

$$\text{Weight of foot and lower leg} \quad = m \times g$$
$$= 4.58 \times 9.81$$
$$= 44.93 \text{ N}$$
$$= \underline{\textbf{45 N}}$$

This, therefore, is the weight of each lower leg and foot combination.

Hence, in order to determine the single joint (over one knee) compressive force acting over the knee joint during standing in a 75 kg athlete it is necessary to take the force value (i.e., the weight) of one lower leg and foot from the force acting over each foot (or under each foot from the ground). This will give the value acting over each knee (the summation of forces in the free body diagram).

It is clear that in a 75 kg athlete, 368 N of force will be acting under each foot (i.e., from the ground on the foot (the ground reaction)). It is also clear that in a 75 kg athlete the single compressive force acting over each knee joint can be calculated as follows:

Free body diagram ($\sum F = 0$)

1. Force acting on **knee joint** from weight of body through the thigh
i.e. upper leg (**−ve**) = **unknow**

2. Force acting from weight of lower leg and foot
(−ve) W = m × g (4.58 × 9.81)
= **45 N (−ve)**

Summation of external forces

3. Forces from ground acting on body (one leg) = **368 N (+ve)**
or $^1/_2$ body weight

Single joint compressive force acting over each knee

$\sum F = 0$
$368 + (-F_1) + (-45N) = 0$
$368 - 45 = F_1$
$\underline{323\ N = F_1}$ (acting over each knee joint (downward (−ve))

At this stage it is important to identify what is defined by the term compression in this context of knee joint forces. *Fig. C8.3* helps to illustrate this in more detail.

The force that has just been determined is the single joint compressive force acting over the knee during standing. It is the force that will cause the tibia (lower leg) and the femur (upper leg) to be compressed together (i.e., it is the force acting downwards over the joint; because the athlete is standing on the ground there will be a force acting vertically upwards thus causing this compression).

As the knee flexes the line of gravity will fall behind the knee joint line (i.e., behind the point in *Fig. C8.1*) and it will create a moment arm about which the force acts (*Fig. C8.4*). For most of the stance phase during walking the knee will flex through less than 20° of movement. Throughout this action of walking the muscle force in the quadriceps and hamstrings will be continuously changing to accommodate the moment (and the imbalance) caused about the knee joint. Muscle forces are affected by many factors which include friction, the momentum and mass of body and the velocity of the movement. Therefore, in order to understand statics it would be useful if we could first determine what muscle force would be required to hold the body in a position of flexion in a static posture.

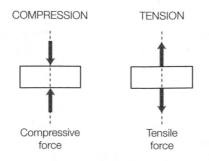

COMPRESSION   TENSION

Compressive force   Tensile force

Fig. C8.3.   Compressive and tensile forces

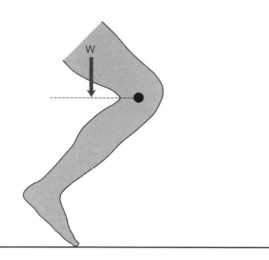

Fig. C8.4.   Moment arm during knee flexion

Muscle forces

Q. What is the muscle force required to maintain a static position of 20° of knee flexion during a unilateral stance (on one leg)?

Taking the previous example of the 75 kg athlete we will now try to work out the muscle force in the quadriceps muscle needed to hold a position of standing on one leg when the knee is in 20° flexion. Although during walking the action is dynamic (movement) this specific position will be assumed many times during the gait cycle in walking. In this process we first need to determine the single joint compressive force acting over the knee joint when a moment is created (such as in the case of knee flexion). Remember, however, in this case the athlete is standing on one leg only. *Fig. C8.5* illustrates the position in more detail.

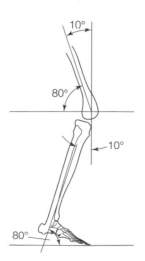

Fig. C8.5.   20° knee flexion position (standing on one leg)

Calculation of super-incumbent weight over one knee

(Super-incumbent means lying on top of and imposing pressure on something else).

$$75 \times 9.81 \quad = \quad 736 \text{ N (weight of athlete)}$$
$$0.061 \times 75 \quad = \quad 4.58 \text{ kg (mass of one lower leg and foot)}$$
$$4.58 \times 9.81 \quad = \quad 45 \text{ N (weight of one lower leg and foot)}$$
$$736 - 45 \quad = \quad \underline{\textbf{691 N}} \text{ (weight acting over one knee – standing on one leg)}$$

In order to use this value to calculate the muscle force needed to hold this static posture it is necessary to determine the perpendicular distance (**dw**) from the joint center (knee) to the line of action of this force (weight) which in this example is caused by the gravity. In addition, it is also necessary to determine the perpendicular distance from the joint center (knee) to the muscle line of pull (**dm**). *Fig. C8.6* illustrates these components in more detail.

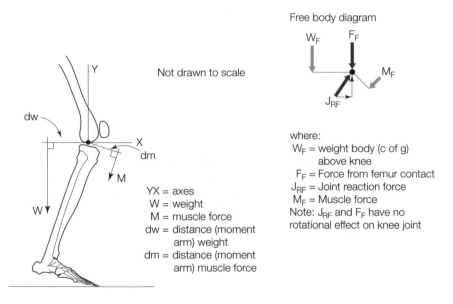

Free body diagram

where:
W_F = weight body (c of g) above knee
F_F = Force from femur contact
J_{RF} = Joint reaction force
M_F = Muscle force
Note: J_{RF} and F_F have no rotational effect on knee joint

Not drawn to scale

YX = axes
W = weight
M = muscle force
dw = distance (moment arm) weight
dm = distance (moment arm) muscle force

Fig. C8.6.   Moments and forces acting about the knee joint

These values (moment arms or perpendicular distances) are usually provided from either kinematic measurements using video digitization techniques (to determine **dw**) or from radiological measurements (to determine **dm**). In this case we can use the following values for **dw** and **dm** respectively 0.064 m and 0.05 m.

$$\textbf{dw} = 0.064 \text{ m (given)}$$
$$\textbf{dm} = 0.05 \text{ m (given)}$$

We now have a force system established in which we can utilize the second condition of equilibrium; which states that the sum of the moments are zero. That is, clockwise moments plus anti-clockwise moments equal zero ($\sum \textbf{M} = \textbf{0}$; no movement, static consideration) to solve the problem for the muscle force. *Fig. C8.7* helps to illustrate this system in more detail.

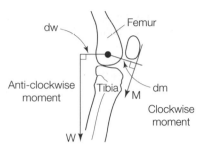

Fig. C8.7. Force/moment system in the knee during standing (one leg-static posture)

Second condition of equilibrium ($\Sigma M = 0$)

$$W.dw + M.dm = 0$$

where **W.dw** equals the anti-clockwise moment (+ve) and **M.dm** equals the clockwise moment (–ve) – remembering that a moment is defined as a force × a perpendicular distance.

Substitute from previous values:

$$W.dw + (-M.dm) = 0$$

$$(691 \times 0.064) + (-M \times 0.05) = 0$$

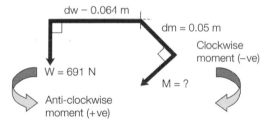

Note the positive sign in this case denotes an anti-clockwise moment.

$(44 \text{ Nm}) + (-M \times 0.05 \text{ Nm}) = 0$
$44 \text{ Nm} = M \times 0.05$
$44 / 0.05 = M$
880 N = M

This represents the muscle force needed in the quadriceps muscle in order to maintain this static position while standing on one leg with 20° of knee flexion as shown in *Fig. C8.5*. Hence, for a 75 kg athlete standing on one leg and holding this position of 20° knee flexion the quadriceps muscle force would be 880 N (1.20 × the subject's body weight).

In the calculation of this force it is important to note several factors.

- That the muscle force throughout walking is actually continually changing and the situation is not really one of a static consideration.
- That other factors play a role in the changing muscle and joint forces such as: friction of the foot and ground; friction in the joint; muscle line of pull; muscle complex arrangement; momentum and movement possessed by the body and equipment/shoes (external forces and torques).

- The problem should not really be considered as a two-dimensional problem (in the sagittal plane as in our calculations) and it is really a three-dimensional problem where all the force vectors are considered.
- Finally, it is important to add that **very rarely** is the problem or consideration of muscle and joint force **purely static**. An example of when the force may be potentially considered as static would be in the case of weightlifting where the athlete squats to the bottom of the weightlifting squat exercise with a bar and weights and then momentarily stops before beginning to rise again. At the moment that the body has stopped its descent (i.e., the vertical velocity downward will be zero) and at the point before it begins (i.e., providing the velocity has stayed zero momentarily) to rise this static calculation may be considered valid to work out muscle and joint force (although there will have been some momentum and velocity possessed by the body directly before this point and also an eccentric–concentric muscle contraction). However, in order to understand the dynamics of the problem it is important to have a good working knowledge of static applications of this method of calculation.

Calculation of joint forces

Fig. C8.8 shows the diagrammatic representation of the muscle force M, which we have just determined for the 75 kg athlete standing on one leg in the static position. The next stage is to determine the joint reaction force that is acting on the knee joint. In order to do this we need to consider both the muscle force M and the ground reaction force G (i.e., which are both acting upward on the lower leg – co-planar non-parallel external forces).

In order to solve a system where there are two non-parallel co-planar forces acting and the system is in equilibrium (i.e., static not moving in this case) we can use the first condition of equilibrium ($\sum F = 0$) to find the resultant of these two forces (the one force that is the equivalent of the two and the force that must be opposing the effect from the two non-parallel co-planar forces in order to maintain equilibrium). This can be achieved by constructing a free body diagram and expressing the two known forces at a point and then resolving for the third.

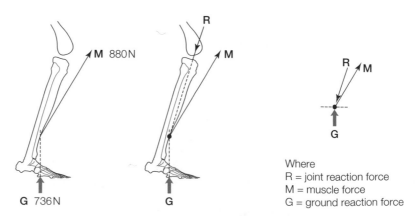

Where
R = joint reaction force
M = muscle force
G = ground reaction force

Fig. C8.8.   Co-planar force system around the knee (standing on one leg)

Resolution of a force system

The question that we need to answer in this example is what force is needed that would have the same effect as the two forces currently known. For example, if you push a box along a table with two forces (say one from each hand) and you push the box in a different direction and with a different force from each hand (independently and simultaneously) the box will move off in a resultant direction by an amount/distance that is dependent on the two applied forces. **What single force and what single direction would be required to have the same effect as the two other forces that you have just applied?** In the case of the knee joint the solution is achieved by using the same resolution of force system seen in Section C5. We currently have two forces that are acting on the joint (namely the muscle force (M) and the ground reaction force (G)) and we wish to find out the effect of these two forces on the joint (namely the joint reaction force (R)). In order to do this we can express the known forces at a point and using the first condition of equilibrium and basic trigonometry we can resolve these two forces into a single force which is the effect or the resultant of the other two.

However, before we can resolve these forces it is important to know at what angle or position the quadriceps muscle force (M) is acting. This can be determined again by either calculated kinematics of body position using video-digitization techniques or by radiological methods (i.e., such as x-ray or ultrasound techniques which, although primarily used to identify bone (x-ray), can be used to determine muscle and tendon line of pull (especially ultrasound techniques)). *Fig. C8.9* helps to illustrate the position and angle of the muscle force in more detail.

From *Fig. C8.9* it is clear that the muscle force needed from the quadriceps to hold this static position will act through the patella tendon. The patella tendon is attached to the quadriceps muscle (together the tendons of the four quadricep muscles form the patella tendon) and it is also attached to the tibial tuberosity (a bony eminence on the anterior part (front) of the lower leg). When the quadriceps muscle contracts it causes a tensile force in the tendon (i.e., because the tendon is attached at the tibia). This force from the quadriceps acts through the patella tendon and it is this position that we use in the resolution of force system. It can be noted from *Fig. C8.9* that the quadriceps tendon (patella tendon) is acting at 60° to the right horizontal in this example.

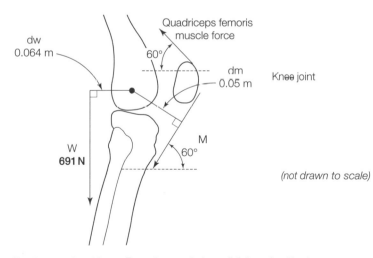

Fig. C8.9.   *Quadriceps Femoris muscle force (M) line of pull/action*

Resolution of forces

Both forces (muscle force (M) and ground reaction force (G)) are now expressed at a point (see *Figs C8.10* and *C8.11*) and it is important to identify that in this case the ground reaction force (which is usually a resultant of three forces) is acting vertically upwards. Normally during movement or dynamic action this would not be the case and the ground reaction force would be acting at an angle as it would be a resultant effect from a vertical, horizontal, and medial or lateral component. However, in this case (no movement static position) the ground reaction force can be expressed as a single force acting vertically.

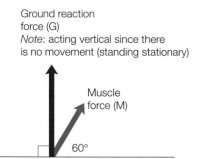

Ground reaction
force (G)
Note: acting vertical since there
is no movement (standing stationary)

Muscle
force (M)

60°

Fig. C8.10.   *Resolution of Muscle and ground reaction forces*

G = **736N**

M = **880N**

60°

Note: that the ground reaction force is now the
whole body weight force (75 × 9.81) because the
athlete is standing on one leg

Fig. C8.11.   *Muscle and ground reaction forces expressed at a point*

Therefore, resolving for two forces at a point:

Sum of vertical components (Fv)

F$\quad$= F sin θ

$\quad$= 736 + 880 sin 60°

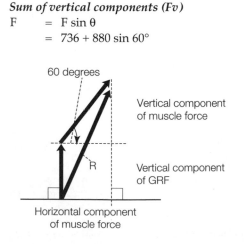

60 degrees

Vertical component
of muscle force

R

Vertical component
of GRF

Horizontal component
of muscle force

(Note: it is not necessary to include the 90° angle for the vertically acting ground reaction force because the sin of 90° is equal to 1.)

$\quad$= 736 + 880 × 0.8660

$\quad$= 736 + 762

Fv$\quad$= **1498 N**

Sum of horizontal components (Fh)

F   = F cos θ
      = 880 cos 60°

(Note: the vertical ground reaction force is acting perfectly vertically and hence it does not have a horizontal component.)

      = 880 × 0.5
Fh   = **440 N**

Magnitude of the resultant

R   = √ Fv² + Fh²   (Pythagoras)

where Fv equals the vertical component calculated previously and Fh equals the horizontal component calculated previously.

R   = √ 1498² + 440²
      = √ 2437604
      = **1561.3 N**

This force represents the resultant of the two forces shown in *Fig. C8.11*. It is the one force that will have the same effect as the two forces expressed in this figure. It is now important to establish the angle (direction) at which this force (which is a vector quantity) is acting.

Determination of angle

Tan θ = Fv
          Fh

       = 1498
          ‾‾‾‾
          440

Tan θ = 3.40
    θ = (inverse tangent) 3.40
    θ = **73.61°**
       = 73° 36′ (expressed as degrees and minutes)

Derived from

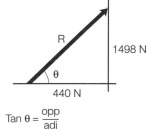

Tan θ = opp
        ‾‾‾
        adj

This is the angle at which the 1561.3 N resultant force is acting. It is now important that we transfer this force and its position to a diagram of the knee in order to understand how the joint is loaded. The joint reaction force which is created at the knee by the application of the ground reaction force and the muscle force will be equal and opposite to this resultant force. *Fig. C8.12* illustrates this in more detail.

From *Fig. C8.12* it is possible to see the joint reaction force in place around the knee joint. The force is 1561.3 N, which is 2.12 × the subject's body weight and it

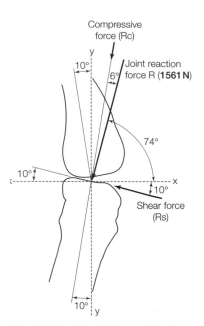

Fig. C8.12.   Knee joint reaction, shear and compressive forces

is acting at an angle of 74° (73.61°) to the right horizontal (in this case). This joint reaction force will cause a shear and compression force that is acting on the actual structures of the knee (such as the ligaments and other soft tissue structures). In order to calculate the shear and compression we use the same method as is applied to determine the joint reaction force.

Determination of knee joint compressive and shear forces

The compressive force is found to be perpendicular to the tibial plateau and is parallel to the long axis of the tibia. It is expressed as **Rc**.

The shear force is found to be parallel to the tibial plateau and it is known as **Rs**.

With the knee in this 20° flexion position, the thigh and lower leg form an angle of 10° with the vertical (see *Fig. C8.5*). Hence, the tibial plateau would also form an angle of 10° with the horizontal (because of the lower leg angle with the vertical). It is important to point out that in the many different angles/positions of knee flexion these angles (formed with the vertical) are not always equal. *Fig. C8.12* helps to show this in particular detail.

The compressive and shear forces are resolved from the joint reaction force expressed at a point, however, it is important to use the correct angles in this interpretation. In order to resolve this single joint reaction force into both the vertical and horizontal components (the other way around from what we did previously) we need to use the angle that is formed between the joint reaction force and the long axis of the tibia. *Fig. C8.13* identifies this angle of 6° (derived from 90° – (74° + 10°)) in more detail.

Calculation of joint compressive force (using $\theta = 6°$)

$\text{Rc} = \text{R} \cos \theta$
$\quad = 1561 \cos 6°$
$\quad = 1561 \times 0.994$
$\quad = \underline{\mathbf{1552\ N}}$

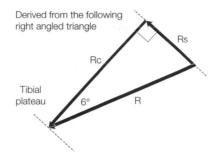

Calculation of joint shear force (using θ = 6°)

Rs = R sin θ
  = 1561 sin 6°
  = 1561 × 0.104
  = **162 N**

In this static calculation we have determined the muscle force needed to hold the limb stationary while standing on one leg for a 75 kg athlete. In addition, we have determined the **joint reaction force** and the **compression** and **shear force** components of the **joint reaction force**. The **shear force** in this example would be **pushing the tibia backwards** (i.e., causing the tibia to translate backward – posteriorly) **with respect to the femur** and therefore it would be stressing (placing a load on) the **posterior cruciate ligament** (*Fig. C8.14*). The **posterior cruciate ligament** is attached from its posterior location on the tibia to an anterior

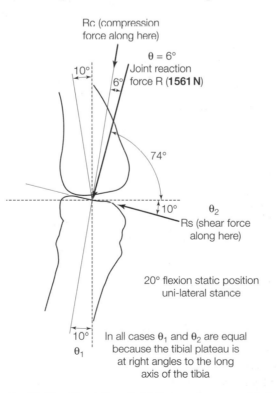

Fig. C8.13.   Resolution of joint shear and compression force

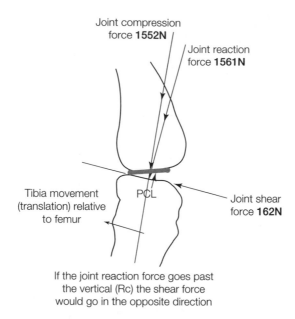

Fig. C8.14.   *Posterior tibial translation relative to femur*

location on the femur and it provides a **passive restraint** to **posterior (backward) movement of the tibia relative to the femur**. However, if the joint reaction force was at such an angle that when it was plotted onto the knee diagram with respect to the horizontal, it went past the vertical compression force which acts down the longitudinal axis of the tibia, the **shear force** created would be going in the **opposite direction** and it would **move the tibia anteriorly (i.e., forward) with respect to the femur**. In this case the motion (of the tibia) would be placing stress on the **anterior cruciate ligament. Ligaments** connect **bone to bone** and they **act passively** to **resist motion in a joint (i.e., provide stability)**. For example, the **passive restraint** (where passive is defined as receiving or being subjected to an action without responding or initiating an action) offered in resistance to **the drawer of the tibia anteriorly** with respect to the femur is provided by the **anterior cruciate ligament**. The cruciate ligaments in the knee are described by their attachments on the tibia. The **anterior cruciate ligament** is attached **anteriorly** at the **front of the tibia** and it extends backward and upward to be attached **posteriorly** on the **femur**. Hence, drawing or moving the tibia forwards relative to the femur (i.e., as in the case of the anterior drawer test used by many clinicians, which is termed the "Lachman" test) will place the anterior cruciate ligament under load (because it will resist this action). The ligaments of the knee provide **passive** support/stability (they are like pieces of string) whereas the muscles surrounding the knee provide **dynamic (active)** support/stability. Within biomechanics this passive and active role of ligaments and muscles can often be misunderstood. Although this application is two-dimensional and is statically determined (i.e., not moving), it is however showing an important mechanism for potential injury and knee ligament rupture.

Summary

In this static, one-legged (unilateral) standing posture we have calculated the following forces acting on and around the knee joint:

- Quadriceps muscle force needed to hold this static position – **880N (1.20 × body weight)**
- Knee joint reaction force – **1561 N (2.12 × body weight)**
- Knee joint compression force – **1552 N (2.10 × body weight)**
- Knee joint shear force – **162 N (0.22 × body weight)**

In addition, we have also seen that some of these forces may be responsible for injury development (such as in the case of the joint shear force where the force is trying to slide the tibia and femur apart). In particular it was clear that the position of this force can change and thus load different ligaments within the knee causing different injury potential mechanisms. Finally, in this single leg example, with no weights the quadriceps muscle force needed to hold this position was 1.20 × the subject's body weight which causes a joint reaction force of 2.12 body weights. This may have important loading implications for injury potential.

Applied example   EX 1   An athlete has a mass of 90 kg and begins to rise from a squatting position with a bar containing 150 kg (composite mass of the weights and bar 170 kg). The mass is distributed equally on both feet with the line of gravity falling 0.30 m behind the knee joint axis. The thigh is horizontal forming a 50° angle with the lower leg. The perpendicular distance from the joint center to the patella tendon line of action is 0.05 m. The patella tendon forms an angle of 35° with the horizontal.

Q.   What is the quadricep muscle force necessary to maintain this position and what are the tibio-femoral and patella-femoral joint reaction forces? Calculate the tibio-femoral shear and compressive forces and express your answers in terms of % body weight and absolute values.

Draw the free body diagrams to illustrate your answer.
Take the quadriceps tendon to be horizontal.

Schematic diagram

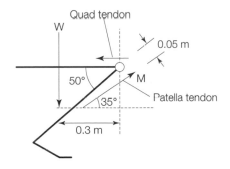

Applied example solution

Solution to question
Mass of athlete 90 kg
Weights and bar 170 kg
Shank and lower leg   0.061 × m (anthropometric data)

Weight over each leg
90 + 170 = 260 kg
260 × 9.81 = 2550.6 N

Divide by 2 per leg (i.e., under each foot)
1275.3 N ground reaction through each leg

Minus weight of 1 lower leg + foot
0.061 × 90 = 5.49 kg
5.49 × 9.81 = 53.86 N
1275.3 – 53.86
= 1221.44 N Acting over each knee
= 1221.44 N

Determination of quadriceps femoris muscle force
2nd condition of equilibrium $\Sigma M = 0$
W.dw + M.dm = 0
1221.44 × 0.30 + M × 0.05 = 0
366.43 + M × 0.05 = 0

$$M = -\frac{366.43}{0.05}$$

M = –7328.6 N

Body weight = 90 × 9.81 = 882.9

% Body weight $= \dfrac{7328.6}{882.9}$

$$= 8.30 \times Bw$$

Calculation of joint reaction force

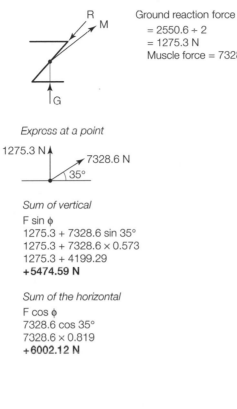

Ground reaction force
= 2550.6 ÷ 2
= 1275.3 N
Muscle force = 7328.6 N

Express at a point

1275.3 N

7328.6 N

35°

Sum of vertical

F sin ϕ
1275.3 + 7328.6 sin 35°
1275.3 + 7328.6 × 0.573
1275.3 + 4199.29
+5474.59 N

Sum of the horizontal

F cos ϕ
7328.6 cos 35°
7328.6 × 0.819
+6002.12 N

$\phi_1 - 42°22'$
$\phi_2 = 90° - (40° - \phi_1)$

Determination of resultant

$R = \sqrt{FV^2 + FH^2}$
$= \sqrt{5474.59^2 + 6002.12^2}$
$= \sqrt{65996580.16}$
= 8123.83 N (9.20 × Bw)

Angle of joint reaction force

$\tan \phi = \dfrac{FV}{FH}$

$= \dfrac{5474.59}{6002.12}$

= 0.9121
ϕ = 42.36°
ϕ = 42°22'

Transfer to diagram

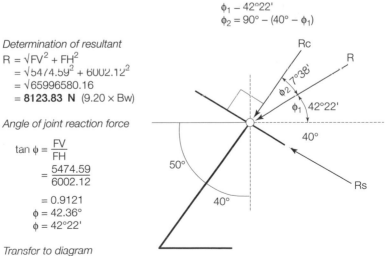

not drawn to scale

Compression force
R cos φ
8123.83 cos 7°38'
8123.83 × 0.9911
8151.52 N   (9.23 × Bw)

Shear force
R sin φ
8123.83 sin 7°38'
8123.83 × 0.1328
1078.84 N   (1.22 × Bw)

Calculation of patella joint reaction force

Quads tendon
7328.6 N (Muscle force)

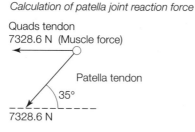

Patella tendon

35°

7328.6 N

Sum of vertical
F sin φ
−7328.6 sin 35°
−7328.6 × 0.5735
−4202.95 N

Sum of horizontal
F cos φ
7328.6 + 7328.6 cos 35°
7328.6 + 7328.6 × 0.8191
7328.6 + 6002.86
+13331.46

Resultant
$R = \sqrt{FV^2 + FH^2}$
$R = \sqrt{4202.95^2 + 13331.46^2}$
$R = \sqrt{17664788.7 + 177727825.7}$
$R = \sqrt{195392614.4}$
R = 13978.29   (15.83 × Bw)

Angle of application
Actually bisects the angle contained between
quadriceps and patella tendons
ie. 35° ÷ 2 = 17.5°

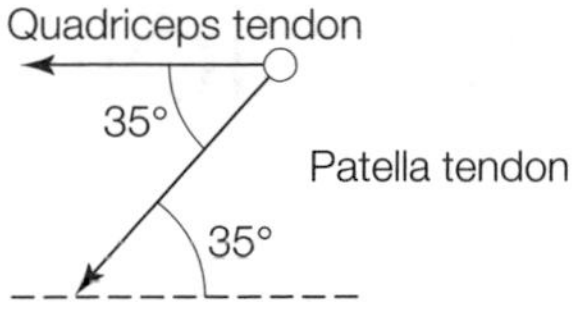

Quadriceps tendon

35°

Patella tendon

35°

alternate angles between
two parallel lines

$$\tan \phi = \frac{FV}{HV}$$

$$= \frac{4202.95}{13331.46}$$

$$= 0.3152$$
$$= 17.49°$$
$$= \mathbf{17° \ 29' \ 24.00''}$$

Summary

Muscle force	7328.6 N (8.30 × Bw)
Joint reaction force	8123.83 N (9.20 × Bw)
Compressive force	8151.52 N (9.23 × Bw)
Shear force	1078.8 N (1.22 × Bw)
Patella femoral force	13978.29 (15.83 × Bw)

C9 THE ESTIMATION OF MUSCLE AND JOINT FORCES – SIMPLE DYNAMIC APPLICATIONS

Key Notes

Introduction

Inverse dynamics calculations form the basis of mathematical approaches that are used in simulation and modeling within biomechanics. Dynamics utilize the acceleration approach to solving problems and they allow the biomechanist to be able to determine realistic muscle and joint forces.

Moment of inertia

Required for calculation of dynamic forces.

The moment of inertia for regular shaped bodies (where the mass is essentially point center located) is determined by:

$$I = \sum m\, r^2$$

The moment of inertia for the human body is determined by using the radius of gyration:

$$I = mk^2$$

Radius of gyration

Required for the calculation of dynamic forces.

The radius of gyration is the distance between the axis of rotation and the point at which the mass of a non-rigid body (i.e., a distributed mass) is considered to be concentrated. It is important to determine about which axis the limb or joint is rotating (proximal or distal). The lower arm segment can rotate about the distal axis of rotation (i.e., the wrist/hand) or it can rotate about the proximal axis of rotation (i.e., the elbow joint). Distal is the point that is furthest away from the attachment of the part to the body and proximal is the point that is nearer to the point of attachment of the part to the body.

For static solutions

Use the first and second conditions of equilibrium.

For dynamic solutions

Combine these with the acceleration approach.

In static situations the acceleration will be equal to zero whereas in dynamic situations there is a potential for acceleration (i.e., there could also be a constant velocity situation where the acceleration would be zero).

Dynamic solution

$$\sum CWM + \sum ACWM = I\alpha$$

where $I\alpha$ = torque and when $\alpha = 0$ the equation becomes $\sum M = 0$ (second condition of equilibrium).

Summary

Inverse dynamics can show that the forces on the joints during movement either increase or decrease (i.e., they can decrease under the effect of gravitational acceleration) when the limb is to be accelerated in a specific direction. This change in velocity (i.e., acceleration) and the potential increase in joint and muscle forces can lead to the possibility of injury. For example, weightlifting, when the athlete has to accelerate the weight and the limbs during an arm curling exercise, can lead to increased muscle and joint forces.

Introduction

Dynamic calculations form the basis of complex inverse dynamics approaches that are used extensively in both modeling and simulation applications in biomechanics. The acceleration approach is used to solve kinetic problems and determine the cause–effect of movement (acceleration) from muscle and joint forces and torques. In static applications the clockwise moments (CWM) or torques are balanced by the anti-clockwise moments (ACWM) resulting in a zero angular acceleration (i.e., no movement). In dynamic applications the net torque produced is not equal to zero and its effect is to produce an angular acceleration. The net torque has the same causal relationship to angular acceleration that net force has to linear acceleration. The resistance to changes in angular velocity is quantified by the moment of inertia and it is an important consideration in these calculations.

Moment of inertia

Moment of inertia is defined as the resistance of an object to start or stop rotating and for bodies where the mass is concentrated at a point, it is determined by how the mass of the object is distributed around the axis of rotation. It is generally defined by:

$$I = \sum m \, r^2 \qquad \text{(C9.1)}$$

where
I = moment of inertia
m = mass
r = distance of mass center from the axis of rotation

If the object is rotated about a different axis or if the mass is redistributed then the moment of inertia changes (as the distance of **r** will change).

The moment of inertia is different for different body shapes and unless a defined axis of rotation is identified, the moment of inertia has little meaning. Generally the moment of inertia is defined by an axis passing through the center of gravity of the object (I_{CofG}) and this provides a reference value for the object from which further calculations can be made. This would be necessary, for example, when there is a change in the axis of rotation from the center of gravity to, say, some other point such as the end of an object (e.g., the handle of a racket or the end of a body segment such as the shank). For a more thorough understanding of this concept the reader is referred to section C3.

Segments of the human body (e.g., shank, thigh, forearm, or head) rotate about axes of rotation at the end of the segment which are referred to as proximal or distal. The proximal end is defined as the point that is nearest to the point of attachment of the limb/segment to the body while the distal end is the point farthest away from the point of attachment of the limb/segment to the body.

Sometimes the body rotates about either proximal or distal ends. For example, in the case of the subject doing a cartwheel the rotation of the body/segments would be about the wrist (the distal end) whereas in the case of doing an arm curl with weights the rotation would be at either the elbow or the shoulder (the proximal end for the lower arm and upper arm segment, respectively).

The reference moment of inertia value (I_{CofG}) can now be revised to include the effect of the new axis of rotation using the parallel axis theorem.

$$I_A = I_{CofG} + md^2 \qquad\qquad \textbf{(C9.2)}$$

where

I_A = moment of inertia of the body or segment about an axis through a point, A

I_{CofG} = the moment of inertia about a parallel axis through the center of gravity of the object

m = mass

d = radius of rotation (the distance from the axis of rotation A to the center of gravity of the object)

Radius of gyration

To simplify this calculation, the new moment of inertia (I_A, equation C9.2) is equated to the general form of the formula as given in equation C9.1

$$I_A = mk^2 \qquad\qquad \textbf{(C9.3)}$$

where **k** is termed the **radius of gyration**. The use of the radius of gyration is helpful in calculations as once a segment mass is known (from segmental data – see *Table C3.2*) the radius of gyration (which is also given as segmental data – see *Table C9.1*) can be used easily to calculate the moment of inertia of a segment without having to perform the larger number of calculations that would be required by equation C9.2.

Table C9.1. Radii of gyration as percentages of segmental lengths (derived from Winter (1990), pp. 56–57)

Segment	From proximal end %	From distal end %
Head, neck, and trunk	83.0	60.7
Arm (upper)	54.2	64.5
Forearm	52.6	64.7
Hand	58.7	57.7
Upper limb	64.5	59.6
Forearm and hand	82.7	56.5
Thigh	54.0	65.3
Leg	52.8	64.3
Foot	69.0	69.0
Lower limb	56.0	65.0
Leg and foot	73.5	57.2

(The values in the table are derived (and reproduced with permission) from data presented in Winter, D. A. (1990) Biomechanics and Motor Control of Human Movement (2nd edition). Wiley-Interscience Publishers. New York. (3rd edition published 2004))

Calculation of muscle and joint forces during a dynamic movement

Consider *Fig. C9.1* and the free body diagram shown in *Fig. C9.2*, which shows the flexion of the elbow (90°) with the forearm in the horizontal position in a 75 kg athlete. The distance of the center of gravity (Fw) of the forearm to the proximal axis of rotation (elbow joint) is 0.154 m and the muscle force (F_M) acts at 80° to the limb and 0.05 m from the proximal joint axis. The joint reaction force (F_J) acts at the proximal joint. A question we may ask is: **What is the muscle force required to maintain this position with the horizontal and what is the muscle force required to accelerate the limb in flexion (i.e., counter clockwise) at 80 rads/s²? In each case the joint reaction force can also be determined.**

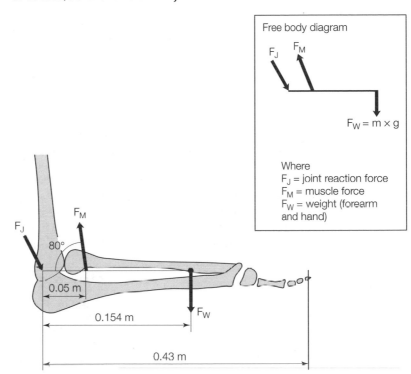

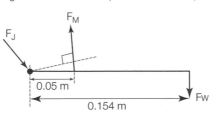

Fig. C9.1. Arm in 90° position of flexion (lower arm horizontal)

Fig. C9.2. Free body diagram for muscle and joint force calculation (static calculations in the first instance)

Case 1. No segment acceleration (static situation)
First, determine the combination of forearm and hand weight.

The anthropometric value for the mass of the forearm and hand is given as the segmental mass ratio × body mass (determined from Table C3.2). Hence

Mass of forearm and hand = 0.022 × body mass = 0.022 × 75 = **1.65 kg**
Weight of forearm and hand **W = m × g** = 1.65 × 9.81 = **16.19 N**

From *Fig. C9.1* we can see that the combined center of mass (forearm and hand) is located 0.154 m from the elbow joint center of rotation. The elbow flexor muscle pulls at an angle of 80° to the left horizontal and it inserts 0.05 m from the joint center of rotation (see free body diagram, *Fig. C9.2*).

We can now solve for the static situation (no movement – to hold the position) Using the second condition of equilibrium and the free body diagram

$$- \sum CWM + \sum ACWM = 0$$
$$- (16.19 \times 0.154) + (F_M \times \sin 80 \times 0.05) = 0$$

Note: the convention is that the anti-clockwise moment is positive and the clockwise moment is negative. The sin 80° (in the anti-clockwise moment) determines the perpendicular distance that the muscle force is acting from the elbow joint center of rotation. Re-arranging and evaluating gives:

$$(F_M \times 0.984 \times 0.05) = (16.19 \times 0.154)$$
$$F_M \times 0.049 = 2.49$$
$$F_M = \frac{2.49}{0.049} = \underline{\mathbf{50.82\ N}}$$

This represents the muscle force that is required in the biceps brachii muscle (elbow flexor) in order to hold the limb in the static position of 90° elbow flexion. In addition, it is now possible (using $\sum F = 0$) to calculate the joint reaction force acting on the ulna/radius (lower arm) from the contact with the humerus (upper arm). This is illustrated in *Fig. C9.3*. However, it is important to remember that this is only a two-dimensional application and for a more detailed understanding of the joint and ligament forces, a three-dimensional approach is needed.

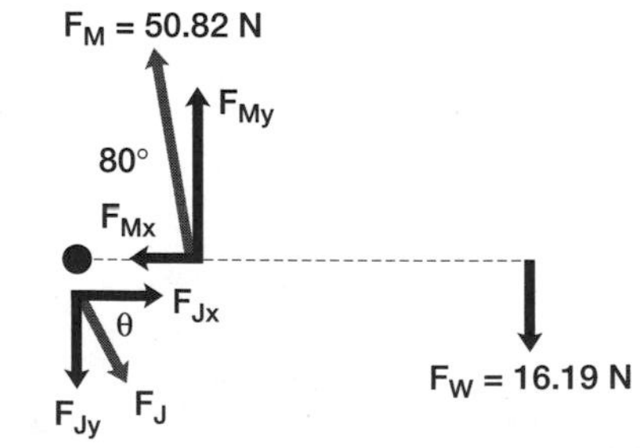

The muscle and joint forces are divided into their vertical and horizontal components respectively ($\mathbf{F_{My}}$, $\mathbf{F_{Mx}}$, $\mathbf{F_{Jy}}$, $\mathbf{F_{Jx}}$)

The vertical component of muscle force is
$\mathbf{F_{My}} = F_M \sin \theta = 50.82 \sin 80 = 50.05$ N
The horizontal component of muscle force is
$\mathbf{F_{Mx}} = F_M \cos \theta = 50.82 \cos 80 = 8.82$ N

Using $\sum F = 0$ (i.e., $\sum Fx = 0$ and $\sum Fy = 0$)

the net vertical force ($\sum Fy = 0$)
$$\sum Fy = \mathbf{F_{My}} - \mathbf{F_{Jy}} - \mathbf{F_W} = 0$$
therefore
$$\mathbf{F_{Jy}} = \mathbf{F_{My}} - \mathbf{F_W}$$
$$= 50.05 - 16.19 = 33.86\ N$$

the net horizontal force ($\sum Fx = 0$)
$$\sum Fx = -\mathbf{F_{Mx}} + \mathbf{F_{Jx}} = 0$$
therefore
$$\mathbf{F_{Jx}} = \mathbf{F_{Mx}} = 8.82\ N$$

$F_M = 50.82$ N

F_{My}

80°

F_{Mx}

F_{Jx}

θ

F_{Jy} F_J

$F_W = 16.19$ N

Note: the joint reaction force F_J is moved slightly for clarity

Therefore the resultant elbow joint force
$= \sqrt{(\mathbf{F_{Jx}}^2 + \mathbf{F_{Jy}}^2)}$ at $\tan^{-1} (\mathbf{F_{Jy}}/\mathbf{F_{Jx}})$
$= \sqrt{(33.86^2 + 8.82^2)}$ at $\tan^{-1} (33.86/8.82)$
$= \mathbf{35\ N\ at\ an\ angle\ \theta = 75.4°}$

Fig. C9.3.   Calculating the joint reaction force (F$_J$) at the elbow

Case 2. Determine the muscle force when the limb is being accelerated counter-clockwise (anti-clockwise) at 80 rads/s^2

As the muscle force increases, the forearm flexes. To achieve a rapid flexion (i.e., with an angular acceleration of 80 rads/s^2) *the muscle force must be quite high.* In order to determine the muscle force required to produce this acceleration it is necessary to combine the second condition of equilibrium with the acceleration approach to produce the following equation:

$$-\sum CWM + \sum ACWM = I\alpha$$

where

I = moment of inertia of segment about a specific axis of rotation
α = angular acceleration (rads/s^2)

Remembering from section C2 that **Iα = torque** and when $\alpha = 0$ (the static case) the second condition of equilibrium is evident ($\sum M = 0$). This equation (shown above) contains the moment of inertia of the limb, so it has to be calculated first. To calculate the moment of inertia of the forearm and hand when it rotates about the elbow (proximal) joint the formula developed at equation **C9.3** is used:

$$I_{elbow} = m_{forearm+hand} * k^2_{elbow} \qquad (C9.3)$$

The mass of the forearm and hand was calculated previously. So it is necessary to calculate the radius of gyration (k$_{elbow}$):

Forearm and hand length = 0.43 m (total length – see *Fig. C9.1*)
Axis of rotation   = elbow (proximal joint)
Radius of gyration (k$_{elbow}$) = 82.7% of segment length (*Table C9.1*, proximal)
                        = 82.7% × 0.43 = 0.356 m from axis of rotation

therefore
Moment of inertia (I$_{elbow}$)   = mk^2 = 1.65 × 0.356^2 =   **0.209 kg.m^2**

Now substituting this into the formula (noting that the clockwise moment is negative):

$$-\sum CWM + \sum ACWM \qquad = I\alpha$$
$$-(16.19 \times 0.154) + F_M \times 0.049 \qquad = 0.209 \times 80$$

Note: the left-hand side is the same as the static case, so evaluating we have the following:

$$-2.49 + F_M \times 0.049 = 16.72$$
$$F_M \qquad = (16.72 + 2.49)/0.049$$
$$F_M \qquad = 19.21/0.049$$
$$F_M \qquad = \textbf{392.0 N}$$

Hence, it can be seen that the muscle force required in the biceps brachii to accelerate the limb counter-clockwise at 80 rads/s^2 is **7.7 × the muscle force** required to keep the limb stationary (50.82 N (stationary) and 392.0 N (moving)). Considering this increasing force, which is evident in the dynamic situation, the importance and significance for injury potential becomes apparent.

Calculation of the joint reaction force
The joint forces can be calculated in a manner similar to the static case illustrated in *Fig. C9.3*, but taking into account the acceleration of the center of mass in the vertical and horizontal direction (i.e., it is necessary to combine the first condition

of equilibrium with the acceleration approach to produce the following equations):

$$\sum Fx = m.a_x$$
$$\sum Fy = m.a_y$$

where a_x and a_y are the linear accelerations in the horizontal (x) and vertical (y) directions. In this example the angular acceleration is chosen to be constant, but as the limb rotates about the joint the linear accelerations will vary depending on the angle made by the limb. It is possible to calculate the accelerations for any angle of the limb, using the relationships covered in section A4 on linear-angular motion. The accelerations are given as:

$$a_x = -r.\alpha.\sin\phi$$
$$a_y = r.\alpha.\cos\phi$$

where **r** is the radius of rotation of the limb's center of mass about the axis of rotation (0.154 m), α is the angular acceleration (80 rads/s²), and ϕ is the angle of the limb to the horizontal.

In this example we shall calculate the joint reaction force for the limb's starting position where $\phi = 0$. This means that $a_x = 0$ and $a_y = 12.32$ m/s². The revised free body diagram for an accelerating system can now be seen in *Fig. C9.4*.

It should be noted in this example some real characteristics of the joints are ignored because they are generally considered to be small and have little influence on the calculations. For example, the friction occurring at the joint is ignored as the synovial fluid between the joint surfaces reduces this to a negligible amount.

The muscle and joint forces are divided into their vertical and horizontal components respectively (F_{My}, F_{Mx}, F_{Jy}, F_{Jx})

The vertical component of muscle force is
$F_{My} = F_M \sin\theta = 392.0 \sin 80 = 386.0$ N
The horizontal component of muscle force is
$F_{Mx} = F_M \cos\theta = 392.0 \cos 80 = 68.1$ N

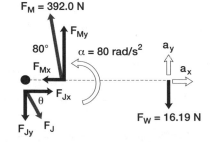

$F_M = 392.0$ N

Note: the joint reaction force F_J is moved slightly for clarity

Using $\sum F = ma$ (i.e., $\sum Fx = ma_x$ and $\sum Fy = ma_y$)

the net vertical force ($\sum Fy = ma_y$)
$$\sum Fy = F_{My} - F_{Jy} - F_W = ma_y$$
therefore
$$F_{Jy} = F_{My} - F_W - ma_y$$
$$= 386.0 - 16.19 - (1.65)(12.32)$$
$$= 349.5 \text{ N}$$

the net horizontal force ($\sum Fx = ma_x = 0$)
$$\sum Fx = -F_{Mx} + F_{Jx} = 0$$
therefore
$$F_{Jx} = F_{Mx} = 68.1 \text{ N}$$

Therefore the resultant elbow joint force
$$= \sqrt{(F_{Jx}{}^2 + F_{Jy}{}^2)} \text{ at } \tan^{-1}(F_{Jy}/F_{Jx})$$
$$= 356.1 \text{ N at an angle } \theta = 79.0°$$

Fig. C9.4.   Calculating the joint reaction force (F_J) at the elbow when accelerating

Applied question   The following problem tries to calculate the muscle force required to accelerate the limb in flexion (i.e., anti-clockwise) with a 10 kg mass held in the hand (as in the case of an arm curl during weightlifting). The following question addresses this issue using a typical weightlifting example and requires you to calculate the muscle and joint forces using the method shown previously.

Q. *Figs C9.5 and C9.6 (free body diagram) identify an athlete holding a 10 kg weight (mass) in the hand 0.35 m from the elbow joint axis. What is the muscular force required to maintain this static position with the horizontal when the athlete is holding this 10 kg weight stationary and what is the muscle force required when the limb is being accelerated counter-clockwise at 80 rads/s²? The elbow joint reaction forces are present in each case.*

The athlete has a mass of 75 kg (as in the previous example shown in this section). Use the anthropometric, radii of gyration and inertia data given in the text thus far (i.e., this section and section C8). The distance from the elbow joint center to the hand is considered to be 0.43 m (total length), which is used for the calculation of the radii of gyration (even though the weight is at a position 0.35m from the elbow joint center of rotation).

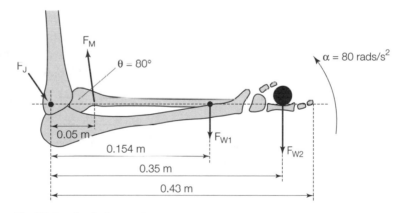

Fig. C9.5.   Applied example

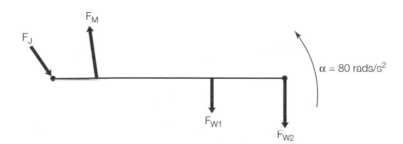

Fig. C9.6.   Free body diagram of accelerated arm shown in Fig. C9.5

Static case

The combination of forearm and hand weight = **16.19 N** as calculated in the previous case. The 10 kg weight (mass) is located 0.35 m from the elbow joint (see *Fig. C9.5*). Using the second condition of equilibrium and the free body diagram:

$$- \sum CWM + \sum ACWM = 0$$
$$- (16.19 \times 0.154) - (10 \times 9.81 \times 0.35) + (F_M \times \sin80 \times 0.05) = 0$$

Rearranging $F_M = (2.49 + 34.3) / 0.049$
and evaluating $F_M = \textbf{750.8 N}$

The muscle and joint forces are divided into their vertical and horizontal components respectively (F_{My}, F_{Mx}, F_{Jy}, F_{Jx}).
The vertical component of muscle force is

$$F_{My} = F_M \sin\theta = 750.8 \; \sin80 = 739.4 \text{ N}$$

The horizontal component of muscle force is

$$F_{Mx} = F_M \cos\theta = 750.8 \; \cos80 = 130.4 \text{ N}$$

Using $\sum F = 0$ (i.e., $\sum Fx = 0$ and $\sum Fy = 0$)

Net vertical force $\sum Fy = F_{My} - F_{Jy} - F_{W1} - F_{W2} = 0$
Therefore $F_{Jy} = F_{My} - F_{W1} - F_{W2}$
                                                  $= 739.4 - 16.19 - 98.1 = 625.1 \text{ N}$
Net horizontal force $\sum Fx = - F_{Mx} + F_{Jx} = 0$
Therefore $F_{Jx} = F_{Mx} = 130.4 \text{ N}$

Therefore the resultant elbow joint force $= \surd\,(F_{Jx}{}^2 + F_{Jy}{}^2)$ at $\tan^{-1}(F_{Jy}/F_{Jx})$
                                                  $= \surd(625.1^2 + 130.4^2)$ at $\tan^{-1}(625.1/130.4)$
                                                  $= \textbf{638.6 N at an angle } \theta = \textbf{78.2°}$

Dynamic case

The moment of inertia of the system (forearm plus weight) around the elbow is

$$I_{elbow} = m_{forearm} k^2_{forearm} + m_{weight}.r^2_{weight}$$
$$= 1.65 \times (82.7\% \times 0.43)^2 + 10 \times (0.35)^2$$
$$= 0.209 + 1.225 = 1.434 \text{ kg.m}^2$$

The location of the forearm plus weight centre of mass from the elbow axis of rotation, D, is given by

$$(m_{forearm} + m_{weight}).D = m_{forearm}(0.154) + m_{weight}.(0.35)$$
therefore $D = 0.322 \text{ m}$

Now using the second condition of equilibrium and the free body diagram for the dynamic case:

$$-\sum CWM + \sum ACWM = I\alpha$$
$$- (16.19 \times 0.154) - (10 \times 9.81 \times 0.35) + (F_M \times \sin80 \times 0.05) = 1.434 \times 80$$

Rearranging $F_M = (2.49 + 34.3 + 114.72) / 0.049$
and evaluating $F_M = \textbf{3092 N}$

The joint forces can be calculated in a manner similar to the static case and with the arm in the starting position (i.e., $\phi = 0$), the arm centre of mass acceleration is $a_x = 0$ and $a_y = \alpha D = 25.76 \text{ m/s}^2$.

The muscle and joint forces are divided into their vertical and horizontal components respectively ($\mathbf{F_{My}}$, $\mathbf{F_{Mx}}$, $\mathbf{F_{Jy}}$, $\mathbf{F_{Jx}}$)

The vertical component of muscle force is:

$$\mathbf{F_{My}} = F_M \sin\theta = 3092 \sin80 = 3045N$$

The horizontal component of muscle force is:

$$\mathbf{F_{Mx}} = F_M \cos\theta = 3092 \cos80 = 536.9N$$

Using $\sum F = ma$ (i.e., $\sum Fx = ma_x$ and $\sum Fy = ma_y$), the net vertical force ($\sum Fy = ma_y$) is

$$\sum Fy = \mathbf{F_{My}} - \mathbf{F_{Jy}} - \mathbf{F_{W1}} - \mathbf{F_{W2}} = ma_y$$

Therefore
$$\begin{aligned}\mathbf{F_{Jy}} &= \mathbf{F_{My}} - \mathbf{F_{W1}} - \mathbf{F_{W2}} - ma_y \\ &= 3045 - 16.19 - 98.1 - (1.65 + 10)(25.76) \\ &= 2631N\end{aligned}$$

The net horizontal force ($\sum Fx = ma_x = 0$)

$$\sum Fx = -\mathbf{F_{Mx}} + \mathbf{F_{Jx}} = 0$$

Therefore
$$\mathbf{F_{Jx}} = \mathbf{F_{Mx}} = 536.9N$$

Therefore the resultant elbow joint force $= \sqrt{(\mathbf{F_{Jy}}^2 + \mathbf{F_{Jx}}^2)}$ at $\tan^{-1}(\mathbf{F_{Jy}}/\mathbf{F_{Jx}})$
$$= \sqrt{(2631^2 + 536.9^2)} \text{ at } \tan^{-1}(2631/536.9)$$
$$= \textbf{2685 N at an angle } \theta = \textbf{78.5}°$$

D1 Work, power, and energy

Key Notes

Work	Work refers to overcoming resistance by the application of a force. Evidence that the resistance has been overcome is seen from the movement of the point of application of the force. Thus, the work done (W) by a force is defined as the product of the force (F) applied to an object and the amount of displacement (d) of the object in the direction of the force and is given by the equation W = F.d. The units of work are Joules (J).
Positive and negative work	When the direction of the force is in the same direction as the motion of its point of application, positive work is being done (e.g., lifting a barbell from the ground). For humans to do positive work they need to expend chemical energy through their muscles. When the direction of the force is in the opposite direction to the motion of its point of application, negative work is being done (e.g., lowering a barbell from the ground). For humans to do negative work they also need to expend chemical energy through their muscles, but often some of this work can be stored as strain energy in the body's tendons.
Power	Power (P) is a term used to describe the rate at which work is being done. For example, lifting a barbell slowly is different from lifting it rapidly, even though the final outcome in terms of the height lifted is the same. When the movement is completed more rapidly, greater power is needed. Power is defined as the rate at which work is being done and if the work is done (W) in a time interval t, then an equation for power can be given as P = W/t. The units of power are Watts (W) – note this has the same **symbol** as work. The former is a unit while the latter is a mechanical concept.
Energy	Energy is defined as the capacity to do work or to perform some action and can be considered a something that is "stored" or "possessed". The units of energy are Joules, which is the same as for work. There are different forms of energy. The most important to sport and exercise biomechanics are potential energy (gravitational and strain) and kinetic energy (linear and angular).
Work and energy	Work and energy have the same units and are closely related. Energy can be stored but work cannot. Essentially work is energy in motion. Energy changes from one form to another by the process of doing work.

Work

The term **work** is commonly used in everyday language loosely to refer to the effort or exertion expended in performing a task. However, when used in a scientific context the term work takes on a specific meaning and refers to the movement of an object by the application of a force. Thus, the **mechanical work** (W) done by a force is defined as the product of the **force** (F) applied to the object and the amount of **displacement** (d) of the object in the direction of the force. This can be written as:

$$\text{mechanical work} = \text{force . displacement}$$
$$W = F. d \qquad\qquad (D1.1)$$

If the force is measured in Newtons and the displacement in meters, then the units of work are **Joules** (J).

A question that is often asked is "if work is being done where does it go?" Work is not a quantity that can be seen, but its effects can be. Usually the effects of doing work are to see a change in the position of an object, often by being moved upward against gravity, or being deformed, or by increasing its velocity (see Examples in *Figs D1.1* and *D1.2*).

Problem

A person is performing a bench press and trying to move a load of 60 kg. With great effort the barbell slowly moves from the chest to a full lock out position, a distance of 40 cm.
How much work is done? Where does this work go?

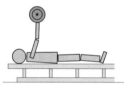

Solution

The work done is given by the relationship W = F.d
The force applied to the barbell is equal to the gravitational load of the weights (as the load is moved slowly, i.e., is not accelerated) given by F = 60. g Newtons.
The displacement of the force is given by the distance moved by the hands placed on the barbell (40 cm) which is 0.4 meters.
Therefore the work done is given by W = (60. g).(0.4) = **235.4 Joules**.
Where does the work go? It goes into raising the barbell against gravity.

Fig. D1.1.   Example 1

Problem

A force of 1000 N is applied to a soccer ball during a kick and it deforms 10 cm
How much work has been done on the ball? Where does this work go?

Solution

The work done is given by the relationship W = F.d.
The force applied to the ball is given as 1000 Newtons.
The displacement of the force is given by the deformation of the ball (10 cm) which is 0.1 meters.
Therefore the work done is given by W = (1000).(0.1) = **100 Joules**.
Where does the work go? It goes into deforming the ball.

Fig. D1.2.   Example 2

Positive and negative work

An important implication of the above definition of work is that if a force is acting but does not move the object, that is, it does not move its point of application, then no work is done. Thus, if a person tries to lift a barbell but is unable to do so, then no work is done even though a lot of effort has been expended. The muscles of the body have contracted and much muscular effort has been expended but this has not resulted in the object being overcome, so from the point of view of the barbell no work has been done on it.

The work done can be classified as either **positive work** or **negative work** depending on whether the force moves in the direction of movement of the object or in the opposite direction. In the case of the bench press example, lifting the barbell upwards results in positive work being done on the barbell as the force of the hand applied to the barbell is in the same direction as the movement. The outcome of this positive work is that the barbell has changed its position against gravity. On the downward movement the force applied by the hand is in the same upward direction, but the direction of movement of the barbell is downwards. This introduces a negative sign into equation D1.1 and so the work done becomes negative. This means that when the barbell is being lowered, the barbell is doing work on the person, rather than when it is being lifted where the person is doing work on the barbell. Where does this negative work go? Often it is dissipated as heat in the muscles so it is lost. Of interest in sports performance is the fact that some of this work can be used to deform structures of the body (most notably the muscles and tendons) which, given the right technique, can be recovered again during their shortening. This phenomenon can explain a range of observations in sport from why people "cheat" when doing tasks such as a barbell bench press (by bouncing the bar off the chest) to the greater efficiency of running compared with walking at certain speeds (due to the stretching and recoil of the tendons of the leg).

Power

Power (P) is a term used to describe the rate at which work is being done. For example, lifting a barbell slowly is different from lifting it rapidly, even though the final outcome in terms of the height lifted is the same. A second example would be a cyclist climbing up a hill either slowly or quickly where again the outcome is the same but the effort involved is greater when the hill is climbed more rapidly. The difference between the two movements in each case is the power generated. When the movement is completed more rapidly, a greater power is needed. **Power** is defined as the rate at which work is being done. If the work (W) is done in a time interval t, then an equation for power can be given as:

$$\text{Power} = \text{work} / \text{time}$$
$$P = W / t \qquad \text{(D1.2)}$$

If the work is measured in Joules and the time in seconds, the units of power are **Watts** (W). Note that this has the same symbol as work – they are differentiated by the context they are used in and rarely cause confusion (see *Fig D1.3*).

This equation can be developed by including the expression for work as defined in equation D1.1.

From equation D1.1 $\qquad\qquad$ $W = F.d$

Substituting into equation D1.2 $\qquad$ $P = F.d / t$

as $d / t = v$, then $\qquad\qquad\qquad$ $\mathbf{P = F.v}$ $\qquad\qquad$ **(D1.3)**

This is an extremely useful equation as many biomechanical methods enable both the force and velocity to be measured together (see *Fig. D1.3* and *D1.4*).

Problem

(a) What is the power produced in the bench press example of Fig. D1.1 if it is completed in 2 s?

(b) A cyclist climbs a 50 m high hill in 3 min 45 s. If cyclist and cycle have a combined mass of 100 kg, what power is being developed?

Solution (a)

The power is given by P = W/t
The work done on the barbell has been calculated as 235.4 J
The time taken for completing the action is 2 s
Therefore the power P = 235.4/2 = 117.7 Watts

Solution (b)

The power is given by P = W/t
The work done in climbing the hill is calculated as W = F.d = (100.g).(50) = 49,050 J
The time taken for completing the action is 225 s
Therefore the power P = 49,050/225 = **218 Watts**

Fig. D1.3. Example 3

Energy

Energy is defined as the capacity to do work or to perform some action. Energy can be considered a something that is "stored" or "possessed" so it is possible to account for it in different ways. The units of energy are **Joules**, which is the same as for work, so there are important links between energy and work.

There are many different forms of energy, including the chemical energy that enables muscles to contract, but there are two **mechanical** forms of energy that are relevant to biomechanics. These are **potential** energy which relates to the energy associated with position or deformation, and **kinetic** energy which relates to the energy associated with motion.

Potential energy has two forms. One is **gravitational potential energy** (E_{GPE}) which is the energy that is stored as a result of position in a gravitational field. If an object is at a height (h) above the ground then its gravitational potential energy is given by the equation:

$$E_{GPE} = m.g.h \tag{D1.4}$$

In figure D1.1 the barbell was lifted 0.4 m above its resting position and so it now has an energy of m.g.h = 60. g. 0.4 = 235.4 J more than it had in its resting position. This is the same as the work done in moving the barbell to this position and illustrates the close relationship between the work done and the energy stored. It is worth noting that the energy that enabled the work done in the first place came from the chemical energy sources available in the muscle. This example also illustrates the way that energy can be converted from one form (chemical) into another (gravitational potential energy).

The second form of potential energy is **strain energy** (E_{SE}) which is the energy that is stored due to the deformation of a material. It is dependent on the amount

Problem

What is the power output generated during a counter movement vertical jump?

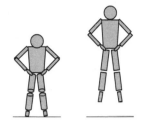

Solution

A force record needs to be obtained from a force platform as the jump takes place. The force record is shown on the graph as the solid line. From an integration of the net force (ground reaction force minus body weight) the acceleration and velocity of the center of mass can be computed (see section A6). The power (dash curve) is given as the product of force and velocity following equation D1.3. Note that for a short period the instantaneous power reaches over 6000 Watts.

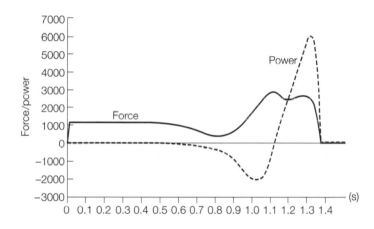

Fig. D1.4. Example 4

of deformation (d) and the stiffness (k) of the material being deformed and is given by the equation:

$$E_{SE} = \frac{1}{2}\,k \cdot d^2 \tag{D1.5}$$

The stiffness is one of the mechanical properties of a material (see section D3) and is determined by physical size, material composition, and construction. In example of figure D1.2, the stiffness of the ball is determined by the pressure of the ball. If the pressure is low the ball is easy to deform and appears "soft", conversely if the pressure is high the ball is difficult to deform and appears "hard". Equation D1.5 is non-linear due to the d^2 term. This means that as the deformation increases more and more energy is stored and so more and more work has to be done. A good example of this is spring-loaded exercise equipment. Springs are easy to extend at the start of their extension but become harder and harder to stretch as their length increases and more energy has to be stored for small additional displacements.

Kinetic energy also has two forms which are related to an object's linear and angular velocity. The first form is termed **linear kinetic energy** (E_{LKE}) and is determined by the linear velocity (v) and the mass (m) of an object. It is given by the equation:

$$E_{LKE} = \tfrac{1}{2}\, m \,.\, v^2 \qquad\qquad (D1.6)$$

Linear kinetic energy expresses the work which has to be done to for an object to move and so reflects the energy stored in the object due to its motion. For example, in the sprint start the sprinter has to supply muscle energy on each stride to increase the body's velocity. The energy is similar for each leg on each drive but the effect on the increase in the sprinter's velocity diminishes as speed increases (see *Fig. D1.5*). This is due to the non-linear term v^2 in equation D1.6. As speed increases it becomes increasingly difficult to increase speed further. It requires four times more energy to increase the sprinter's speed from 2.3 to 4.6 m/s than it does from 0 to 2.3 m/s (see *Fig. D1.5*) even though the change in speed is the same in each case. Another issue worth noting is that an object which has some linear kinetic energy needs to have that energy dissipated in order to stop (i.e., the linear kinetic energy needs to be reduced to zero). To do this we adopt special techniques. In running for example, we brake our forward motion by extending out a leg; when landing from a jump we flex the ankle, knee, and hip joints; when catching a ball we stretch out to meet the ball then absorb the ball into the body. All of these actions are designed to reduce energy in a controlled manner. Usually muscular chemical energy is expended in performing these actions, so muscular chemical energy is required to both increase and reduce the kinetic energy. Thus, movements that involve a lot of starting and stopping (like games play or racket sports, for example) also require high levels of chemical energy expenditure. Equation D1.6 also indicates that linear kinetic energy is related to body mass and so in these examples the heavier person will have a more difficult job to stop. Heavier people are generally considered to be less agile.

Problem
Calculate the velocity of a 75 kg sprinter over the first 5 strides and the change in velocity per stride if the energy delivered by each drive of the legs is 200 J.

Solution
The linear knetic energy of the sprinter is given by $E_{LKE} = \tfrac{1}{2}\, m.v^2$ so the velocity of the sprinter is given by $v = \sqrt{(2 . E_{LKE}/m)}$

Stride	Energy (J)	Velocity (m/s)	Change in velocity
0	0	0	–
1	200	2.31	2.31
2	400	3.27	0.96
3	600	4.00	0.73
4	800	4.62	0.62
5	1000	5.16	0.54

The greatest change in velocity is on the first stride. Even though the energy applied to each stride is the same, the benefit (indicated by the change in velocity) reduces as speed increases.

Fig. D1.5. Example 5

The second form of kinetic energy is termed **rotational (or angular) kinetic energy** (E_{RKE}) and is determined by the angular velocity (ω) and moment of inertia (I) of an object and is given by the equation:

$$E_{RKE} = \tfrac{1}{2}\,I\,.\,\omega^2 \tag{D1.7}$$

Most sports actions involve rotation of the limbs about a joint and so during these actions energy is contained in the rotation of the limbs. As joints flex and extend (e.g., the knee joint) the limb segments move forward and backward, changing their direction on each cycle. Muscular chemical energy is required to increase the angular velocity of the limbs, but also to slow them down and to change their direction. Thus, actions that require a lot of limb movement (for example, sprinting) require high levels of chemical energy expenditure.

Work and energy

It has already been noted that work and energy are closely related and that they have the units of Joules. Energy can be stored, work cannot. In essence work is the process of changing energy from one form to another and that enables relationships between work and energy to be defined. The basic relationship is that the work done (W) equals the change in energy (ΔE) and is given by the equation:

$$\textbf{work done} = \Delta E = E_{final} - E_{initial} \tag{D1.8}$$

where the change in energy is defined by the energy value at the start of when work is being done until its end. In the example of *Fig. D1.5* the energy change between strides is 200 J which is due to the work done during the drive on each stride. *Fig. D1.6* provides another example.

Problem
A high jumper of mass 70 kg applies an average force of 2000 N over a distance of 0.4 m. Calculate the jumper's velocity at take-off.

Solution
The high jumper has a zero vertical velocity at the lowest point of the jump (initial) and maximum vertical velocity at the moment of take-off (final).

Relationship	Work done – change in kinetic energy
Formula	$F.d = [\tfrac{1}{2}\,m.v^2]_{final} - [\tfrac{1}{2}\,m.v^2]_{initial}$
As initial KE = 0	$F.d = [\tfrac{1}{2}\,m.v^2]_{final}$
	$2000 \times 0.4 = \tfrac{1}{2}\,70.v^2$
therefore	$v^2 = 22.85$
and	$v^2 = 4.78$ m/s **ANSWER**

Fig. D1.6. Example 6

D2 THE CONSERVATION OF ENERGY

Key Notes

Law of conservation of energy	The law of conservation of energy states that energy can be neither created nor destroyed and expresses the fact that the total amount of energy remains constant as it changes from one form to another. This law is one of the cornerstones of science and helps us to develop a better understanding of the world around us. Although this law applies to the energy exchanges that occur in sports and exercise in practice its application is rather limited because the possible energy combinations are too numerous, but a more restricted form of the law can be identified that does have value.
Conservation of mechanical energy	The conservation of mechanical energy refers to the specific form of the law of conservation of energy which is of value in sport and exercise science as it uses only mechanical forms of energy. It refers to exchanges between just two types of energy: the gravitational potential energy, and linear and angular kinetic energy. In general the conservation of mechanical energy applies to projectile flight where air resistance can be neglected. It cannot be applied where there are obvious energy losses due to friction or other resistances.

Law of conservation of energy

The **law of conservation of energy** states that **energy can be neither created nor destroyed** and expresses the fact that the total amount of energy remains constant as it changes from one form to another. This law is one of the corner-stones of science and helps us to develop a better understanding of the world around us. Although this law applies to the energy exchanges that occur in sports and exercise in practice its application is rather limited, but a more restricted form of the law can be identified which does have value.

In section D1 several forms of **mechanical energy** were identified, principally **gravitational potential energy**, **strain potential energy**, **linear and angular kinetic energy**. The examples used the idea that chemical energy is used by the muscles to generate muscle tension. The **muscle** is essentially a device which converts **chemical to mechanical energy**. When energy changes from chemical to mechanical a certain amount of **heat** is given off. The heat is a by-product of the energy conversion process and while it may have some biological value in main-taining body temperature it does not generally contribute to the performance and so is considered a waste product. Energy conversion processes often produce heat as a by-product. For example, when a ball is dropped it is compressed as it hits the ground and after the recoil never quite reaches the same height from which it was dropped. This failure to regain the original drop height is due to a loss of energy as a result of the compression and is indicative of the efficiency of energy con-version which, if heat is generated, is always less than 100%. If that compression were repeated many times the ball would warm up, a characteristic which is used to good effect in the game of squash where the warm ball rebounds with greater speed than a cold ball.

One energy conversion process, though, is not associated with the generation of heat. The conversion of gravitational potential energy to kinetic energy can be achieved without the production of heat and is 100% efficient. This provides a valuable tool for studying mechanical energy exchanges which is particularly useful in the biomechanical study of sports and exercise as many activities utilize this specific form of energy exchange.

The conservation of mechanical energy

The specific form of the law of conservation of energy which has the property of perfect energy exchange between its components is referred to as the **conservation of mechanical energy**. This refers to exchanges between just two types of energy – the gravitational potential energy (E_{GPE}) and kinetic energy (linear, E_{LKE} and angular, E_{RKE}) and is given by the equation:

$$E_{GPE} + E_{LKE} + E_{RKE} = \textbf{total mechanical energy} \qquad \textbf{(D2.1)}$$

where the total mechanical energy is a constant.

It is important to state the conditions where equation D2.1 **does not apply**. It does not apply to strain potential energy as the process of deformation causes molecules to rub across each other and to lose energy due to the process of friction. It does not apply to other situations in which there is a loss of energy due to friction, such as an object sliding down a surface (e.g., a child's slide, or ski slope). It cannot be used if the influence of air resistance is important, in practice if relative air speeds exceed 5–6 m/s.

Situations in which it **does apply** are mainly to do with projectile flight where the body or a projectile moves slowly in the air. Situations such as athletics jumping, gymnastics, diving, trampolining, throwing actions, and many other activities can all be investigated using this relationship. It can also be used to understand techniques used to play shots in racket sports where in a looped forehand or backhand drive the gravitational force is used to help generate racket head speed. The principle can also explain the manner in which limbs are used in walking and running actions. In short, the conservation of mechanical energy has widespread application to sport and exercise situations.

An application

Consider a trampoline movement in which the trampolinist is just about to leave the bed and perform a straight bounce (i.e., no rotation) (*Fig. D2.1*). The vertical velocity is the greatest at this point. As the trampolinist rises in the air the height increases but the velocity reduces. This continues until the velocity becomes zero and the greatest height is reached. The descent now begins with the velocity increasing in the negative direction and the height reducing until contact is made once again with the trampoline bed.

In this example, the energy according to equation D2.1 remains constant so we can equate the energy conditions at two points (the take-off and the top of flight) to give:

$$[E_{GPE} + E_{LKE}]_{take\text{-}off} = [E_{GPE} + E_{LKE}]_{top\ of\ flight}$$

As the E_{GPE} is zero at the start and E_{LKE} is zero at the top of flight we have:

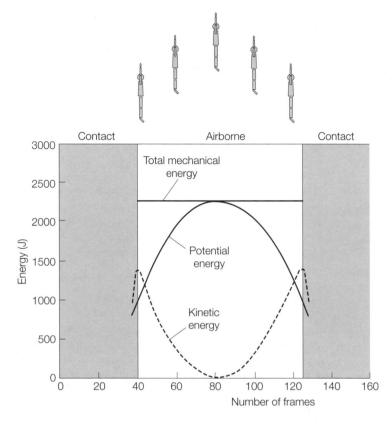

Fig. D2.1.   Energy changes during a trampoline bounce

$$[E_{LKE}]_{\text{take-off}} = [E_{GPE}]_{\text{top of flight}} \qquad \text{(D2.2)}$$

or using the full expressions for each energy term:

$$\tfrac{1}{2}\,m \cdot v^2_{\text{take-off}} = m.g.h_{\text{top of flight}}$$

Canceling the term "m" and re-arranging gives:

$$v = \sqrt{(2.g.h)} \qquad \text{(D2.3)}$$

which is a general expression linking the velocity and height for any projectile motion.

If a high jumper takes off with a velocity v = 4.78 m/s (see example in section D1) the height raised by the center of mass will be ½ v²/ g = 1.16 m. To this, of course, must be added the starting height of the center of mass – probably around 1 m so the maximum possible jump height would be around 2.16 m. Of course the jumper must rotate the body in order to clear the bar.

In a dive from the 10 m board, with what velocity would the diver hit the water? Using equation D2.3 it can be easily calculated that the diver will have a velocity:

$$v = \sqrt{(2.g.h)} = \sqrt{(2.g.10)} = 14.01 \text{ m/s}.$$

In all of the above examples, air resistance has been ignored. As velocity increases this becomes a less reasonable thing to do. Earlier it was stated that speeds greater than 5–6 m/s are likely to have an effect on the motion of an object. The detail of this depends on a number of factors considered in more detail in section D6. For the purposes of using equations D2.1–D2.3 it is sufficient to be aware of the limitation. So, in the above example of the diver, the velocity calculated represents the maximum velocity that would be achieved. In practice, due to the effects of air resistance, the velocity will be a little smaller – but probably unlikely to have an effect on the general point being made – that impact with the water is fast!

Adding rotation

Up to now any rotation that might be present has been ignored. If in the trampoline case the performer also rotated, some of the energy would be taken up with rotation and so less energy would be available for the linear kinetic energy, thus reducing the height which could be reached. Consequently, the more rotation the less height. This is the reason that trampolinists start their routine with a series of straight bounces to gain height, as when they perform their stunts involving rotation they progressively lose height. After two or three movements the trampolinist has lost some height and, to regain it, they have to perform an "easy" bounce so that they can focus on regaining height for the next complex series of stunts in their routine.

Another example, of perhaps novelty value, is to consider the velocity with which a person would hit the ground if they were simply to fall over. The person keeps rigid and rotates about the feet. To solve this problem we need to know a few things about the person such as their mass (70 kg), the location of their center of mass (1.0 m from the ground) and moment of inertia ($I = 80$ kg/m²) about the feet. During the rotational fall the person has only rotational kinetic energy, so equation D2.1 can be developed in the same way as equation D2.2 but using only the rotational kinetic energy rather than the linear kinetic energy. In this case, using the ground level as the reference zero level, the gravitational potential energy at the start is equal to the rotational kinetic energy at the end, in other words:

$$m.g.h = \tfrac{1}{2} I . \omega^2$$

so :

$$\omega = \sqrt{(2.m.g.h / I)}$$

Therefore

$$\omega = \sqrt{2.70 \times 9.81 \times 1/80}$$

$$\omega = \sqrt{1373.4/80}$$

$$\omega = \sqrt{17.16}$$

$$\omega = 4.14 \text{ rad/s}$$

The **unit of angular velocity** here is the **radian per second** (rad/s). The **radian** is a scientific unit for measuring angles and is derived from the number of times the radius of a circle goes into its circumference. Thus, one circle is equal to 360° which is equal to 2π radians, giving 1 rad = 57.3°.

The linear velocity of the center of mass rotating at 1.0 m from the axis of rotation with an angular velocity of 4.14 rad/s is given by $v = r . \omega = 4.14$ m/s.

So the body would make contact with the ground at around 4.14 m/s; the top of the body will make contact with the ground even faster (6–7 m/s). If the hands were placed outwards to support the body as it hit the ground the hands would make contact with a similar velocity. As the arms do not have the same strength as the legs for stopping the body falling, it is unlikely that the arms will provide much protective effect. In fact the impact associated with this type of fall often leads to a collar bone fracture in the young and more serious problems in the elderly.

$$m.g.h = \tfrac{1}{2}\,k.\,\Delta x^2 = \tfrac{1}{2}\,m.v^2$$

D3 THE MECHANICAL CHARACTERISTICS OF MATERIALS

Key Notes

Load and deformation

Material solids can sustain applied loads but they have a tendency to deform. Depending on the load applied their state will be one of tension and have a tendency to extend; compression and tend to shorten; shear and tend to slide; or torsion and tend to twist.

Stress and strain

Stress is defined as the force per unit area and describes the way the force is distributed through the material. Strain is defined as the increase in length divided by the original length and is often expressed as a percentage.

Hooke's law

For many materials, stress is linearly related to strain, and this relationship is known as Hooke's law. This relationship holds until a material reaches its elastic limit or yield point where the material begins to disintegrate.

Elasticity

Elasticity describes the way in which a material deforms and then returns to its original shape. Materials that do this well are called elastic (e.g., an elastic band or spring). Materials that do this poorly are called inelastic (e.g., putty or a deflated soccer ball).

Stiffness and modulus of elasticity

The elasticity of a material can be computed from the way it deforms under load. If the force which causes a deformation is used, their ratio is the stiffness. If the stress (force per unit area) and strain (percentage length change) are used, their ratio is called the modulus of elasticity. The stiffness is more widely used in sport and exercise biomechanics.

Hysteresis

When an object is deformed and then allowed to return to its original state a certain amount of energy is lost. This energy loss is termed hysteresis.

Area elastic and point elastic surfaces

Surfaces in sport can be characterized as area elastic or point elastic depending on how they deform. Area elastic surfaces deform over a large area and floors which are designed to be area elastic are generally referred to as sprung floors. These have advantages in terms of energy return to the player and are generally more comfortable to play on. Point elastic surfaces deform locally and typify playing fields and artificial playing surfaces. These are generally less comfortable to perform on.

Introduction

Materials are classified as either **solids** or **fluids**. The latter will be dealt with more fully in section D6. Material solids have certain mechanical properties that affect their function and determine how they influence performance and injury in sport and exercise.

Load and deformation

The mechanical properties of a material are determined by the way it reacts to a **load**. The applied load can be categorized as a **force** or a **torque** (or twisting moment) or a combination of these. The applied load can either be **gradual** (such as when lifting a barbell), or **impulsive** (such as heel strike impact in running). The applied load can either be applied once (**acute loading**) or several times (**repetitive loading**). These latter two load characteristics are useful when considering the injury effects of loading, as an acute load can lead to a fracture of the bones or a torn tendon, while a repetitive load can lead to an overuse injury.

When an applied load acts on a material it causes the material to **deform**, and the nature of this deformation can be described and related to its function. When the forces applied to the two ends of a material are directed away from each other, the material is said to be in **tension** (*Fig. D3.1a*) and has a tendency to extend. When the forces are directed towards each other the material is said to be in **compression** (*Fig. D3.1b*) and has a tendency to shorten. When the forces are directed along different lines of action (*Fig. D3.1c*) then **shear** is created. When torques (or twisting moments) act at each end of the material in opposite directions then **torsion** is created (*Fig. D3.1d*) causing the material to twist. Combinations of forces and torques lead to more complex types of deformation but these do not need to be considered here.

Examples of common load deformation conditions occur in the snatch event in weightlifting. As the weight is lifted from the ground the arms are in tension. When the weight is supported above the head the arms are in compression. The force of the arms act upwards to support the bar. The force from the weight plates act down due to gravity so a shear force is produced on the bar, in this case causing it to bend.

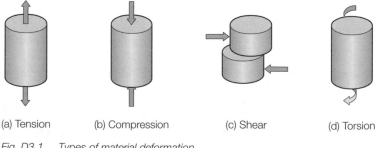

(a) Tension   (b) Compression   (c) Shear   (d) Torsion

Fig. D3.1.   Types of material deformation

Stress and strain

Consider a material that is in tension due to the application of force at each end causing it to extend (*Fig. D3.2*). If the force (F) is applied over an area (A) then material experiences a **stress** which is defined as the **force per unit area** (i.e., F/A) and describes the way the force is distributed through the material. Similarly, the material experiences a deformation (in this case an extension) which is termed the **strain** and is defined as the increase in length divided by the original length. Strain is often expressed as a percentage. For example, one

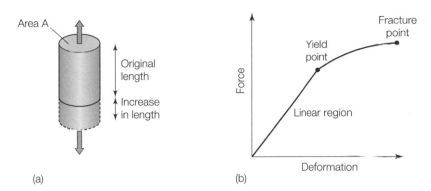

Fig. D3.2.   Stress (force) – Strain (deformation) relationship for a material.

might expect the Achilles' tendon to stretch 3% during the contact phase in running which would be a measure of its strain. For many materials stress is linearly related to strain, and this relationship is known as **Hooke's law** (*Fig. D3.2*). This relationship holds until a material reaches its **elastic limit** or **yield point**, where the material begins to deform easily for a small increase in stress and then finally the **fracture point** where the material fails. For the tendon this will occur at a strain of about 10%.

Elasticity

Elasticity describes the way in which a material deforms and then returns to its original shape. Materials that do this well are called **elastic** (e.g., an elastic band or spring). Materials that do this poorly are called **inelastic** (e.g., putty or a deflated soccer ball). Materials used in sports and exercise have a range of elasticity depending on their function. Materials with good elasticity would be a trampoline bed with springs, or a bow used in archery. Materials with moderate elasticity would be a gymnastic beat board, or fiberglass pole for pole vaulting. Materials with poor elasticity would be a squash ball or the human foot.

The **linear region** of Hooke's law (*Fig. D3.2*) implies that as the force (or stress) increases the deformation (or strain) increases in the same proportion and so the force-to-deformation ratio and the stress-to-strain ratio are constant. This constant is known as the **stiffness** when the force and deformation are used to describe the behavior of the material, and the **modulus of elasticity** when the stress and strain are used. In sport and exercise science it is more common to measure force (F) and deformation (d) so the first term, stiffness (k) is often used and is expressed as:

Force (F) = stiffness (k). deformation (d)

$$F = k.\ d \qquad\qquad (D3.1)$$

As the force is applied it moves its point of application and, following the principles established in section D1, the force does work. The work done on the material is stored as elastic energy (E_{SE}) given previously by equation D1.5.

$$E_{SE} = \tfrac{1}{2}\ k.\ d^2$$

The stored elastic energy is also given by the area under the force deformation graph (shaded area in *Fig. D3.3a*).

When the load is removed the extension is reversed with the subsequent shortening called **restitution**. There is a loss of force during this phase that can be

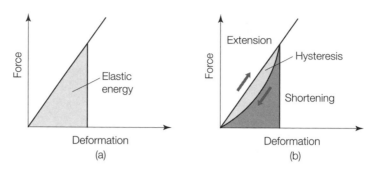

Fig. D3.3.   (a) Loading energy and (b) unloading energy. The difference between the two in (b) gives the hysteresis energy loss

seen when the load deformation during elongation is plotted onto the same graph as the force deformation during shortening (shaded areas in *Fig. D3.3b*). The darker shaded area represents the energy returned during restitution and is a measure of the **resilience** of a material. The lighter shaded area represents the energy lost during the recoil, and is termed **hysteresis**. A good example of hysteresis energy loss is when a ball bounces on the floor. If the ball is dropped from a certain height it will never rebound to exactly the same height. The energy stored during compression allows the ball to bounce back. Some of this energy is lost due to friction between the molecules that develop heat during the compression, and so the recoil energy is never quite enough to get the ball back to the same height from which it was dropped. The same principle applies to the contact between a tennis ball and tennis racket. The tennis racket and tennis ball are poor devices for returning energy, but the strings are excellent, allowing the deficiencies of the ball and racket to be overcome. Gut strings are often preferred by experts as they are more elastic and have better energy return properties, although they are more expensive and have a shorter lifespan.

Other characteristics
A special note should be given to sports surfaces. In sports like gymnastics and tumbling the surfaces are described as **area elastic**, that is they deform over a large area when jumped on and have good elasticity to aid the performer. Wooden gymnasium floors that are "sprung" are also area elastic. Surfaces like real or artificial turf are considered **point elastic**, that is they deform in a localized region when jumped on (*Fig. D3.4*). Generally point elastic surfaces have poor elasticity.

Permanent deformations are referred to as **set**, and describe the **plastic** behavior of materials. Set can be important in some sport materials, for example those used in the midsoles of running shoes. The expanded foam material that is used to provide cushioning as the foot makes contact with the ground gradually permanently deforms through use. This happens because the normally closed cells which make up the foam material gradually fracture and release their internal pressure, and ultimately collapse. When this happens the running shoe becomes thinner and harder. A worn shoe is a known injury risk factor as it is likely to increase the impact force on heel strike. Shoes that show any sign of this type of wear should be replaced.

There are other properties of materials relevant to sport and exercise. The term **hardness** is used to describe how much resistance a material has to penetration,

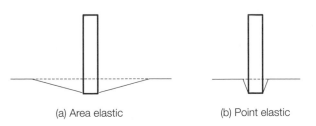

(a) Area elastic   (b) Point elastic

Fig. D3.1.   Point and area elastic surfaces

stretching and wear. A material that is difficult to penetrate (such as the ball bearing in the bearing race of a bicycle) is very hard. Conversely a "crash" mat used in gymnastics is very soft. As there is such a wide range of hardnesses in common materials there are various "scales" used to measure these. For intermediate materials, such as the midsole material of a running shoe, the Shore "A" scale is used. A measurement is made by a device which has a spring loaded blunt pointer which is impressed into the surface (as in *Fig. D3.4b*) . The amount of penetration is recorded and used as the measurement. Clearly, the strength of the spring has to match the general deformability of the surface tested. For very hard surfaces a much stronger spring is used. Running shoe midsole materials range from a Shore "A" value of 20 (hard) to 50 (soft) with 35 being a commonly found value.

D4 IMPACTS

Key Notes

Impacts	Impacts are characterized by large forces over small times of contact. Impacts can be mechanically analyzed using the two impact equations, the conservation of momentum and the coefficient of restitution.
The conservation of momentum	The conservation of momentum is an important law in science that defines how objects behave when they interact and represents the general situation of a collision in sport. The conservation of momentum law states that the momentum in a system before an interaction takes place is the same as that after the interaction, in other words momentum has been conserved. An equation for this can be written and used in the analysis of a problem. The conservation of momentum is the first impact equation.
The coefficient of restitution (e)	The coefficient of restitution (e) is a term which quantifies how the energy stored in a deformed material is returned. It provides a numerical value that reflects the elasticity of a material, or more specifically a material combination as the coefficient of restitution is dependent on both materials in a collision. The smallest value "e" can have is zero, when an object sticks to the floor when dropped. The largest value it can have is 1, when the ball returns to its original height. In practice, this latter situation represents an ideal case and never actually happens. The coefficient of restitution gives the second impact equation.
Collisions and central impact	A collision describes the way in which two objects interact with each other during an impact. There are may ways this can happen but one important class of collision is known as central impact where the velocities of the center of mass of each object are directed towards each other along a line of impact. In other words, a head-on collision.

Impacts

Impacts are defined by **large forces** that act over **short periods of time**. Examples of impacts are propelling a ball with the head in a soccer header, hitting a ball with a racket in tennis, and the contact between club and ball in a golf drive. In all these cases the time of contact between the striking device and object is small ranging from about 20 ms in the case of heading the soccer ball, to 5 ms in tennis and 0.5 ms in golf. In addition, the forces applied are also large, ranging from approximately 1000 N when heading the ball to 10 000 N in golf.

The different contact times and contact forces in these examples indicate that the nature of impact is dependent on both objects involved in the collision. Generally speaking, the softer the object the longer the impact takes and the lower the force generated. In order to understand the specific nature of impact it is necessary to consider the **conservation of momentum** and the **coefficient of restitution** relationships. These lead to two equations which are known as the **impact equations**.

The conservation of momentum

The **conservation of momentum** (see also section B4) is an important **law** in science that defines how objects behave when they interact. The law applies to a system that may contain many objects but for the purpose of this text the interaction between just two objects will be considered. These two objects can represent, for example, the head and ball in soccer heading, or the ball and racket in tennis, or the club and ball in golf. In other words, they represent the general situation of a collision in sport.

Consider two masses (m_A and m_B) each moving with an initial velocity v_A and v_B (*Fig. D4.1*). If these two objects collide then after the collision their velocities are found to be v'_A and v'_B . The **conservation of momentum law** states that the momentum of a system (of two objects) before an interaction takes place is the same as that after the interaction, in other words momentum has been conserved. This can be expressed in equation form as:

$$m_A \cdot v_A + m_B \cdot v_B = m_A \cdot v'_A + m_B \cdot v'_B \qquad \text{(D4.1)}$$

system total momentum before = system total momentum after

This rather complex looking equation is the **first impact equation** and an example of its application will be given after the next topic.

Before impact   After impact

Fig. D4.1.   An illustration of the conservation of momentum

Coefficient of restitution

The **coefficient of restitution (e)** is a term which quantifies how the energy stored in a deformed material is returned. It provides a numerical value which reflects the **elasticity** previously described in section D3. Elasticity was defined in general terms and an example was given of a ball bouncing from the floor after having been dropped from a certain height. If the ball was dropped from a height H_{drop} and it rebounded to a height $H_{rebound}$ then the coefficient of restitution is given as:

$$e = \sqrt{\frac{H_{rebound}}{H_{drop}}} \qquad \text{(D4.2)}$$

It can be appreciated from this equation that the smallest value "e" can have is zero, when the ball sticks to the floor and $H_{rebound}$ is zero. The largest value it can have is 1, when the ball returns to its original height. In practice, this latter situation represents an ideal case and never actually happens, even though with a "super ball" the rebound height can come close to the original drop height.

Recall the expression linking height of drop and velocity of impact from section D2.

$$v = \sqrt{(2.g.h)} \qquad \text{(D2.3)}$$

If this expression is substituted into equation D4.2 then the expression for "e" is made a little simpler:

$$e = \frac{V_{rebound}}{V_{drop}}$$

In this equation, it should be remembered that the direction of the velocity when dropping is opposite to that when rebounding, so the directions of motion need to be taken into account where necessary. This relationship can be applied to the more complex situation in which two objects collide, such as the situation that led to equation D4.1. When such an impact takes place, momentum is conserved and there is a relationship between the masses and velocities before and after impact. If the velocity terms used in equation D4.1 are used in the expression for "e" previously, then the more general form of the coefficient of restitution equation can be written as:

$$e = \frac{v'_A - v'_B}{v_B - v_A} \qquad \text{(D4.3)}$$

Equation D4.3 is the **second impact equation**. This equation takes into account the relative velocities before impact ($v_A - v_B$) and the relative velocities after impact ($v'_A - v'_B$) and, through the use of a negative sign (taken into account by reversing v_A and v_B on the denominator), the direction of drop, which is opposite to the direction of rebound. In fact, if the rebound situation described in figure D4.1 is applied to equation D4.3, then, if the first mass represents the ball (m_A) and the second mass the floor (m_B), whose velocity is zero both before ($v_B = 0$) and after ($v'_B = 0$), equation D4.3 becomes:

$$e = \frac{v'_A}{-v_A}$$

with the negative sign representing the change in direction of the ball after impact.

It is relatively easy to undertake experiments in which the drop and rebound height of a ball is measured. These experiments show that the coefficient of restitution for various sports ball which are dropped onto a concrete floor range from 0.75 (basketball and soccer ball) through to 0.67 (tennis ball) to 0.32 (cricket ball). When the ball is dropped onto a softer surface (such as a wooden or grass floor) the coefficient of restitution is found to be smaller. This finding indicates that the coefficient of restitution is not a fixed value for a sports ball, but is dependent on the nature of both it and the impacting surface. The softer the surface, the lower will be the coefficient of restitution, and hence the lower the rebound height. Similarly, the softer the ball or the lower its pressure, the lower the coefficient of restitution will be.

Collisions

When two objects collide they may do so in two main ways, described as central impact or oblique impact. The latter type of impact will be considered in the next section. In **central impact** the velocities of the center of mass of each object are directed towards each other, along the **line of impact** (*Fig. D4.2*). This represents a **head-on collision**. Given some information about the objects involved in the collision, it is possible to find out further information by using the two

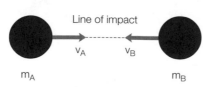

Fig. D4.2.   Line of impact

impact equations (equations D4.1 and D4.3). In these equations there are seven terms (m_A, m_B, v_A, v_B, v'_A, v'_B and e) . As there are two impact equations, these can be used to find a maximum of two unknowns. In other words, in a collision five further pieces of information are needed for a complete solution. An example is given in *Fig. D4.3.*

Problem

A 60 kg rugby wing is moving at a speed of 8 m/s to the right and collides with a 100 kg forward who is at rest. If after collision the rugby forward is also observed to move to the right at a speed of 3.6 m/s determine the velocity of the wing after collision and the coefficient of restitution between the two players. Assume no interaction with the ground.

Solution

Known variables $m_A = 60$, $m_B = 100$, $v_A = +8$, $v_B = 0$, $v'_A = ?$, $v'_B = +3.6 = ?$

Using the conservation of momentum equation

$$m_A.v_A + m_B.v_B = m_A.v'_A + m_B.v'_B$$
$$60(8) + 100(0) = 60v'_A + 100(3.6)$$

therefore $v'_A = 2.0$ m/s

Using the coefficient of restitution equation

$$e = (v'_A - v'_B)/(v_B - v_A)$$
$$e = (3.6 - 2.0)/(8.0 - 0)$$

therefore $e = 0.2$

Fig. D4.3.   Example

Application to soccer kicking

An interesting fact about some forms of impact is that the mass being hit can have a velocity after impact greater than the hitting mass. In other words, there seems to be a velocity gain. This is true in soccer when the foot has an impact velocity of around 20 m/s while the ball is propelled with a velocity of 25 m/s or more. Why does that occur and is it possible to get something for nothing?

The two impact equations can be rearranged so that they are expressed in terms of the ball velocity and foot velocity. In the two impact equations, consider the mass A to be the foot and the mass B to be the ball and there is no influence of the leg on the foot. Remember that when kicking a stationary ball, the initial velocity of the ball (v_{ball}) is zero. These equations now become:

$$m_{foot} \cdot v_{foot} = m_{foot} \cdot v'_{foot} + m_{ball} \cdot v'_{ball}$$

and

$$e = \frac{v'_{foot} - v'_{ball}}{-v_{foot}}$$

It is possible to re-arrange these equations in order to isolate the velocity of the ball after impact (v'_{ball}) and express this in terms of the velocity of the foot before impact (v'_{foot}), as:

$$v'_{ball} = v_{foot}\{[m_{foot} / (m_{foot} + m_{ball})] \times [1 + e]\}$$

The term $[m_{foot} / (m_{foot} + m_{ball})]$ represents the mass proportions of the foot and ball and for typical values for an adult male foot and a soccer ball, the term has a value of around 0.8. The term $[1 + e]$ represents the effectiveness of the impact due to the hardness of the ball (as a result of its pressure) but also the rigidity of the foot (due to its tendency to deform and flex at the ankle). A typical value for this term is around 1.5. (Note: the coefficient of restitution for a soccer ball on the foot is lower then when dropped onto concrete.) Substituting these values into the above equations:

$$v'_{ball} = 1.2 \, v_{foot}$$

This relationship suggest that in a typical kick (maximal instep kick by a competent player) the ball should travel about 20% faster than the foot travels. This speed gain is the result of the greater mass of the foot compared with the ball. If a player can increase the mass of the foot (by a heavier boot) or can increase the quality of impact (by having a more rigid foot) then the ball should fly off even faster. A low percentage gain would indicate a poor skill level.

D5 OBLIQUE IMPACTS

Key Notes

Oblique impact	Oblique impact is a class of collision where the velocities of the two objects are not directed along the line of impact. Oblique impact is a more general case of central impact. It is important to appreciate that the same two impact equations as used in central impact (the conservation of momentum and the coefficient of restitution) apply to oblique impact. As each object in a two object collision has its own direction of travel before collision and after collision, there are a further four variables that make up an oblique impact problem. In order to solve this with just the two impact equations, quite a lot of information about the collision is required.
An assumption of oblique impact	For the mechanical analysis illustrated in this section it is assumed that there is no frictional interaction between the two objects that collide. This condition is referred to as a "smooth" interaction and means that the velocity of the object perpendicular to their line of impact remains unchanged as no friction force acts to slow it down. Thus, when applied to each object two further equations are obtained, making an analysis of a collision problem easier to complete.
Contact with a surface	When one of the objects involved in a collision is a surface, the problem becomes easier to solve. A further advantage of this type of analysis is that it gives insight into how balls may bounce off surfaces in sports like tennis, table tennis, and soccer. It should be noted that in the real case, surface friction has to be included and the spin of the ball needs to be taken into account. These are quite complex issues, which are outside of the scope of this text.

Oblique impact

In **central impact** the velocities of the center of mass of two objects (A and B) are directed towards each other along the **line of impact**. In **oblique impact** the velocities of the center of mass of each object are directed towards each other at an angle (θ_A and θ_B) to the line of impact as illustrated in *Fig. D5.1*. Immediately after impact the objects move away from each other again but with different velocities and different angles (θ'_A and θ'_B). Compared with the problem of central impact in which there were **seven terms** (m_A, m_B, v_A, v_B, v'_A, v'_B and e) in oblique impact there are **11 terms** (additionally θ_A, θ'_A, θ_B and θ'_B).

In order to solve this problem it is necessary to divide the velocities into components, one **along the line of impact** and one **perpendicular to the line of impact** as illustrated in *Fig. D5.2*.

In the direction **along the line of impact** the collision is a central impact and can be dealt with by resolving along the line of impact. In the conservation of momentum (equation D4.1):

$$m_A \cdot v_A + m_B \cdot v_B = m_A \cdot v'_A + m_B \cdot v'_B \tag{D4.1}$$

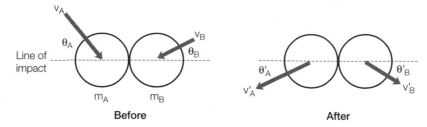

Fig. D5.1.   An illustration of oblique impact in the instant before contact and the instant after contact

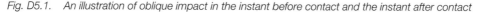
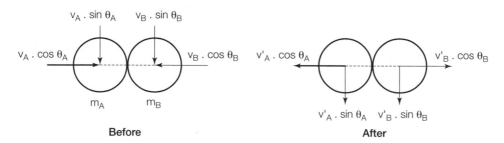

Fig. D5.2.   Velocity components for oblique impact in the instant before contact and the instant after contact

the velocity components terms in *Fig. D5.2* are substituted to give:

$$m_A \cdot v_A.cos\theta_A + m_B \cdot v_B .cos\theta_B = m_A \cdot v'_A.cos\theta'_A + m_B \cdot v'_B .cos\theta'_B \quad \textbf{(D5.1)}$$

Similarly, using the coefficient of restitution (equation D4.3) and substituting the velocity components of *Fig. D5.2* gives:

$$e = \frac{v'_A.cos\theta'_A - v'_B.cos\theta'_B}{v_B.cos\theta_B - v_A.cos\theta_B} \quad \textbf{(D5.2)}$$

In the direction **perpendicular to the line of impact**, the velocity is not affected as the interaction between the two objects in this direction is considered frictionless so there is no interacting force to slow the velocities in this direction. In other words, their momentum in this direction is conserved so the following can be written:

$$v_A.sin\theta_A = v'_A.sin\theta'_A \quad \textbf{(D5.3)}$$

$$v_B.sin\theta_B = v_A.sin\theta_A \quad \textbf{(D5.4)}$$

Equations D5.1–D5.4 enable four unknowns to be calculated provided the other seven variables (of the 11 which make up these problems) are known. An example is given in *Fig. D5.3*.

Contact with a surface

When an object such as a ball makes **contact with a surface**, the ball will bounce off with a **reduced angle** due to the loss of vertical velocity through energy loss, while the horizontal velocity, in the ideal case, remains unchanged (*Fig. D5.4*). The same principles as discussed above are used to solve these types of problems. An example is given in *Fig. D5.5* for a squash ball making an impact with a vertical wall.

Problem

Two identical smooth balls collide with velocities and direction as shown. If the coefficient of restitution $e = 0.9$ determine the magnitude and direction of each ball after impact.

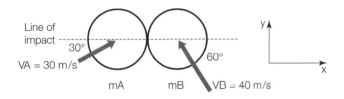

Solution

First obtain the component velocities before impact

$(V_A)_x = V_A \cos 30 = 26.0 \text{ m/s}$ $(V_B)_x = -V_B \cos 60 = -20.0 \text{ m/s}$

$(V_A)_y = V_A \sin 30 = 15.0 \text{ m/s}$ $(V_B)_y = V_B \sin 60 = 34.6 \text{ m/s}$

(i) Consider velocities **perpendicular** to the line of impact (i.e., y direction) after impact. Since no force acts (smooth balls) then these remain unchanged, in other words

$(V'_A)_y = 15.0 \text{ m/s}$ $(V'_B)_y = 34.6 \text{ m/s}$

(ii) Consider velocities **parallel** to the line of impact (i.e., x direction). This is governed by the conservation of momentum and coefficient of restitution equations.

(a) Conservation of momentum gives

$$m_A (V_A)_x + m_B (V_B)_x = m_A (V'_A)_x + m_B (V'_B)_x$$

as masses are equal, these cancel, and substituting values for $(V_A)_x$ and $(V_B)_x$

gives $(V'_A)_x + (V'_B)_x = 6.0 \text{ m/s}$ (1)

(b) The coefficient of restitution equation gives

$$e = \frac{(V'_A)_x - (V'_B)_x}{(V_B)_x - (V_A)_x}$$

therefore $(V'_A)_x - (V'_B)_x = -41.4 \text{ m/s}$ (2)

Solving equations (1) and (2) simultaneously gives $(V'_A)_x = -17.7$, $(V'_B)_x = 23.7 \text{ m/s}$

Using the values for the velocities in the y direction, the vector velocities after impact are

$$V'_A = \sqrt{[(V'_A)_x^2 + V'_A{}_y^2]} = 23.2 \text{ m/s}, \quad V'_B = 41.9 \text{ m/s}$$

at angles $\theta_A = \tan^{-1} [(V'_A)_y/(V'_A)_x] = 40.3°$, $\theta_B = 55.6°$

Fig. D5.3.   Example 1

In the real case, the condition of a frictionless impact does not apply. The ball will have a tendency to slide over the floor during the time of contact. This can be seen with the skid marks left by a tennis ball on a clay court. As the ball skids, a friction force acts on it to slow down its horizontal velocity. Thus, the ball will lose speed in both the perpendicular and parallel directions with the surface. This could lead to the ball rebounding at a higher angle than the incident angle although its velocity will be considerably reduced. This tendency to "sit up" on contact is a feature of some tennis courts (mainly shale and clay type surfaces and are described as "slow"). Other tennis courts, such as grass, do not slow the ball down as much as the ball slips more easily on the grass surface, particularly when damp. The ball has a tendency to come off "low" (as in the example in *Fig. D5.4*)

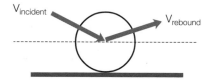

Fig. D5.4.   Oblique impact of a ball with a surface

Problem

A squash ball is hit against a smooth vertical wall with velocity V = 20 m/s at an angle of 60°. If the coefficient of restitution e = 0.4 determine the magnitude and direction of the ball's velocity after impact.

Solution

Resolve the initial velocity of the ball into components

Horizontal x   Vx direction = 20.cos 60 = 10 m/s
Vertical y   Vy direction = 20.sin 60 = 17.3 m/s

Vertical motion after impact:
since the wall is smooth, there is no change in the vertical motion

therefore   V'y = 17.3 m/s

Horizontal motion after impact:
this is covered by the coefficient of restitution applied to the velocities normal to the wall (x direction). Note that the wall has velocity = 0

$$e = \frac{V'x - 0}{0 - Vx} =$$

therefore   V'x = e.Vx = (0.4) × (10) = 3 m/s

The resultant velocity is given by adding the two velocity components

$$V' = \sqrt{[(V'x)^2 + (V'y)^2]} = 17.8 \text{ m/s},$$

and the angle of rebound

$$\theta = \tan^{-1}(V'y/V'x) = 77.0°$$

The squash ball rebounds with a lower speed and an angle closer to the wall.

Fig. D5.5.   Example 2

and as the velocity is substantially maintained these courts are described as "fast". The large differences between tennis court surfaces is a considerable skill challenge to players, some of whom do well only on one type of surface. It is because of these differences that the International Tennis Federation have recently introduced different types of balls, which have different rebound qualities.

It should also be noted that when the ball skids on the surface the friction force which slows down its velocity parallel to the surface, also has a tendency to impart top spin. This spin may also be important in games, such as tennis and table tennis. In particular it can add to the spin that the ball already has in order to affect its total spin. Further, if the ball already has spin, then this spin will influence the ball's interaction with the surface and influence its final direction of travel. In games such as tennis, table tennis, and cricket the "break" of the ball after hitting the surface is a major tactical aspect of the game. The detailed mechanics of these situations is complex and beyond the scope of this text.

D6 FORCES IN A FLUID

Key Notes

Fluids	A fluid is defined by the particles that make up the fluid. It has no fixed shape and distorts under the action of a shear force. Fluids can be categorized as liquids and gases.
Properties of a fluid	The two main properties of a fluid are its density, given by the ratio of its mass and volume, and its viscosity, which describes how easily the fluid flows.
Fluid flow	Fluid flow is either laminar (where the fluid flows as if it were in sheets) or turbulent (where the particles of the fluid move perpendicular to the direction of flow). Turbulent flow is also associated with the formation of eddy currents.
Buoyancy (U)	Buoyancy, or upthrust, is a force that acts perpendicular to a fluid surface, which normally means that it acts vertically upwards. Its value is given by Archimedes' principle and is equal to the weight of the fluid displaced. The buoyancy force acts at the center of buoyancy, which is the geometrical center of the submerged volume. Thus, the center of buoyancy will change its location as a function of the submerged volume.
Flotation and stability	When an object floats the buoyancy force must equal the gravitational force. If the gravitational force is greater than the maximum buoyancy force (i.e., when the object is fully submerged) the object will sink. The stability of a floating object is dependent on the location of the center of gravity and center of buoyancy of an object, and how the buoyancy and gravitational forces interact.
Bernoulli's equation	Bernoulli's equation links the velocity and pressure of a fluid together. Essentially it explains why the pressure of a fluid drops as the velocity of a fluid increases. Knowing this it is possible to appreciate why pressure differences occur around an object helping to understand why the drag and lift forces occur.
The fluid drag force	The fluid drag force is a force that is developed when an object moves relative to a fluid (either the object can be stationary and the fluid flows, or the fluid can be stationary and the object moves) and is directed opposite to the direction of motion of the fluid. In air this is termed air resistance, while in water this is termed hydrodynamic resistance. The fluid drag force is dependent on the coefficient of drag (a term describing how streamlined the object is), the fluid density, the cross-sectional area of the object in the direction of motion and the square of the velocity. A formula for fluid drag force can be given which links these variables together.

The fluid lift force	The fluid lift force is a force that is developed when a fluid flows around an object in such a way as to cause a pressure difference perpendicular to the direction of fluid flow. The lift force is directed perpendicular to the direction of motion of the fluid. This can occur due to: 1) inclination of a plate shape to the direction of flow so that the fluid is deflected away from the direction of flow; 2) an aerofoil (or hydrofoil) where the fluid has asymmetrical flow around the surface creating a pressure differential; 3) spinning ball creating a pressure differential – this is called the Magnus effect; and 4) unevenness of surfaces on one side of a ball compared with the other and which applies specifically to the swing of a cricket ball. The fluid lift force is dependent on the coefficient of lift (a term describing how effective the object is at creating lift), the fluid density, the cross-sectional area of the object in the direction of motion and the square of the velocity. A formula for fluid lift force can be given that links these variables together.

Fluids

A simple distinction between solids and fluids is that **solids** have a fixed shape, and individual particles are arranged in a fixed structure, while **fluids** have no fixed shape and flow freely, so individual particles have no fixed relationship with each other. Fluids can be subdivided into **liquids** and **gases**. A liquid will change shape but retain the same volume, while a gas will expand to fill the available volume (i.e., its density is not fixed). In sport and exercise science the main liquid of interest is water while the main gas of interest is air.

Properties of a fluid

An important characteristic of a fluid is density (ρ) and is defined as the mass (m) per unit volume (V) of that substance, in other words,

$$\rho = \frac{m}{V} \quad kg/m^3 \tag{D6.1}$$

In **liquids** the density decreases with increasing temperature. This will affect buoyancy. Density is increased by mineral impurities, for example a 1% salt concentration leads to a 2.3% increase in density. Density is little affected by pressure, and so a liquid is known as an incompressible fluid. A typical value for water density = 1000 kg/m³.

In **gases** the density decreases with increasing temperature, but increases with increasing pressure. Therefore, a gas is known as a compressible fluid. Compressibility is important to the air we breathe, which is compressed at depths below sea level, and expands above sea level. A typical value for air density = 1.2 kg/m³.

Fluid flow

The main feature of a fluid is that it will distort under the action of even a very small **shear force**. In a solid, a shear force is a force that tends to produce twisting or rotation but in a fluid it causes it to flow (*Fig. D6.1a*). For example, a shear force may act due to gravity when the fluid is allowed to flow down a slope. This ability for fluids to distort under the action of a force provides a varied environment for the performance of sports in air and water.

Fluid flow is either **laminar** or **turbulent**. The feature of laminar flow is that the fluid flows, as if it were, in sheets, one sliding on top of the other. In turbulent

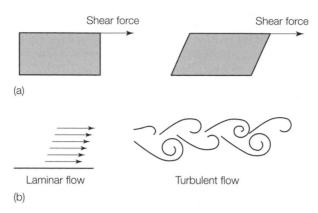

(a)

(b)

Fig. D6.1.   Shear forces cause a fluid to flow

flow particles of the fluid move perpendicular to the main direction of flow. This often occurs through the formation of eddy currents as depicted in *Fig. D6.1b*.

The **viscosity** of a fluid describes how easily or not a fluid flows. A fluid like syrup flows very slowly and is said to have high viscosity. A fluid like alcohol flows freely and is said to have low viscosity. In effect, the viscosity of a fluid describes the interaction between layers of the fluid as they slide over each other, and is best thought of as the friction between these layers. A typical value for the viscosity of air is $= 1.8 \times 10^{-5}$ Pa.s and for water is $= 1.00 \times 10^{-3}$ Pa.s (Pascals.seconds).

Buoyancy

The **hydrostatic pressure** in a fluid increases with depth. This can be experienced during a dive to retrieve an object on the bottom of a swimming pool. The pressure on the ears and lungs is felt to increase as depth increases, as the air inside the ears and lungs is compressed. This pressure acts over the surface of any object under water and creates a force or upthrust known as the **buoyancy force (U)** given by:

$$U = V. \rho.g \text{ Newtons} \qquad\qquad (D6.2)$$

An expression for the buoyancy force is also obtained from "Archimedes principle", which states that **the buoyancy force or upthrust (U) acting on an object submerged in a fluid is equal to the weight (W = m.g) of the fluid displaced**. As the weight of the fluid is given by W = m.g, and the mass from Equation D6.1, then the upthrust is equal to U = W = m.g = V.ρ.g.

The buoyancy force is a **fluid static force**. If the weight of the object submerged is greater than the upthrust then the object will sink. If the weight is less than the upthrust, the object will rise in the water until the upthrust is equal to the weight. This condition describes **floating**.

As an example, consider the upthrust acting on a beach ball of radius 15 cm (volume = 0.014 m³) submerged in water which can be calculated as U = 0.014 × 1000 × 9.81 = 137 N. This is quite a high value and some effort is needed to keep the ball submerged.

There are some interesting applications in sport and exercise.

Floating
Some people have great difficulty in floating in fresh water because their density is too high. This can happen because of low body fat and high bone mineral density. Breathing in and out can have a major effect on buoyancy. Those who have little buoyancy will have greater difficulty in learning to swim. Also the

body floats higher in sea water than in fresh water due to the higher density of sea water and the greater buoyancy force.

Scuba diving

The wet suit contains bubbles of air both within its construction and between the suit and the body. As the depth of dive increases these bubbles compress and reduce buoyancy. When this happens the diver has to get rid of some ballast, which has been used initially to enable him/her to descend in the water, otherwise the diver will continue to sink.

Airborne objects

Objects in any fluid have a buoyancy force acting on them, even objects in air, although this force is quite small. An example is the hot air balloon, which rises due to the volume of hot air that is less dense than the colder surrounding air.

Flotation and stability

The buoyancy force acts at the center of buoyancy (C o B), is located at the center of the geometric area submerged, and it is directed vertically upwards. It should be noted that the center of buoyancy and the center of gravity (C o G) are not in the same place. This leads to some interesting situations in sport and exercise due to the interaction between the gravitational and buoyancy forces and this affects an object's floating stability. If the downward force (G) acting on the center of gravity is below the upward force (U) acting on the center of buoyancy the object is stable. In *Fig. D6.2* the hull of a ballasted keel yacht is stable because

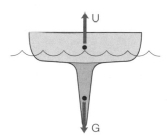

Stable – as G is below U
due to ballasted hull

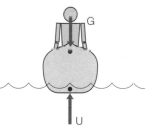

Unstable – as G is above U
due to body mass of canoeist

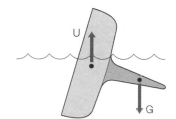

Self-righting boat – when capsized the turning
moment due to the gravitational force (G)
and the buoyancy force (U) forces causes
the boat deck to return to the surface

Fig. D6.2. Stability of yachts and canoes

the two forces act in this way. Conversely, in canoeing the gravitational force can be above the center of buoyancy (in some canoes) and so is inherently unstable as when the canoe tilts the force acts to capsize the canoe.

When a person floats in the water the buoyancy force (U) and gravitational force (G) also interact to influence the way the body lies in the water (*Fig. D6.3*). The buoyancy force acts at a location closer to the head than the gravitational force because of the lungs which make the upper body less dense. As the buoyancy force acts at a higher point in the body than the gravitational force the feet tend to sink. This is counteracted by the use of leg kick to keep the feet close to the surface.

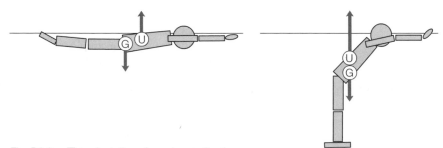

Fig. D6.3.   The orientation of a swimmer floating

The fluid drag force

When an object moves through a fluid or a fluid flows past an object a force is experienced. This force is a **dynamic force**, due to the movement of the object relative to the fluid, and is generally referred to as the **drag force**. In air, this force is usually termed **air resistance** while in water it is termed **hydrodynamic resistance**. The drag force (F_{drag}) is dependent on a number of factors and is written in the following form:

$$F_{drag} = C_D. \ 0.5. \ \rho. \ v^2. \ A \qquad\qquad (D6.3)$$

where ρ = fluid density, v = fluid velocity relative to the object, A = cross-sectional area, and C_D = coefficient of drag (which relates to the shape of an object). These factors illustrate what contributes to air resistance and water resistance. This equation applies to both air and water and, as the density of water is about 1000 times that of air, the drag force in water is about 1000 times that of air.

The drag force can be controlled by controlling the terms in equation D6.3. For example, if it is necessary to reduce the drag force in cycling, the bicycle and cyclist could be streamlined thereby reducing the coefficient of drag (by using tri-spoke wheels, aero helmet, and lycra clothing, for example). The cyclist could also use drop handlebars to reduce the cross-sectional area in the direction of travel. Little can be done about the air density (except to cycle at altitude where some world records have been broken) and of course the velocity needs to maximized so it is not possible to reduce that, although when fatigued, cyclists do reduce their speed to reduce the resistive forces they have to overcome.

Streamlining is an effective way to reduce the drag force. Well designed objects have a C_D around 0.1. Sports balls may have a C_D around 0.5 although this changes as a function of speed and other fluid flow factors. Poorly designed objects will have a C_D greater than unity.

How the drag force is caused

The drag force occurs because of a difference in pressure between the front and back of the object as it moves through a fluid. To understand how this happens it is necessary to consider some important concepts.

Flow around an object – the boundary layer
When a fluid flows in a laminar manner over a surface there is a part of the fluid
that sticks to the surface due to the viscosity of the fluid (see *Fig. D6.4*). This leads
to a region of fluid flow that is called the boundary layer. The flow of one layer of
fluid over another causes an energy loss due to the friction generated between the
layers.

**Bernoulli's
equation**

Bernoulli's equation – pressure and velocity
It is found that as the **fluid velocity** increases its **pressure** drops and **Bernoulli's
equation** is useful for describing the relationship between fluid velocity and
pressure. Specifically, as a fluid flows around a sphere (*Fig. D6.5*) there is a
region of very high pressure at the front of the sphere as the fluid impacts the
sphere. The fluid is forced around the outside of the sphere and as it does so its
speed increases. As a result of this, the pressure drops (low pressure L in *Fig.
D6.5*). As the fluid moves to the rear of the object it tries to regain its natural
flow lines. In doing so the fluid velocity reduces and its pressure increases. This
creates a region of higher pressure, but this pressure is not as high as it was at
the front of the sphere. The fluid fails to reach its original flow lines due to the
energy lost on its diversion around the sphere and the fluid breaks off and
leaves a **turbulent wake**. This creates a pressure differential between the front
and back of the sphere. Bernoulli's equation gives a specific link between fluid
velocity and pressure and takes the form:

$$P + 0.5\,\rho.v^2 + \rho.g.h = \text{constant} \tag{D6.3}$$

where P = external compressive pressure, the term $0.5\,\rho.v^2$ is called the dynamic
pressure due to the motion of the fluid, and the term $\rho.g.h$ is the hydrostatic
pressure.

In the case of the fluid flowing around a sphere, the hydrostatic pressure can be
considered constant, therefore the equation becomes $P + 0.5\,\rho.v^2 =$ constant. This

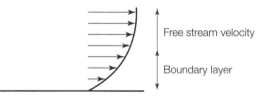

Free stream velocity

Boundary layer

Fig. D6.4.   The boundary layer

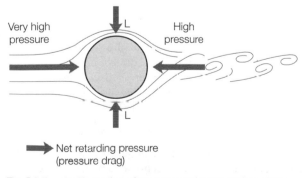

Very high
pressure

L

High
pressure

L

Net retarding pressure
(pressure drag)

Fig. D6.5.   An illustration of pressure acting around a sphere

means that as the velocity (v) increases, the compressive pressure must reduce to keep the sum of the terms constant.

The difference in pressure between the front and back of the sphere creates a **pressure drag** which is the cause of the **drag force**. If fluid can be encouraged to flow around the sphere more efficiently (i.e., with less energy loss) then the fluid has a better chance of regaining its original flow state and produce a smaller turbulent wake. Thus, streamlining encourages the fluid to pass over the object with minimal energy loss, so producing a smaller drag force.

Turbulence in the boundary layer

An interesting phenomena in sport is that sports balls can sometimes be made to travel faster and farther than they would normally. This effect can be explained by **turbulence in the boundary layer**. If the fluid flow is very fast, or if the surface is rough, then the laminar flow, which makes up the boundary layer, becomes **turbulent** (*Fig. D6.6*). This is actually an advantage, as energy from the free stream velocity can enter into the boundary layer region, giving it more energy than it would otherwise have so enabling it to flow better around the sphere. The important consequence is that as the turbulent wake reduces, the pressure drag also reduces, in other words there is a **lower drag force**. Whether turbulence in the boundary layer will occur or not for a sports ball depends on the size of the ball and its surface roughness. A soccer ball will easily go "turbulent" and this enables goalkeepers to make a kick from goal to almost the other end of the pitch. In table tennis the ball is too small and smooth ever to be able to generate turbulent flow and take advantage of a reduced drag force. In cricket the ball can be made to go turbulent by a fast bowler as the ball becomes roughened during play. The dimples of a golf ball help it to go turbulent and increase the distance of a drive, although the dimples also have an important influence on flight due to the ball's spin (see below).

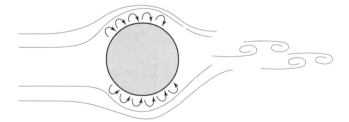

Fig. D6.6.   Turbulence in the boundary layer

The fluid lift force   Under certain circumstances the fluid flow can cause a force to be generated **perpendicular to the direction of flow**, and this force is termed the **lift force**. The lift force can act in any direction, not just upward. The lift force is responsible for a range of effects seen in sport from the aquaplaning of water skis, to the flight of aircraft, to the slice and hook experienced in golf. The lift force (F_{lift}) has the same general form as the drag force, in other words,

$$F_{lift} = C_L \cdot 0.5 \cdot \rho \cdot v^2 \cdot A \qquad (D6.4)$$

where C_L = coefficient of lift (dependent on the shape), ρ = fluid density, v = relative velocity, A = cross-sectional area, and these are similar to those used for the drag force. The different causes of the lift force are detailed.

Deflection of fluid from the main direction of flow
When an object is angled to the direction of flow, some of the fluid is forced away from the direction of motion, causing the object to be forced in the opposite direction (*Fig. D6.7*). This type of lift is created when a plate travels over the surface of water as in water skiing or when a boat "planes". If an object is immersed in a fluid (either air or water) then the effect can be maximized by careful design of the shape (*Fig. D6.8*). Fluid traveling over the top surface has a greater distance to go than that traveling over the bottom surface. As a result, there is greater velocity on the top surface, and, according to Bernoulli's equation, there is lower pressure. This creates a pressure differential causing lift. The aerofoil is a key shape in modern life enabling flight. The same applies in water with the hydrofoil, enabling boats to travel faster and more efficiently.

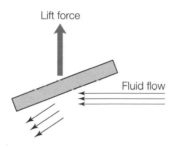

Fig. D6.7.   Fluid hitting a plate cause lift

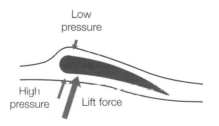

Fig. D6.8.   The lift force on an aerofoil

The spin of an object – the Magnus force
When a sphere moves through a fluid and it **spins**, it increases the speed of fluid on one side, and reduces it on the other (*Fig. D6.9*). As the speed of a fluid is inversely related to its pressure (from Bernoulli's equation) there is a pressure difference at right angles to the direction of flow causing a lift force known as the **Magnus force**. The Magnus force is found to increase non-linearly with the angular velocity of the ball (typically it is related to the square of the spin). The Magnus force acts in the direction of spin where there is the lowest pressure (i.e., or highest fluid velocity).

The Magnus force explains the motion of sports balls as they spin. For example, the topspin and backspin in tennis and table tennis; the hook and slice in golf; the torpedo swerve in rugby; the spin swing in cricket; the swerve of a soccer ball or volleyball.

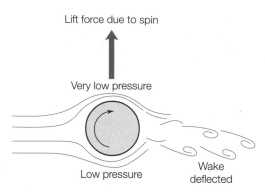

Fig. D6.9.   *The Magus effect: the lift force created due to spin*

The variation in surface roughness – the "swing" of a cricket ball

A cricket ball is observed to "swing" at high speeds. This is quite a crucial aspect of the game. It is found in practice that as a new ball begins to deteriorate through use, it has a tendency to "swing" more. As it further deteriorates, its ability to swing is lost.

The "swing" of a cricket ball can be explained by the asymmetrical roughness on one side of the ball. The bowler has to keep one side smooth and allow one side to become rough. There are various legal (and non-legal) ways of doing this. On the smooth side, the fluid flow is laminar while on the rough side it can become turbulent. Turbulence is encouraged by the presence of the seam (*Fig. D6.10*). Fluid flows more easily over the turbulent side (higher velocity) leading to a lower pressure (from Bernoulli's equation). The ball therefore swings in the direction of the seam. As the ball ages it becomes rough on both sides and turbulence occurs on both sides of the ball. When this happens the asymmetry is destroyed and the ball no longer swings.

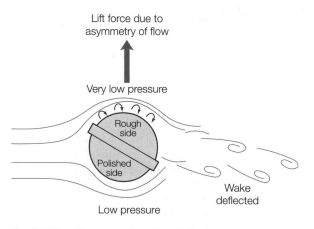

Fig. D6.10.   *The swing of a cricket ball*

E1 BIOMECHANICAL CHARACTERISTICS OF WALKING

Key Notes

The gait cycle	Describes the style of locomotion. For walking this is divided into support (when the foot is on the ground) and swing (when the foot is off the ground) phases. The support phase contains periods of single support (one foot on the ground) and double support (both feet on the ground).
Stride	The movement from when one foot touches the floor to when the same foot next touches the floor. Each stride is made up from two steps.
Speed	The speed of walking is found from the stride frequency (number of strides per second) multiplied by the length of each stride.
Forces during walking	The vertical ground reaction force during walking typically peaks at a little above body weight. The force rises relatively slowly as the load is transferred from one foot to the other during the periods of double support (when both feet are on the floor). The horizontal force is initially negative, indicating that it acts in the opposite direction to the movement and serves as a braking action. During the latter half of the support phase the horizontal force becomes positive to propel the body forward into the next step.
Upper body movement in walking	During walking the arms swing in the opposite direction to the legs such that when the left leg is forward the left arm is back. This movement helps to overcome the angular momentum of the lower body and to reduce the energy cost of walking.

The gait cycle

In the analysis of any skill it is important to understand the role of the various joint movements and body segments involved. Walking is no exception. The joint and segmental interactions involved in walking are so complex that it takes most humans a year to be able to "toddle" and a further 3 to 4 years to perfect walking.

Gait (the style of locomotion) is defined according to the sequence of **swing** and **support** phases of the legs when the foot is either in the air (swing) or in contact with the floor (support or stance). Walking is characterized by the occurrence of a period of double support with both feet in contact with the ground, separating periods of single support when the other leg is swung forwards to make the next step. There is no time at which both feet break contact with the ground at the same time, in other words no flight phase.

Stride

One complete gait cycle, for example, from right heel strike (when the heel of the right foot contacts the floor) to the next right heel strike, is known as a **stride** (*Fig. E1.1*). Each stride is made up from two **steps**, each step covering the period from

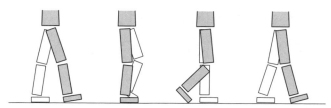

Fig. E1.1a. The phases of the walking gait cycle. One full stride of the right leg is shown from heel strike to heel strike

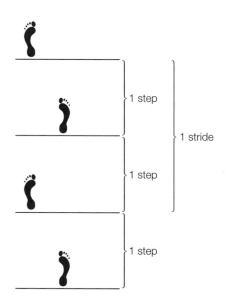

Fig. E1.1b. The composition of the gait cycle into steps and strides. One stride is made up from two consecutive steps

one heel strike to heel strike of the contralateral limb (other leg). The distance covered in a stride is known as the stride length, the rate at which strides are taken is referred to as the stride frequency or cadence, measured as the number of strides per unit of time (strides per second). Based on the stride length and frequency it is possible to calculate the velocity at which someone is walking:

Stride length × Stride frequency = Velocity

Example
If the stride length was 1.2 m and the stride frequency 1.5 Hz (1.5 strides per second or 3 strides every 2 seconds) then the velocity could be found by:

$$1.2 \text{ m} \times 1.5 \text{ Hz} = 1.8 \text{ m.s}^{-1} \text{ (~4 miles per hour)}$$

Speed

An increase in the speed of walking must result from an increase in the stride length and/or an increase in stride frequency.

Gait cycle

To try to understand human gait it is important to perform a kinesiological analysis of the movements. This involves the description of movement in terms of the sequencing and range of joint movements and the muscle actions involved. To assist this form of analysis it is normal to break the action down into phases and consider it joint by joint.

Before starting to describe the movements of the gait cycle it is necessary to break the action down into phases (smaller segments), each phase needs to have a clear start and end point, and the phases need to fit together to give a continuous sequence of movement.

In gait the first division of the movement is to separate the cycle into **swing** and **stance** phases, in other words, the periods when the foot is either in the air or in contact with the ground. The ratio of stance to swing times is a useful measure for quantifying normal and abnormal gaits. The normal ratio of stance to swing during walking is 60% stance/40% swing.

However, these phases are too long to allow us to perform a useful analysis so these are further divided to sub-phases. A common division of the gait cycle is into five phases, these same phases can be used to describe both walking and running actions *(Fig. E1.2)*.

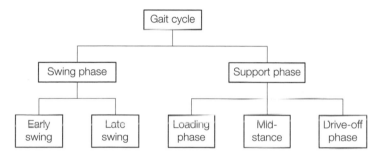

Fig. E1.2.   The five phases of the gait cycle

These can be defined with the following start and end points:

Swing phase	From	Toe-off
	To	Foot strike
Early swing	From	Toe-off
	To	The start of knee extension
Late swing	From	The start of knee extension
	To	Foot strike
Stance phase	From	Foot strike
	To	Toe-off
Loading phase	From	Foot strike
	To	Foot flat
Mid-stance	From	Foot flat
	To	Heel raise
Drive off	From	Heel raise
	To	Toe-off

The movements associated with each of these phases are summarized in the following tables.

Swing phase

Early swing

Joint	Movement	Range of motion
Hip	Flexion	9° ext–30° flex
Knee	Flexion	30° flex–60° flex
Ankle	Dorsi-flexion	5° PF–0° DF

Late swing

Joint	Movement	Range of motion
Hip	Extension	30° flex–25° flex
Knee	Extension	60° flex–10° flex
Ankle	Dorsi-flexion	0° DF–5° DF

Stance phase

Loading phase

Joint	Movement	Range of motion
Hip	Flexion	25° flex–30° flex
Knee	Flexion	10° flex–20° flex
Ankle	Plantar-flexion	5° DF–10° PF

Mid-stance

Joint	Movement	Range of motion
Hip	Extension	30° flex–0° ext
Knee	Extension	20° flex–5° flex
Ankle	Dorsi-flexion	10° PF–20° DF

Drive-off

Joint	Movement	Range of motion
Hip	Extension	0° ext–9° ext
Knee	Flexion	5° flex–30° flex
Ankle	Plantar-flexion	20° DF–5° PF

DF = Dorsi-flexion
PF = Plantar-flexion

These movements of the lower limb are representative of normal walking. However, it should be noted that whilst the general pattern of walking is quite characteristic there will be notable inter-individual variability in the absolute joint angles moved through. The values in the table above are indicative of the pattern of movement and should not be interpreted as representing a set of normative data.

Movements of the limbs are powered by the contractions of the musculature. *Fig. E1.3* summarizes the muscle activity associated with walking. The muscles are mainly involved in the initiation and cessation of limb movements. Much of the action for swinging the leg is achieved by the pendulum effect of the gravity and does not require a significant muscular effort.

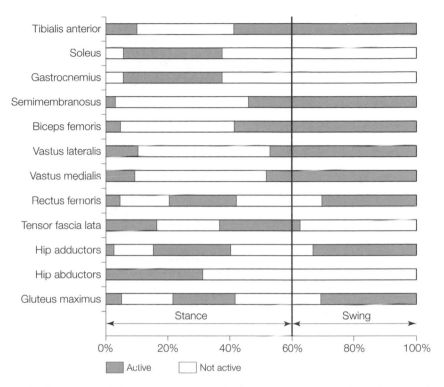

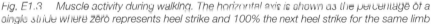

Fig. E1.3   Muscle activity during walking. The horizontal axis is shown as the percentage of a single stride where zero represents heel strike and 100% the next heel strike for the same limb

With the average person taking nearly 2000 steps per mile while walking, it is important to understand the processes involved in the contact between the body and the ground. Application of Newton's first law states that a force must be acting in order for the movements associated with walking to take place. The study of the forces associated with gait therefore forms an important part of any gait analysis.

Fig. E1.4 shows a typical ground reaction force trace for normal walking. The trace shows the periods of single and double support characteristic of walking and the way that the forces are transferred from one foot to the other. The double support phase, where the load is transferred from one foot to the other, allows the loading rate (slope of the curve) to be controlled and therefore kept relatively low. The vertical force remains relatively close to body weight throughout the periods of single support, with peak forces at impact and drive-off of only slightly above body weight. These peaks result from the body decelerating at impact (heel strike) and then accelerating at toe-off; during mid-stance the trough results from a net downwards acceleration of the center of mass as it passes over the foot.

The anterior–posterior forces (*Fig. E1.5*) show the forces acting along the direction of movement, these are known as either braking or driving forces dependent upon their direction. At impact the force is acting in the opposite direction to the movement and is therefore a braking force. The magnitude of the braking force will fluctuate as the gait style changes. As the center of mass passes over the foot and the forward drive begins, so the force becomes positive and acts as a driving force. The point at which the force changes from braking to driving is

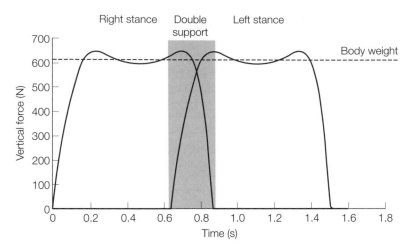

Fig. E1.4. *Typical vertical forces during normal walking*

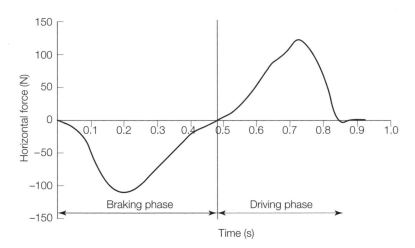

Fig. E1.5. *Typical anterior–posterior forces during normal walking*

normally between 45% and 50% of the total stance time. Variations from this normal pattern will be suggestive of an abnormal gait.

The role of the upper body during gait is to act as a stabilizer to limit the changes in the angular momentum of the body and to thus conserve energy. If there were no upper body then there would be large positive and negative swings in the angular momentum of the body as the legs rotated forward and backward.

The arms swing in a contralateral fashion in exact anti-phase to the legs, in other words the right arm reaches maximum shoulder and elbow flexion when the right leg reaches toe-off and maximum shoulder and elbow extension at right heel strike. Thus the upper body is being rotated in the opposite direction to the pelvis. This out of phase action of the arms generates an opposite angular momentum value to the legs and so reduces the change in the angular momentum of the whole body.

Note: despite the greater mass of the legs the arms are able to generate a momentum almost equal in magnitude to that of the legs. This is possible because the arms are positioned further from the mid-line of the body and so require less mass to achieve the same moment of inertia.

Angular momentum (H) = Moment of inertia (I) × Angular velocity (ω)

In the horizontal plane, the arms have no effect as they act in opposite directions, in other words one forward and one back. Vertically the arms contribute approximately less than 5% to the total lift of the body.

E2 BIOMECHANICAL CHARACTERISTICS OF RUNNING

Key Notes

The gait cycle

Describes the style of locomotion. For running this is divided into support (when one foot is on the ground) and swing (when the foot is off the ground) phases. There are also periods of flight when both feet are off the ground simultaneously.

Running speed

Running speed is the product of stride length and stride frequency. Increases in speed are normally achieved by increasing stride length up to approximately 7 ms^{-1} after which changes in stride frequency are required. Changes in stride frequency are normally accompanied by an increase in the energy cost per unit distance covered (economy).

Pronation

During running the subtalar joint (the joint between the talus and the calcaneus at the bottom of the leg) experiences a pattern of movement known as pronation and supination. Pronation involves the flattening of the foot to the floor by the combination of eversion, adduction and dorsiflexion. Supination raises the arch off the floor by inversion, adduction and plantar-flexion.

The typical range of motion during running is from 10° supination at foot strike to 10° pronation by mid-stance.

Forces during running

The peak vertical impact force during running is typically about 2–2.5 times body weight. The size of the impact force varies with body weight and the speed of running. The force rises rapidly and reaches a peak within the first 50–100 ms after foot strike.

Running on soft surfaces or in cushioned shoes generally leads to a reduction in the vertical impact forces.

The horizontal force initially acts as a braking force slowing the body. At about 50% of the support phase the force becomes positive and serves to accelerate the body into the next flight phase.

Foot strike

Running gait is often described according to which part of the foot makes first contact with the ground. In most runners the first contact is in the rear third of the foot and these runners are described as "heel strikers". A "mid-foot striker" makes first contact in the middle third of the foot and a "toe striker" in the front third.

Mid-foot and toe strikers typically produce a vertical force trace without an obvious impact peak. Rather the initial impact is absorbed by the muscular structures of the lower limb.

Gait cycle

As with walking, the running action is made up from a series of steps in which forward progress is made by sequentially planting the left and right feet on the ground. The style of movement is described as the **gait** and can be considered to be a cyclic (repeated) movement and is thus referred to as the **gait cycle**. The running gait cycle is divided according to the sequence of swing (when the foot is in the air) and support/stance (when the foot is on the ground) phases of the legs. During running the periods of single support (where only one foot is on the ground) are separated by a flight phase in which there is no ground contact and there is no double support phase (no periods when both feet are on the floor at the same time). In running the ratio of stance to swing rises to approximately 40% stance and 60% swing. The exact ratio depends upon speed, with the relative duration of stance decreasing as speed increases such that in maximal sprinting stance occupies only about 20% of the gait cycle.

Running speed

To increase running speed requires an increase in the **stride length** (the distance covered in each step) and/or an increase in **stride frequency** (the number of strides taken in each unit of time). Up to a speed of approximately 7 m.s^{-1} most of the increase is achieved through increasing the stride length while maintaining a nearly constant stride frequency. Above this speed the stride frequency increases. It is suggested that the reason why stride length is usually increased first with changes in stride frequency reserved until higher velocities is because there is an optimum stride frequency at which the energy cost of running is least. If the stride frequency is changed, then the energy cost per unit distance increases making the athlete less energy efficient.

Despite the obvious logic, there is only a low correlation between the stride length of an athlete at a given speed and the anthropometric measures of that person. There is a stronger relationship with factors such as strength and flexibility.

The running gait cycle can be broken into similar phases as walking to facilitate kinesiological analysis. The following is a summary of the typical movements seen in running. Note that the range of motion indicates the start and end points of the joint in each phase and that the absolute ranges of motion are dependent upon speed and generally increase as speed increases.

Swing phase

Early swing

Joint	Movement	Range of motion
Hip	Flexion	9° ext–55° flex
Knee	Flexion	25° flex–90° flex
Ankle	Dorsi-flexion	20° PF–10° DF

Late swing

Joint	Movement	Range of motion
Hip	Extension	55° flex–45° flex
Knee	Extension	90° flex–20° flex
Ankle	Plantar-flexion	10° DF–5° DF

Stance phase

Loading phase

Joint	Movement	Range of motion
Hip	Flexion	45° flex–50° flex
Knee	Flexion	20° flex–40° flex
Ankle	Dorsi-flexion	5° DF–20° DF

Mid-stance

Joint	Movement	Range of motion
Hip	Extension	50° flex–15° flex
Knee	Flexion	40° flex–40° flex
Ankle	Dorsi-flexion	20° DF–30° DF

Drive-off

Joint	Movement	Range of motion
Hip	Extension	15° flex–9° ext
Knee	Extension	40° flex–25° flex
Ankle	Plantar-flexion	30° DF–20° PF

DF = Dorsi-flexion
PF = Plantar-flexion

Although they share certain similarities the movements involved in running differ from walking in a number of ways. The major differences are described below.

Hip: at foot strike the hip is flexed to approximately 45° (greater than in walking where this is typically 30°), this angle is maintained during early stance by knee flexion. During drive-off the hip extends to approximately 9° at toe-off (the same as walking). Flexion during the swing phase reaches about 55° (only reaching 25° during walking).

Knee: At heel strike the knee is flexed to an angle of approximately 25°; it is never straight at impact (knee flexion of 10° during walking). The knee flexes to about 40° by mid-stance (only 20° during walking). From mid-stance the knee extends to toe-off. During swing the maximum knee flexion reaches 90° (maximum flexion of 60° during walking).

Ankle: the ankle reaches maximum dorsiflexion of about 30° by mid-stance. Plantar flexion at toe-off is significantly greater than in walking reaching 20° compared with only 5° in walking.

As speed increases, the flexion of hip and knee joints during the swing phase increases, this serves to reduce the moment of inertia of the limb, thus allowing for a faster swing. There may also be a slight increase in the degree of knee flexion at impact.

The phasing of muscle activity in running is shown in *Fig. E2.1*. In general, muscles are most active in anticipation for and just after initial contact. Muscle contraction is more important at this time than it is for the preparation for and the act of leaving the ground. As the speed of gait increases so the degree of muscle activity increases. The swing phase becomes more active with a greater muscular contribution to the movement. The increased ranges of joint movement result in a greater period of muscle activity during all phases of the cycle.

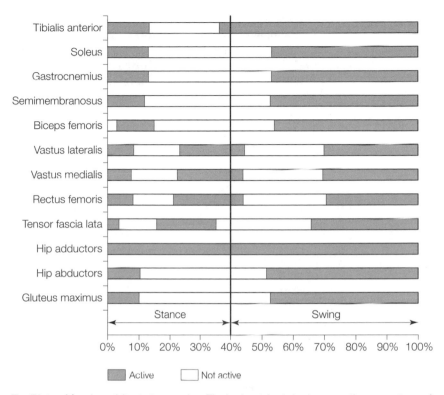

Fig. E2.1. Muscle activity during running. The horizontal axis is shown as the percentage of a single stride where zero represents heel strike and 100% the next heel strike for the same limb

So far, only the major movements of the lower limb in the sagittal plane (side view) have been considered. While these are clearly the largest movements that occur there are other, smaller movements that are of equal or greater importance to a consideration of injury.

The movements of the subtalar joint have been the focus of much attention in the research literature due to their suspected role in the etiology of injury. The subtalar joint lies just below the ankle joint and is formed by the talus above and calcaneus (heel bone) below. It is at this joint that the movements of **inversion** (turning the sole of the foot inwards) and **eversion** (turning the sole of the foot outward) mainly occur.

Pronation

During any form of gait the motions of the subtalar joint and the other plane joints in the foot and ankle act to serve a shock absorbing function. Just prior to impact the foot is positioned in a supinated position (inverted, adducted and plantar flexion) such that the outside portion of the heel makes first contact with the floor. Immediately following impact the foot flattens as the whole of the foot is placed onto the floor and the subtalar joint moves from inversion to eversion. This movement is known as pronation. As this happens the plantar fascia (ligaments and tendons of the sole of the foot) becomes stretched and the supporting musculature works eccentrically to resist this flattening. These actions help to reduce the impact force by effectively softening the impact and slowing the descent of the body more gradually.

Once in mid-stance the foot remains in the pronated position as the body weight moves over the foot, before re-supinating to form a rigid lever for toe-off. *Fig. E2.2* shows a "typical" trace for the rear foot (pronation/supination) angle during a single stance phase. Note that most of the pronation happens during the first 0.05–0.1 s after foot strike. It is therefore a very rapid movement. The normal range of motion is from ~10° supination to ~10° pronation.

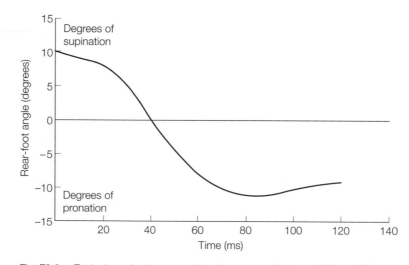

Fig. E2.2.   Typical rear-foot movement during running. Data end during mid-stance as it is not possible accurately to measure the rear foot motion in two dimensions after the heel lifts from the floor, hence the data do not show the return of the foot to a supinated position for toe-off

If the foot strikes the ground in a pronated position, or if the muscles and plantar fascia are loose then there will be little resistance to the normal range of pronation and the shock absorbing function of the foot will be lost. The nature of the subtalar joint means that if the foot pronates then there must be an accompanying rotation of the tibia. Pronation leads to an internal rotation of the tibia while supination leads to an external rotation. Excessive pronation >20° will lead to an excessive internal rotation of the tibia. This will disrupt the normal loading of the ankle joint and may lead to malalignment problems at the knee leading to anterior knee pain.

There are a number of factors which have been suggested to be related to the amount and the rate of pronation during gait. The greater the body weight of an individual, the greater will be the load during stance. Excessive body weight will tend to lead to a more flattened arch position and either greater or more rapid pronation of the foot. As running speed increases, the foot strikes the ground in a more supinated position whilst pronation ends at more-or-less the same final angle at all speeds. This will give a greater range of motion and, because stance time decreases with running speed, the rate of pronation is greater at faster speeds.

Forces during running

During running, the vertical movement of the body is greater than in walking as a consequence of there being a flight phase. As the body will be falling from a greater height it will have a greater vertical velocity at foot strike. The slope of the force–time curve (loading rate) is also greater, reaching a peak after only about 0.05 s (whereas in walking this peak does not occur until about 0.15 s).

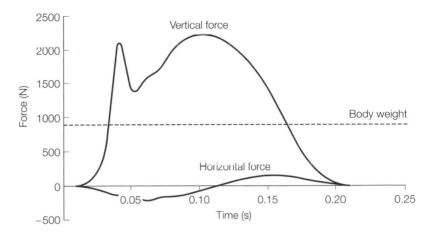

Fig. E2.3.   Typical vertical and horizontal forces during heel to toe running

This will mean that the tissues of the body are loaded more rapidly and a greater stress placed upon them. During running the impact is onto a single leg rather than a gradual transfer from one leg to the other as seen during the double support phase of walking.

The vertical forces experienced during running are directly related to body weight. This is logical, as Newton's second law states that the force is proportional to the mass of the body. Typically vertical ground reaction forces in running are of the order 2–2.5 times body weight. Similarly, as the speed increases so the impact peak force increases, with an accompanying rise in the rate of force loading. The vertical ground reaction impact force increases by about one times body weight (from ~2 × body weight to ~ 3 × body weight) when running speed is increased from 3 to 6 m.s^{-1} (9 minute mile pace to 4.5 minute mile pace). The vertical drive-off force is unaffected by the increase in running speed. As the contact time with the ground decreases as speed increases there is also a significant rise in the rate at which the force is applied to the body.

Foot strike

The "normal" force trace presented in *Fig. E2.3* showed the characteristics of a heel to toe runner, this is the style used by about 80% of runners. Running style is classified according to which part of the foot makes first contact with the ground. If the point of first contact lies in the rear third of the foot then the runner is classified as a rear foot striker; first impact in the mid third of the foot is classified as a mid-foot strike and first contact in the forward third is classified as a fore-foot strike. Runners with a mid- or fore-foot impact will tend to experience a lower impact force as the loading is cushioned by the active contraction of the calf muscles.

It is widely accepted that modern running shoes act to reduce the impact forces seen during walking and running. The various cushioning devices found in the shoes function to prolong the impact and reduce the stress on the musculoskeletal system. The choice of running surface will also have a significant effect on the magnitude of the forces. The harder a surface the greater will be the forces experienced. More compliant surfaces give a greater time to stop the downward motion of the body and so reduce the impact force. However, running on a soft surface uses up more energy and will lead to a slower pace or earlier fatigue.

As the body rises and falls and speeds up and slows down during the gait cycle there are changes in the **potential** and **kinetic** energies in the body. Potential energy describes the energy due to a body's position and is related to its weight and height above the ground (mass × gravity × height). Kinetic energy is the energy due to the body's motion and is related to the mass and the velocity of the body (1/2 mass × velocity^2). In running the changes in these two energies are in phase, such that when potential energy is high so is kinetic energy. Running has been likened to an individual on a pogo stick, propelling the body from a low point during the middle portion of stance (stance phase reversal) to a peak during the flight phase. To reduce the total energetic cost of running two mechanisms are used:

1. the storage and later return of elastic potential energy by the stretch of elastic structures;
2. the passive transfer of energy from one body segment to another.

These seek to ensure that the amount of metabolic energy required to run is minimized. During the initial period of ground contact the body's energy decreases as the velocity slows during the braking phase and the mass center is lowered by hip and knee flexion. Some of this energy can be stored in the tissues of the lower limb, for example, the Achilles' tendon in the form of elastic potential energy. Later, during the drive-off phase this energy can be returned to contribute to the increase in center of mass height and velocity. In so doing this storage and reuse of energy reduces the requirement for the work to be performed by active muscular contraction.

E3 BIOMECHANICS OF JUMPING

Key Notes

The counter-movement

Most forms of jumping are initiated with a downward movement of the body referred to as a "counter-movement". This action serves to increase by about 10% the distance which can be achieved in the jump. This increased performance is attributed to a greater range of movement during the propulsive phase and the use of the stretch–shorten cycle.

The stretch–shorten cycle

The stretch–shorten cycle describes the sequence of movement whereby an active muscle is first stretched by an eccentric contraction and then shortened by concentric action. The stretching phase results in a more forceful shortening of the muscle than if there had been no pre-stretch.

Jump height

In a standing vertical jump the height achieved is directly proportional to the vertical velocity at take-off such that:

$$\text{Jump height} = \frac{\text{Take-off velocity}^2}{}$$

The vertical velocity is related to the vertical impulse generated during the contact period such that:

$$\text{Take-off velocity} = \frac{\text{Force} \times \text{Time}}{\text{Mass}}$$

The impulse can be determined by recording the vertical ground reaction force and integrating this.

Arm swing in jumping

Swinging the arms is believed to add between 10% and 20% to the height or distance achieved in a jump. This is related to both a direct contribution to mass center momentum and to the creation of greater ground reaction forces. The mass center momentum is a summation of the individual segmental momenta. If the arms are being swung upwards and forwards they will contribute to the mass center's forward and upward motion.

Control of angular momentum

During the flight phase of a jump the body will tend to rotate (somersault) in accordance with the body's angular momentum at take-off.

$$\text{Angular momentum} = \text{Moment of inertia} \times \text{Angular velocity}$$

In order to control the rate of rotation jumpers manipulate their moment of inertia by changing the orientation of their limbs. Adopting an elongated body position increases the moment of inertia and hence slows the rate of rotation. Alternatively rotating the arms and legs in a forwards direction (a hitch-kick) will slow the rate of forward body rotation by use of a transfer of momentum.

Jumping is a movement fundamental to a large number of sports. The goal of which may be to try to achieve the greatest possible distance, either horizontally or vertically, as in the long-jump or high-jump events; or the jump may be performed in order to intercept an object, for example to head a soccer ball. Jumps can be performed either from a standing position or following a run-up and might involve a take-off from one or both feet. As a consequence there is no single description for the movements involved in jumping. However, there are sufficient similarities in the movements to allow a separate analysis of two-footed and single-footed jumps.

Standing two-footed jumps are the simplest form of jump to analyze. The basic movements are the same whether the objective is for maximum vertical height or horizontal distance. The movement can be broken down into the following phases:

Counter-movement	From	First movement
	To	Maximum knee flexion
Propulsion	From	Maximum knee flexion
	To	Take-off
Flight	From	Take-off
	To	Landing
Landing	From	Landing
	To	End of the movement

The counter-movement

During the counter-movement phase the hip, knee, and ankle joints all undergo a period of flexion. The amplitude of the flexion will depend upon the demands of the task and the specific situation in which the jump is performed. However, generally for greater jump heights there will be a greater range of hip flexion. The amount of knee and ankle flexion remains more-or-less constant.

The counter-movement serves two purposes: the first is to move the body into a better position to start the propulsive phase and the second is to stimulate the stretch–shorten effect. If the start position is upright standing it is obvious that there is very limited potential for the performer to generate lift from this position as the joints of the lower limb are at, or close to, their end of range of movement. To jump from this position can only really be achieved by ankle plantar flexion. By performing a counter-movement the joints are initially flexed, thus permitting a greater range of movement for the propulsive phase.

The greater range of motion allows for the creation of a larger impulse during the propulsive phase as the force can be applied for a greater time. The impulse (force × time) is directly related to the change in velocity of the body. Thus a greater impulse will result in a higher take-off velocity and thus a greater jump distance.

Stretch–shorten cycle

During the counter-movement the main muscular actions are the eccentric contraction of the hip, knee, and ankle extensors. These muscles work to resist the flexion at the joints which will occur as a natural consequence of the gravitational force. Hence, the muscles are producing a resistive force whilst their length is increased. This eccentric contraction is often referred to as a pre-stretch, as the stretch of the muscles precedes the use of the same muscles in the following propulsive phase. Pre-stretching a muscle before it is shortened leads to an increase in the force that the muscle can produce. This increase in force following the pre-stretch is known as the **stretch–shorten cycle**. The faster the stretch and the shorter the delay between the stretch and the shortening the greater will be the enhancement in the muscle force produced. It is important that the counter-movement is performed rapidly, and that there is a minimal delay between the end of the counter-movement and start of the propulsive phase.

Jump height

If the body is lowered slowly, into a position comparable to that achieved at the end of the counter-movement, and this position is held prior to the upward movement, the jump is described as a squat jump; as the jump begins from a squat position. The jump distance achieved in squat jumps is usually about 10% less than that achieved from a comparable counter-movement jump. Thus suggesting that the counter-movement adds about 10% to the distance that can be achieved.

During the propulsive phase, the joints of the lower limb (hip, knee, and ankle) undergo extension. The hip is always the first joint to start to extend, accelerating the large, heavy trunk segment. Extension at the knee and ankle joints follows after a short delay, the initiation of knee and ankle extensions may happen either simultaneously or in sequence (knee then ankle, or ankle then knee). However, there is no clear evidence to suggest that performance is better with any particular sequence for the initiation of knee and ankle movements.

Arm swing in jumping

In addition to the movements of the lower limb, the arms play an important part in the performance of standing jumps. During the counter-movement phase the arms are swung downward and backward before swinging forwards and upwards in the propulsive phase. Arm swing has been shown to add between 10% and 20% to the distance achieved in a jump. To be effective, the arm swing must be timed appropriately, such that take-off occurs as the point when the arms are at, or close to, maximum velocity.

The exact mechanism through which the arm swing contributes to jumping performance has not been determined. However, it is believed to be related to both a direct contribution to mass center momentum and to the creation of greater ground reaction forces. If the body is considered as a series of individual segments, then it is possible to calculate the velocity and hence momentum (mass × velocity) of each of these. The mass center velocity, and hence momentum, is a consequence of the summation of these individual segmental velocities and momenta. If the arms are being swung upwards and forwards then they will contribute to the mass center's forward and upward motion.

Standing vertical jumps are often used as an assessment of athletic performance. The height of a standing vertical jump is determined by the vertical velocity at take-off which is, in turn, related to the vertical impulse such that:

$$\text{Impulse} = \text{Change in momentum}$$

$$\textbf{Force (N)} \times \textbf{Time (s)} = \textbf{Mass (kg)} \times \textbf{Change in velocity (m.s}^{-1}\textbf{)}$$

In a standing jump the initial velocity can be considered to be zero, as the performer starts from a stationary position, and so the change in velocity is in fact equal to the take-off velocity. Thus:

$$\textbf{Force} \times \textbf{Time} = \textbf{Mass} \times \textbf{Take-off velocity}$$

And therefore:

Take-off velocity = Force × Time / Mass

After take-off the body will experience a negative acceleration due to gravity (−g) which will cause the body to slow and come to rest at the apex of the jump. At this moment the vertical velocity will once again be zero. It is possible to calculate the height of the jump based on the take-off velocity using the equations of uniformly accelerated motion:

$$V^2 = U^2 + 2 \times a \times S$$

where V = final velocity, U = initial velocity, a = acceleration and S = displacement

This can be rearranged to find S:

$$S = (V^2 - U^2) / (2 \times a)$$

In this case, V will be the velocity at the apex of the jump and is thus zero; U is equal to the take-off velocity; a is the acceleration due to gravity (−g) and S is the height of the jump. Therefore:

$$S = U^2 / 2g$$

Jump height = Take-off velocity^2 / 2 × gravity

The most accurate method for obtaining measures of the take-off velocity is to use a force platform to record the vertical ground reaction force. From the force-time data it is thus possible to determine the impulse as the start point for the calculation above. Another common method used to determine the height of a vertical jump is to measure the flight time and use this to calculate the height achieved. This calculation also makes use of the equations of uniformly accelerated motion.

$$S = U \times T + \tfrac{1}{2} a \times T^2$$

where S = displacement, T = time, a = acceleration

If it is assumed that the take-off and landing are performed at the same relative height, then the apex of the jump will occur at exactly half the flight-time at which point the velocity will be zero, and the acceleration will be due to gravity (−g). Therefore, at the apex of the jump the body will have a zero velocity and, from this point to landing, the body will experience a displacement equal to the height jumped in a time of ½ T. Therefore:

U = 0, T = ½ flight time, a = g (note the minus sign has been dropped as the direction of the displacement is not important)

$$S = 0 \times t + \tfrac{1}{2} \times g \times (\tfrac{1}{2} T)^2$$

$$S = \tfrac{1}{2} \times g \times (\tfrac{1}{2} T)^2$$

In many athletic events and other sports, jumping actions are performed from a single leg and following a run-up. Performance in running jumps, is also determined by the velocity of the body at take-off. Unlike in standing jumps, the body will posses an initial velocity as a consequence of the run-up and thus the take-off velocity will be determined by the combination of the run-up velocity and the change in velocity due to the take-off impulse. In the case of horizontal jumps, for example, the long jump, the athlete will attempt to generate a large initial velocity during the run-up. Correlation coefficients of 0.8–0.9 are found between run-up speed and jump distance indicating that generally jump distance increases with increasing run-up speed.

Despite the strength of the relationship between run-up speed and distance it is worthy of note that athletes do not achieve maximum running speed in the run-up. Each individual will have an optimal run-up speed somewhere below their maximum running speed. Increasing the run-up speed above this leads to a decrease in performance as the athlete is unable to generate sufficient impulse during the take-off phase.

The actions of the take-off leg in single-legged jumps are similar to those in the two-legged jumps; after heel strike the hip, knee, and ankle of the take-off leg experience an initial flexion followed by extension. These movements give rise to what are commonly described as compression and drive-off phases. The compression phase is comparable with the counter-movement and involves similar mechanisms, although the range of joint motion is much less than seen in two-legged jumps.

In addition to the action of the take-off leg, the contralateral (other) leg also makes an important contribution to the take-off velocity. The non-take-off leg is often referred to as the "free leg" as it is free to swing in space. The motion of the free leg and the arms contribute to the take-off velocity in a similar way to that described for the arms in standing jumps. The momentum within these segments increases mass center velocity by 10–15% provided that the movements are timed correctly.

Control of angular momentum

During the take-off phase of a jump the body experiences a combination of horizontal and vertical forces. As the line of action of these does not always pass through the center of mass there is a resultant moment about the mass center tending to cause rotation. Generally the ground reaction forces act to create a forward, somersaulting moment. Therefore, during take-off, the body is subjected to a torque impulse (Torque × Time) which leads to a change in its angular momentum (Moment of inertia × Angular velocity). Once in the air, the body will tend to rotate forwards and the performer has to take action to control the consequences of this rotation on their performance.

In the long and triple jump events, the athlete is required to control the rate of forward rotation to allow them to achieve an optimal landing position. As the amount of angular momentum is constant during flight (there is no possibility to apply a corrective torque whilst the body is in the air) there are two methods by which the rotation can be controlled. The first requires a simple consideration of the definition of angular momentum:

$$\text{Angular momentum (kg.m}^2\text{.s}^{-1}) = \text{Moment of inertia (kg.m}^2) \times$$
$$\text{Angular velocity (rad.s}^{-1})$$

As the angular momentum is constant the rotation of the body during flight will depend upon the moment of inertia and hence the body position. If the moment of inertia is maximized by adopting an elongated body position in flight the angular velocity will be minimized and only a small degree of rotation will occur during the flight. This is seen in the long jump by athletes adopting a "hang" technique where the arms are extended above the head and the legs extended.

The other method for controlling angular motion during flight requires an analysis of the segmental contributions to whole body angular momentum. If the body is considered as a series of individual segments, the angular momentum of each segment can be calculated and used to determine the whole body angular momentum. If one group of segments are rotated in such a way as to generate an angular momentum equal to the whole body angular momentum there would be no net rotation about the mass center. This is what happens in the "hitch-kick" technique where the athlete performs a running leg action in the air. The angular momentum generated by rotating the legs makes a sufficient contribution to whole body angular momentum to prevent the forwards rotation of the body.

E4 MECHANICAL CHARACTERISTICS OF THROWING

Key Notes

The phases of a throw

The throwing action can be broken down to preparation, pulling/driving and follow-through phases. It is common to divide the pulling phase to early and late pull. The time duration and range of motion in these phases will vary according to the purpose of the throw.

The preparation phase

The preparation phase provides a longer pulling path for accelerating the arm and serves to pre-stretch the musculature. Both of these allow a greater impulse (force × time) to be developed during the pulling phase.

During this phase the muscles of the anterior shoulder region become stretched by the abduction and horizontal extension of the shoulder. These eccentric contractions facilitate the use of the stretch shortening cycle to enhance the force of the early preparation phase and thus increase the velocity of the movement.

The pulling phase

The pulling phase is where the velocity of the throw is developed. Initial pelvic and then trunk rotations accelerate the shoulder axis in the horizontal plane and cause the flexed lower arm to lag behind by external rotation of the shoulder. The shoulder then internally rotates and the elbow extends in the late pulling phase.

The pulling phase is the primary phase for accelerating the motion of the upper limb. There is a sequential acceleration of the joints and a transfer of momentum from proximal to distal segments.

During the early pull those muscles stretched during the preparation phase overcome the external force and begin to contract concentrically to rotate and flex the trunk.

The follow through

The follow through acts to bring the throwing action to a controlled stop. Muscle actions in this phase are mainly eccentric. The shorter the follow through the more forcefully the muscles have to contract.

Many sports involve the use of some form of overarm throwing or striking action. There are many variants to the throwing action depending upon the object used and the requirements of the skill. It would be unrealistic to try to cover all of these, therefore this section will focus upon the general movement patterns and highlight how these can be varied to achieve different goals.

Throwing is considered to be an **open chain** movement. Open chain movements are those where the distal end of the moving segment is free to move in space. In the case of throwing the hand is the distal segment and can be moved freely to any position. This is in contrast to **closed chain** movements where the distal segment is constrained such that it is not free to move relative to the other parts of the body, for example, the foot during weight bearing. The differentiation of movement into "open" and "closed" chains is somewhat contrived and the terms are more commonly used in the context of physiotherapy than biomechanics.

The phases of a throw

The throwing action can be broken down into preparation, pulling/driving and follow-through phases. As the movements in the throwing action are rather complex it is possible further to divide these phases. A common division of the pulling phase is into early and late pull.

Preparation	From	First backward movement of the hand
	To	Maximum horizontal extension of the shoulder
Pulling phase	From	Maximum horizontal extension of the shoulder
	To	Release of object
Follow through	From	Release of object
	To	Maximum shoulder extension

The pulling phase can be further divided into:

Early Pull	From	Maximum horizontal extension of the shoulder
	To	Maximum external rotation of the shoulder
Late Pull	From	Maximum external rotation of the shoulder
	To	Release of object

Each of the different variants of the throwing action will have different relative lengths of these phases. As with most open chain exercises, the freedom of the distal segment allows substantial variation in the performance of the skill. During the analysis of the movement and muscle actions of throwing it is important to consider the role of each phase to be able to understand why different throwing and striking actions utilize different relative phase lengths.

Table E4.1.

Preparation phase

Joint	Movement
Trunk	Lateral flexion (to left)
	Rotation (to right)
	Hyper-extension
Shoulder	Horizontal extension
	Abduction
Elbow	Flexion
Wrist	Extension

Early pull phase

Joint	Movement
Trunk	Rotation (to left)
	Flexion
Shoulder	Horizontal flexion
	External rotation
Elbow	No movement
Wrist	No movement

Late pull phase

Joint	Movement
Trunk	Rotation (to left)
	Flexion
Shoulder	Internal rotation
Elbow	Extension
Wrist	Flexion

Follow-through phase

Joint	Movement
Trunk	Rotation (to left)
	Flexion
Shoulder	Adduction
	Internal rotation
	Extension
Elbow	Flexion
Wrist	Flexion
	Pronation

These movements are common to all of the variations of the throwing action. The techniques differ in the degree of motion at each joint dependent upon the goal of the action. The functions for each movement phase are described below.

The preparation phase provides a longer pulling path for accelerating the arm and serves to pre-stretch the musculature.

The pulling phase is where the velocity of the limb is developed. This involves a sequential movement of the trunk and upper limb. Initial pelvic and then trunk rotations accelerate the shoulder axis in the horizontal plane and cause the flexed lower arm to lag behind, thus inducing greater external rotation of the shoulder. Many coaches teach that the thrower should "lead with the elbow". This in reality does not happen. The elbow remains behind the shoulder axis throughout the throw and it is the initial trunk rotation that generates the external rotation and the lagging behind of the lower arm. The degree of external rotation at the shoulder is also related to the elbow angle. If the elbow is allowed to flex beyond the 90° position the moment of inertia of the limb will decrease and a smaller rotational torque will be exerted on the shoulder. It is therefore important that the elbow angle can be maintained during the early preparation phase.

The late pulling phase demonstrates a rapid elbow extension as the radius of the arm is increased to generate maximum linear velocity in the distal segment.

During follow-through the rapid arm movements are gradually slowed. The longer this phase the lower the force that is required to slow the limb.

The movements of the preparation phase are initiated by a forceful contraction of the prime movers for each action (concentric contraction). If a run-up is used before the throw this will help to create momentum within the body and make the preparatory movements faster and require less muscular effort to initiate. As the body reaches the end position of the preparation phase the rotator muscles of the trunk become stretched and the stretch reflex stimulates an eccentric contraction. Similarly, the muscles of the anterior shoulder region become stretched by the abduction and horizontal extension of the shoulder. These eccentric contractions facilitate the use of the stretch-shortening cycle to enhance the force of the early preparation phase and thus increase the velocity of the movement.

The pulling phase is the primary phase for accelerating the motion of the upper limb. There is a sequential acceleration of the joints and a consequent transfer of momentum from proximal to distal segments. During the early pull those muscles stretched during the preparation phase overcome the external force and begin to contract concentrically to rotate and flex the trunk. The muscles of the anterior shoulder region continue to work eccentrically as the forward rotation of the trunk tends to leave the arm lagging behind. During this early preparation phase the triceps reach their peak activity, although no movement is seen at the elbow. They contract in an isometric/eccentric fashion to resist the flexion of the elbow during this phase. It is important that the elbow remains at about 90° to maintain the moment of inertia of the lower arm and to promote external rotation of the shoulder.

As the upper arm reaches its peak velocity in the late pulling phase the elbow rapidly extends. This motion does not involve a muscular action of the triceps; the elbow would extend at the same time and with the same velocity even if the triceps had not been functioning. The movement is performed by the transfer of momentum from the trunk and upper arm to the lower arm segment. If the muscle were to be active it would be unable to generate much force due to the high velocity of the movement.

The gradual slowing of the movements requires eccentric contractions of the antagonistic muscles. If the movements are brought to an abrupt halt then the tension developed in the antagonists will be great and the risk of injury greater. A long gradual follow-through is the most desirable, but is not always practical within the sporting context.

There are many different variations to the throwing action, which are distinguished by small changes in the length of the movement phases and the orientation of the various segments. These variations give rise to techniques such as:

- overarm throwing
- round arm throwing
- bowling
- overhead striking/serving/smashing

Within each of these different techniques there are many more variations that make the task of defining them all impossible. Essentially the techniques are differed by the orientation of the trunk and the degree of abduction of the shoulder. These differences in limb orientations will lead to some differences in the prime mover muscles involved in the action. However, the general phasing and the nature of the muscle actions will be consistent across all of the variants of the throwing action.

To calculate the average force acting on the object (ball) during a throw it is necessary to apply the impulse momentum relationship. For example, a ball of mass 0.5 kg, change in ball speed of 30 m.s^{-1} is achieved with a pulling phase lasting about 0.1 s.

The impulse momentum relationship gives:

Force × Time (impulse) = Mass × Change in velocity (change in momentum)

So in this example:

$$Force = (0.5 \times 30) / 0.1$$

$$Force = \underline{\textbf{150 N}}$$

It is possible to determine the relative importance of each of the joint movements involved in a throw by calculating the degree to which they lead to the development of ball velocity. This can be achieved by analyzing both the angular velocity at the joint and the perpendicular distance between the axis of rotation and ball. The product of angular velocity and radius gives the linear velocity:

$$\mathbf{V = \omega.r}$$

So for the upper limb summing the linear velocity contributions from each joint would give the final velocity of the ball such that :

$$\mathbf{V\ release = V_{shoulder} + V_{humerus} + V_{forearm} + V_{hand}}$$

$V_{shoulder}$ is considered to be the velocity of shoulder segment relative to the ground due to the run-up and movements of the lower body and trunk.

For the remaining segments their contribution to the linear velocity (V) will be related to the angular velocity (ω) and radius (r):

$$V_{hand} = \omega(rad/ulna).r + \omega(flex/ext).r$$

$$V_{forearm} = \omega\ (pro/sup).r + \omega(flex/ext).r$$

$$V_{humerus} = \omega\ (int/ext\ rot).r + \omega\ (flex/ext).r + \omega\ (abd/add).r$$

When applied to overarm throwing the following contributions have been reported:

For a release speed of 28 m.s^{-1}
Internal rotation of humerus = 8 m.s^{-1}
Wrist flexion = 7 m.s^{-1}
Horizontal flexion of humerus = 6.5 m.s^{-1}
Forearm pronation = 4 m.s^{-1}
Forward motion of shoulder = 2.5 m.s^{-1}

Shoulder flex/ext, ulna deviation and elbow extension do not make a significant contribution to release speed. They may, however, be important to ensuring that the release of the object is optimal in terms of angle, orientation, or spin.

E5 PROPULSION THROUGH A FLUID

Key Notes

Propulsion through a fluid	Within biomechanics propulsion through a fluid can involve the movement of a body or object through air or through water. The activity of swimming will be used to explain concepts regarding movement through fluids. Some of the principles described will also apply to the movement of other objects through other types of fluid (i.e., such as a discus through air).
Swimming velocity	This is defined as the speed at which a swimmer is able to achieve through the water. The swimming velocity of a swimmer depends on the stroke rate (**SR**) and the distance per stroke or stroke length (**DPS**). Hence swimming velocity (**V**) is given as **SR** × **DPS**. Stroke rate is determined from the time it takes the swimmer to complete the pulling and recovery phase of the arm stroke. Distance per stroke is governed by the propulsive and resistive forces that act on the swimmer.
Resistive forces	These are a direct consequence of drag and they can be classified as **form**, **wave**, and **surface** drag. **Form drag** is related to the cross-sectional area of the body that is exposed to the water, the shape of the body and the relative velocity of the fluid flow. Form drag can be reduced by adopting a more streamlined position in the water. This type of drag is probably the most significant in terms of resistive forces offered to a swimmer's progression through the water. **Wave drag** is related to the waves that are created at the interface between the swimmer and the water. Large bow waves (a v-shaped wave caused by an object moving through a fluid) act to drag the swimmer backwards. Any fast movements, such as arm recovery, need to be performed in the air rather than in the water. Many swimmers now adopt as much of the race as they can underwater in order to reduce such bow waves. **Surface drag** is related to the amount of surface area actually in contact with the water during swimming. The wearing of the fast-skin shark suits now seen in many competitions is designed to reduce surface drag. These suits claim to create eddy currents of water around the body that cause a water on water interaction rather than a swimmer on water interaction (i.e., less friction).
Propulsive forces	Propulsive forces take the form of **drag** and **lift** propulsion. **Drag propulsion** through the water is achieved by pushing the water directly backwards (i.e., the swimmer moves forward) whereas **lift propulsion** utilizes the same principle of lift force that is used to cause airplanes to fly. Swimmers use a combination of both drag and lift propulsion to propel them through the water. Modern techniques utilize complex underwater pull patterns that optimize the maximum amount of propulsion that can be achieved through these two methods. Many modern elite swimmers in freestyle now adopt a pronounced bent elbow pull pattern that is like the action used to climb a ladder.

Propulsion through a fluid

In this section the activity of swimming will be used to explain concepts regarding propulsion through a fluid. Although this is specifically related to movement through water many of the principles will apply to movement through other fluids such as air.

Swimming velocity

Swimming velocity can be defined as the speed which a swimmer is able to achieve through the water by movements of the body. This velocity is dependent upon two factors: **stroke rate** (stroke frequency) and **distance per stroke** (stroke length).

Swimming velocity (V) = Stroke rate (SR) × Distance per stroke (DPS)

$$V = SR \times DPS$$

The **stroke rate** of a swimmer is determined by the time it takes to complete both the pulling and the recovery phase of the stroke. **Distance per stroke** is governed by the **propulsive** and **resistive forces** that act on the swimmer as they move through the water.

In order to increase swimming velocity an athlete can increase either or both of the components described previously (stroke rate or distance per stroke). However, an increase in one component should not be achieved at a loss or detriment in the other.

Stroke rate (frequency) can be improved by increasing the number of strokes for each length (lap) of the pool (cadence). However, one of the main drawbacks to increasing swimming velocity by this method is that the more hurriedly the swimmer tries to swim the more likely there will be deterioration in the swimmer's technique.

Distance per stroke can be achieved by increasing the **propulsive forces** while reducing the effect of the **resistive forces**. Hence, it is critical to understand these two types of forces in more detail.

Resistive forces

Resistive forces are a direct consequence of drag which can be classified into three types: **form, wave, and surface drag**.

Form drag is concerned with the cross-sectional area of the body that is exposed to the oncoming flow of water, the shape of the body, and the relative velocity of the fluid flow. **Wave drag** is involved when the swimmer moves at the interface between the air and the water. During wave drag some of the energy of the swimmer is transformed into wave motion. This wave motion also acts to drag the swimmer backward. **Surface drag** is concerned with the amount of body surface area, the smoothness of the body's surface, and the relative velocity of the oncoming flow (frictional drag). *Figs E5.1, E5.2*, and *E5.3* illustrate these three forms of drag in diagrammatic form.

Form drag can be reduced by adopting a more streamlined shape as in the case of the swimmer in Example A in *Fig. E5.1*. Reduced form drag would be achieved by a more streamlined body shape in the water. In Example A the athlete is lying almost flat in the water (level with the surface of the water). This would create less frontal resistance to the oncoming flow of water (note: the athlete actually moves forward and the oncoming water is stationary). Form drag is probably the largest resistive force in swimming and the most effective way to reduce form drag is to try and adopt a more streamlined (hydrodynamic) body position. The ability to be more streamlined is, however, closely related to the amount of buoyancy (ability to float) possessed by the swimmer. The more buoyant a swimmer the

Sagittal plane view: Example (A)

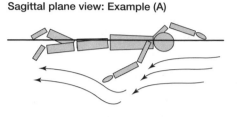

Reduced form drag
(streamlined shape)

Sagittal plane view: Example (B)

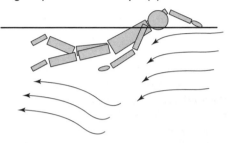

Increased form drag
(non-streamlined shape)

Fig. E5.1.   Drag forces in swimming (form drag)

Sagittal plane view

Bow wave

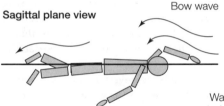

Waves formed above the water and at each
side of the swimmer, which are caused by
excessive movements (splashing) of the arms
(inefficient technique) and the forward motion
of the swimmer through the water

Transverse plane view

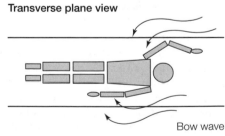

Bow wave

Fig. E5.2.   Drag forces in swimming (wave drag)

easier it is to maintain a more streamlined body shape. In the context of form drag
it may be important to point out that some elite athletes do not swim in the most
streamlined position. Elite swimmers who have a very powerful leg kick will
adopt a more angled swimming position (higher upper body and lower leg position) in order to utilize the leg kick more effectively.

   Wave drag probably accounts for the next most significant resistive force in
swimming. The method used to reduce this type of drag would be to reduce the
size of the bow wave (a V-shaped wave created by an object moving across a

Sagittal plane views

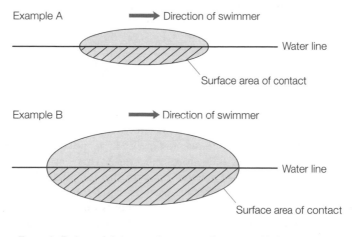

Example B shows a larger **surface area** of contact with the water.
This example would have an increased amount of surface drag.

Fig. E5.3.   Drag forces in swimming (surface drag)

liquid surface) created by the swimmer. Any fast movements of the body, such as arm recovery, should be performed in the air rather than while in contact with the water. Large bow waves in swimming result from excessive vertical and lateral movements of the body. Crashing the arms and hands into the water during entry can create large bow waves. The hands should slip and glide into the water. In freestyle (which is the most common swimming stroke) the hand should be initially placed side on and the arm should glide into the stroke through the same hole (path) made in the water by the hand entry. Often, large amounts of wave drag are created by the side to side movements through the water (like a snake type movement) which is a common fault seen in adult swimmers. The reduction in wave drag and the effect on performance can be seen by observing that many elite swimmers now spend as much of the race as possible, within the defined rules, under the water (from the start and at each turn). Indeed many modern competitive strokes now utilize more underwater actions, for example, modern breaststroke which now uses undulating under water movements that resemble the butterfly stroke.

Surface drag is created by the amount of surface area actually in contact with the water during swimming (*Fig. E5.3*). In this context very little can be done to reduce surface drag during swimming apart from the wearing of the friction reducing shark suits developed by sports companies such as Speedo and Adidas. These suits aim to create a surface that causes eddy currents to form around the swimmer as he/she moves through the water. These eddy currents reduce the friction by allowing water on water interaction (i.e., the eddy currents around the suit and the surrounding water act on each other reducing frictional drag). The result is that the swimmer is able to slip, slide, and glide through the water more effectively. Other methods, such as wearing swimming caps and body shaving, are also used to reduce surface drag. However, these techniques are probably only really helpful at the elite level of the sport. In swimming 90% of the drag comes from the shape of the swimmer as he/she moves through the water (form drag) and only 10% is attributed to the friction caused between the skin, the costume, and the water. However, at the elite level this 10% would mean the difference between breaking a world record and not. In 1875 Mathew Webb, while swimming the English Channel, wore a swimming costume that weighed (mass) 4.5 kg (10 lb). In Athens, at the 2004 Olympic Games, the body suits weighed (mass) only a few ounces (0.09 kg) saving over 98% in weight (mass) since the original swimming costume used in 1875. It is speculated that there is an 8% reduction in drag resistance while wearing these suits, and they are even better than swimming with no suit on at all.

Propulsion forces   Prior to the 1970s, propulsive forces in swimming were thought to be due entirely to the action–reaction method (i.e., push backward in the water and you moved forward – Newton's third law). This was termed **drag propulsion** as it relied on the large surface area of the hand to push the water backward (like the paddle wheel propulsion used in small boats). However, in the years that followed the 1970s, the term **lift propulsion** was introduced. This was primarily attributed to the work of James Counsilman in the USA, and it involved both lateral and vertical movements of the hand through the water. This technique is still used by many elite level swimmers today.

The term **lift** gives a slightly false impression as to how the principle works in swimming. It implies that the force is always directed upward (i.e., to lift the body). In swimming this is not necessarily the case and the lift force can act in almost any direction. Therefore it is more accurate to indicate that the lift force acts at right-angles (at 90°) to the direction of movement of the object (or fluid flow) that is causing the lift force to be created. Since it is the hand that would cause the lift force to be created in swimming it is clear that this force can occur in any direction.

Lift is based around **Bernoulli's principle** of fluid dynamics, which is more commonly seen applied to aerodynamics (movement through air) and the movement or flight of aeroplanes. *Fig. E5.4* helps to explain this principle in more detail.

Fig. E5.4 shows the cross-sectional view of an aeroplane wing. When the wing moves forward (propelled by the jet engines of the plane) the layers of air that are oncoming to the wing separate. Some travel over the top and others below the wing. Due to the shape of the wing the path over the top of the wing is longer than the path underneath the wing. The shape of the wing and its inclination causes the air over the top of the wing to travel faster than the air underneath the wing (it also has a greater distance to travel). The result is that this difference in the speed

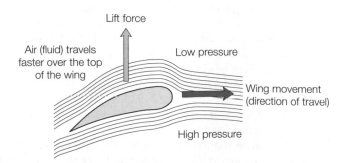

Fig. E5.4.   Propulsive forces (lift) aeroplane wing

of air travel causes a pressure differential to occur. A high pressure is created below the wing and a low pressure is created above the wing. According to Bernoulli's principle, which relates to the pressure differential and the association of lower pressure with a faster fluid velocity, the result is a "lift" force that acts at right-angles to the direction of fluid flow or in this case the direction of the aeroplane wing (since it is traveling forward). As we have seen the wing is traveling forward and here the lift force is acting upwards (at 90° to the forward motion). It is this lift force that enables the aeroplane to take off the ground and fly through the air (i.e., the lifting the wings). This continues to occur in flight as the wing is still being propelled through the air (fluid) by the engines of the aeroplane.

In swimming the hand performs the same function as the aeroplane wing and it can also create a lift force if it is moved through a fluid (water) in a certain manner (i.e., shape and angle). As an example the next time you are a passenger in a moving car try carefully holding your hand out of the car window in a shape like an aeroplane wing (cup your hand to create the shape seen in *Fig. E5.4*). The result will be that your hand will lift upwards as the car travels forwards. In this case the hand is stationary and the car is traveling forward which provides the flow of air over your hand. If you angle your hand at different positions to the oncoming air flow you will see the effect of different lift forces. At some positions and angles of the hand you can even create a lift force that is directed downward and not upward. This downwardly directed lift force is often used on racing cars by having the spoiler on the back of the car angled and positioned in a certain way. This creates a downward lift force that helps to keep the car firmly attached to the ground offering better traction when it is traveling at high speeds.

In the case of the swimmer, although the hand is moved through the water at much slower speeds it can still create enough lift force to propel the body forward through the water. As a practical example position yourself vertically in the water and use the horizontal sculling action of your hands to keep you afloat. This method of flotation is primarily dependent upon the lift forces that you create while your hands are sculling horizontally under the water. In the freestyle stroke, the hand is angled and moved in an elliptical pull pattern throughout the arm pull phase of the stroke. Depending on the hand position as you pull it through the water it will create different amounts and directions of lift force. If the hand movement is in the appropriate direction the lift forces that are created can be directed either to help keep you in a flat horizontal position or indeed propel you forwards through the water. *Fig. E5.5* shows the lift forces created by the positioning of the hand as it is moved through the water (which is analogous to the aeroplane wing).

During swimming, the advantages of this application are significant. In the old action–reaction method of propulsion (drag propulsion) half of the stroke was classified as recovery because you could not always be pushing directly backward at every point throughout the pull phase. However, by combining the lift (lift propulsion) and drag method of propulsion (drag propulsion) the whole of the pulling phase can be utilized to propel the body forward through the water. In the modern swimming stroke, athletes use lift and drag propulsion to swim more effectively.

The direction in which the hand is inclined in the water is termed the **angle of attack** and increasing the angle of attack will increase the lift forces. However, with increasing lift forces there will also be increased drag forces that will act against the movement of the hand through the water. The angle of attack position can reach a limit before there would be less lift force created. In aerodynamics the

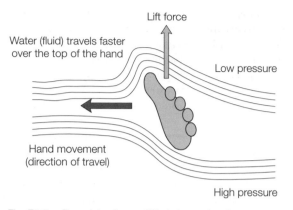

Fig. E5.5.   Propulsive forces (lift) during swimming

optimum lift to drag ratio can be achieved with an angle of attack of between 4°
and 15°. However, in swimming the angle of attack is at an optimum between 30°
and 50°. This occurs because the swimmer is prepared to generate greater lift at
the cost of trying to overcome the larger drag forces. Distance swimmers trade off
some propulsive lift by having a smaller angle of attack of the hand (i.e., less lift
force). This has the benefit of reducing the energy cost of the stroke because the
swimmer does not have to overcome so much drag force during the movements
of the arms/hands. Sprinters, on the other hand, use a larger angle of attack
position because the race is much shorter and the increased lift is imperative for
greater propulsion (i.e., faster progression). Each swimmer has to develop a
"feel" for the water and it is important to note that the drag force does not act
against the direction of the swimmer but acts against the direction of the movement of the hand or object that is creating it. *Fig. E5.6* helps to show the lift and
drag force ratios during the different pitches of the hand or aeroplane wing.

In modern swimming the lift and drag propulsion method is used throughout
all the four competitive strokes. Athletes adopt complex pull patterns under the

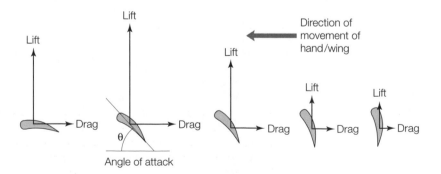

As the "angle of attack" is increased the amount of lift increases. However
this increased lift force is only apparent up to a point and beyond this point
the lift force reduces as the angle of attack increases.
Note the associated drag force with each lift force component.

Fig. E5.6.   Lift to drag ratios during hand/wing pitch "angle of attack"

water to utilize these principles more effectively. Today, in freestyle, many elite swimmers will use a pronounced bent arm pattern throughout the arm pulling phase. This is a similar action to that used in climbing a ladder. Distance swimmers in particular will be seen utilizing the bent arm pull pattern. However, most strokes will still have common features for effective propulsion through water. These are summarized as follows: curved elliptical pull patterns; an extend – "catch the water" – flex – extend pull pattern; a high elbow position (bent arm); hand entry to create minimum splash; utilization of both lift and drag propulsion; and streamlined body alignments (with the exception of the strong powerful leg kicking athletes). Two objectives are apparent for effective propulsion in swimming: 1) to propel the body forwards with respect to the hands (using an optimum combination of lift and drag propulsion); and 2) to minimize resistance to the propulsion of the body (reduce drag and maintain optimum body alignment). *Fig. E5.7* shows the pull patterns of the four modern competitive swimming strokes.

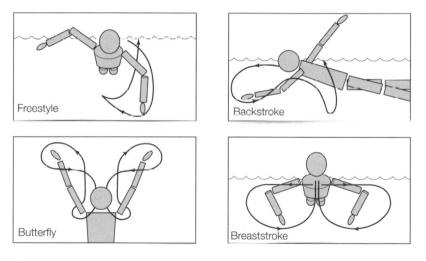

Elliptical (curved) pull patterns of the various swimming strokes utilizing the lift and drag propulsion.

Note the lateral (sideways) and vertical directions of the pull patterns in the different strokes.

Fig. E5.7. Underwater elliptical pulling path of modern swimming strokes

In swimming it is speculated that 85% of the propulsion comes from the arm movement through the water. However, there are many that would argue that the legs are a far more significant contributor to propulsion than only 15%. As already mentioned, many athletes with powerful leg kicks will angle their body downward to utilize this propulsive force provided by the legs. Throughout the leg kick the same principle for lift and drag propulsion also applies (depending on the angle, position, and direction of movement of the foot). Arm action in swimming should not be totally classified as either lift or drag propulsive. When the hand moves predominantly backward it is likely that the majority of propulsion would be drag propulsion. Conversely, when the hand is moved laterally and vertically lift propulsion would be more prominent. An effective coach would be tolerant of different techniques and should always be prepared to change a swimmer's technique if inefficiencies are detected. However, this can only be achieved with a good working knowledge of the biomechanics of effective propulsion through a fluid such as water.

E6 MECHANISMS OF INJURY

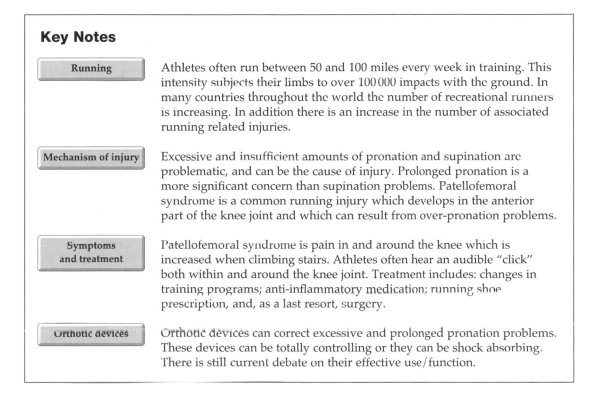

Key Notes

Running

Athletes often run between 50 and 100 miles every week in training. This intensity subjects their limbs to over 100 000 impacts with the ground. In many countries throughout the world the number of recreational runners is increasing. In addition there is an increase in the number of associated running related injuries.

Mechanism of injury

Excessive and insufficient amounts of pronation and supination are problematic, and can be the cause of injury. Prolonged pronation is a more significant concern than supination problems. Patellofemoral syndrome is a common running injury which develops in the anterior part of the knee joint and which can result from over-pronation problems.

Symptoms and treatment

Patellofemoral syndrome is pain in and around the knee which is increased when climbing stairs. Athletes often hear an audible "click" both within and around the knee joint. Treatment includes: changes in training programs; anti-inflammatory medication; running shoe prescription; and, as a last resort, surgery.

Orthotic devices

Orthotic devices can correct excessive and prolonged pronation problems. These devices can be totally controlling or they can be shock absorbing. There is still current debate on their effective use/function.

This section presents a biomechanical consideration of sports injury. By the end of the section it is expected that you will have an understanding of the basic biomechanics associated with some of the different types of injury in sports, (i.e., possible biomechanical mechanisms and preventative measures).

Patellofemoral syndrome during running

Running

Runners subject their bodies to a considerable amount of loading during the course of their running career. Many athletes run between 50 and 100 miles (approximately 80–160 km) every week. Consider the fact that much of this running is carried out on concrete surfaces and aspects of injury become evident very quickly: these athletes could be subjecting their limbs to over 100 000 weekly impacts with the ground. This loading effect, over time, could be responsible for a variety of problems. Involvement in running does not seem to be diminishing – the London Marathon, for example, regularly attracts over 80 000 applications every year. In recent years the entry to this event has been capped for safety reasons (i.e., 46 500 runners).

Search the Internet to see if you can find out how many athletes completed the London Marathon in 2006 and indeed how many actually applied to take part?

Middle distance runners usually contact the ground with the heel and then adopt a heel–mid-foot–toe stance pattern. Most runners (80%) will be heel strikers and they will land on the lateral (outside) edge of the heel. Once the foot has made contact with the ground (in a supinated position) the foot is required to pronate (this occurs at the subtalar joint in the foot). This pronation allows for a shock absorption process. After the foot has reached the maximum pronation point it then undergoes supination in which it forms a rigid lever for toe-off. This is needed so the athlete can push off the ground. This is a normal component of foot function, whether in walking or running. The foot initially pronates then it is required to supinate. This specific biomechanical detail has been identified in section E2 (biomechanical characteristics of running).

Mechanism of injury

Excessive pronation can be a problem for runners, as can insufficient amounts of pronation. Similarly, excessive and insufficient amounts of supination can also be a serious concern. However, it is important to point out that these excessive or insufficient components of pronation and supination can also be the direct effect from another problem regarding the runner's gait. This is an important consideration for clinicians.

The relationship with pronation and supination and rotation of other structures is shown in some detail in *Table E6.1*. It is important to note that this is a ratio representation and that, for example, for every degree of pronation that occurs there would be 2.5° of internal rotation of the tibia and fibula complex (1:2.5 ratio). By comparison for every degree of supination there is only 0.5° of tibia and fibula external rotation (2:1 ratio). This is one of the many reasons why over-pronation is more of a problem for injury development in running than over-supination. For example, for 10° of pronation the tibia and fibula complex would internally rotate approximately 25°.

Excessive and insufficient amounts of both pronation and supination can be a problem for runners. Problems can be associated with the shins, knees, hips, and even the back. The amounts of pronation and supination are affected by the type of running shoe the athlete is wearing, the surface they are running on, their running style and the type and intensity of training they undertake. For example, running on a beach or cross-country will affect the runner's foot movement. Similarly, and adding more complexity to the problem, excessive and insufficient amounts of pronation and supination can manifest from an injury to another structure. Therefore the athlete may be excessively pronating by necessity in order to keep running without pain.

Table E6.1 Pronation and supination relationship with other structures of the leg

SUPINATION (External rotation)	2:1 ratio with fibula and tibia rotation
	1:1.5 ratio with femoral rotation
	1:1 ratio with pelvic rotation
PRONATION (Internal rotation)	1:2.5 ratio with fibula and tibia rotation
	1:1.5 ratio with femoral rotation
	2:1 ratio with pelvic rotation

Correcting the degree of over-pronation may indeed not be the correct solution for the medical practitioner. It becomes a careful balance and interpretation of the exact cause of the excessive or insufficient pronation and supination.

Many injuries result from or cause excessive or insufficient amounts of pronation and supination. An excessive pronator may land on the ground in a rolled over or pronated position (i.e., on the medial (inside) edge of the heel or mid-foot) and then continue to pronate too much and for far too long into the stance phase. On the other hand, an excessive supinator may land on the lateral edge of the heel and then not pronate at all. This athlete may roll outwards on the outer edge of the heel from heel strike all the way through to toe-off. These are two extreme cases of over-pronation and over-supination. Injuries such as patella tendinitis, plantar fasciitis, shin splints, illio-tibial band friction syndrome, and patellofemoral syndrome are just a few of the many that can manifest from pronation and supination concerns. However, one of the more problematic injuries, and one that is often seen in many runners, is that of patellofemoral syndrome (in the anterior part of the knee joint).

When the foot moves from heel strike to mid-stance the foot normally undergoes a pronation movement. The ankle dorsi-flexes, the calcaneus everts and the forefoot abducts causing the tibia and fibula complex (lower leg) to rotate internally (*Fig. E6.1*). When the foot pronates past the point of mid-stance and, indeed when the foot pronates too much (usually measured by the amount of eversion of the calcaneus), the lower leg is internally rotating excessively and for too long. This pronation continues into the stance phase and past the point of mid-stance. The leg (knee) reaches a point of maximum knee flexion and the quadriceps cause a pull on the patella that attempts to move this bone laterally (towards the outside – away from the body mid-line). However, because the lower leg is still internally rotated and the foot is still pronated, this lateral pull causes the patella to laterally track over the lateral femoral condyle of the knee. Normally the lower leg would be externally rotating and the foot supinating, which would mean that the patella could be pulled naturally within the groove contained between the femoral condyles.

To add to this problem, as the knee is flexed during mid-stance the ankle normally undergoes a degree of dorsi-flexion. This movement is decelerated by the gastrocnemius and soleus muscle complex. If the athlete has a tight gastrocnemius–soleus muscle complex then normal amounts of ankle dorsi-

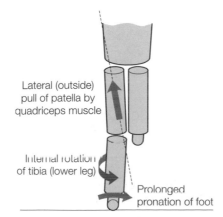

Lateral (outside) pull of patella by quadriceps muscle

Internal rotation of tibia (lower leg)

Prolonged pronation of foot

Fig. E6.1.   Lower leg and foot movement

flexion are prevented. This has the result of making the athlete increase the amount of knee flexion, which further forces the patella down onto the femur. This creates additional aggravated knee pain. This problem is commonly referred to as having a "tight heel cord" (see *Fig. E6.2*).

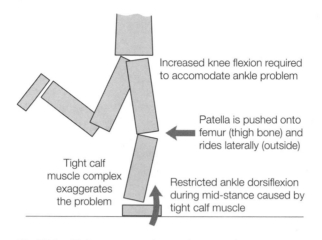

Increased knee flexion required to accomodate ankle problem

Patella is pushed onto femur (thigh bone) and rides laterally (outside)

Tight calf muscle complex exaggerates the problem

Restricted ankle dorsiflexion during mid-stance caused by tight calf muscle

Fig. E6.2.   Tight gastrocnemius – soleus muscle complex

Symptoms and treatment

Symptoms of patellofemoral syndrome are generally described as pain in and around the knee joint. This pain is increased when the athlete climbs stairs, walks, or even just sits. Often an audible "click" may be heard as the knee moves. In addition to this, continued patellofemoral syndrome may lead to an inflammation of the bursa underneath the patella and a painful knee swelling. If not treated this can also produce a degeneration of the patella bone. Finally, it is important to remember that patellofemoral syndrome may result from a problem that is evident in another structure of the body (e.g., in the back) and indeed may be a symptom caused by another injury.

The treatment rationale for this problem consists of recommending changes to training programs and potentially training on more cushioned surfaces. Often anti-inflammatory medicine is applied. Sometimes it may be necessary to consider surgery, for if the patella is constantly being pulled laterally, it may need re-attaching in a more biomechanically optimum position to minimize this problem. However, this is rarely recommended as routine and is a last resort. Other non-invasive methods include the use of recommended running shoes and/or the prescription of orthotic devices to control the excessive and prolonged amounts of pronation.

Some of the aspects/components of the running shoe that help to reduce the onset or condition of patellofemoral syndrome include: an extended medial support that aims to prevent excessive inward rolling (calcaneal eversion); increasing the density of the mid-sole in the shoe, again to try and control the excessive inward roll; and prescribing a shoe with little or negative heel flare. The heel flare is the angle made by the sole component of the shoe when viewed from the rear. Older shoe models, such as the *Brooks Rage* for women (which are no longer manufactured) possessed a specific roll bar at the mid-foot of the shoe. This was really effective in controling prolonged and pronounced pronation.

Using the Internet, search for specific types of anti-pronating shoes that are currently available to athletes.

Orthotic devices

Orthotic devices are another method for controling excessive pronation and prolonged pronation (note: they can also be designed to correct for supination problems). They are a type of insole, usually made by a podiatrist, that is placed inside the shoe. Orthotics can take on many different forms and can range from rigid orthotics with limited shock absorbing capacity to soft orthotics that are shock absorbing. The type of orthotic device depends very much on the problem and the type of foot strike of the athlete. Currently, there is a scientific debate as to whether orthotic devices are actually a form of management or a form of treatment. For example, if the orthotic devices are taken away from the athlete after a period of use, will the foot continue to function as though the orthotics were still present?

Search the Internet for information on orthotic devices for running shoes and see whether you can find any information or research that may help answer this question.

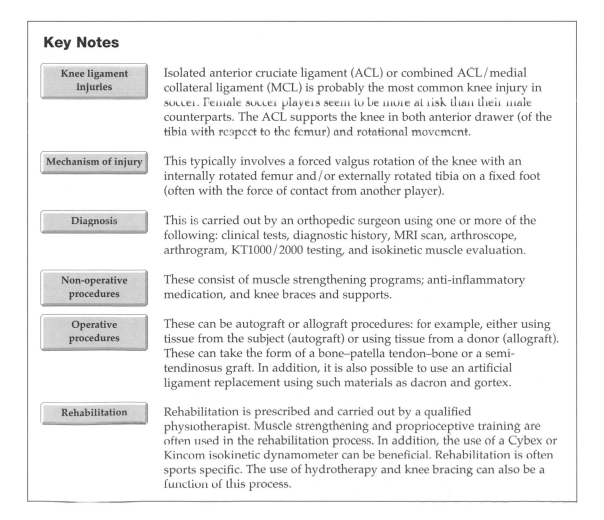

Key Notes

Knee ligament injuries

Isolated anterior cruciate ligament (ACL) or combined ACL/medial collateral ligament (MCL) is probably the most common knee injury in soccer. Female soccer players seem to be more at risk than their male counterparts. The ACL supports the knee in both anterior drawer (of the tibia with respect to the femur) and rotational movement.

Mechanism of injury

This typically involves a forced valgus rotation of the knee with an internally rotated femur and/or externally rotated tibia on a fixed foot (often with the force of contact from another player).

Diagnosis

This is carried out by an orthopedic surgeon using one or more of the following: clinical tests, diagnostic history, MRI scan, arthroscope, arthrogram, KT1000/2000 testing, and isokinetic muscle evaluation.

Non-operative procedures

These consist of muscle strengthening programs; anti-inflammatory medication, and knee braces and supports.

Operative procedures

These can be autograft or allograft procedures: for example, either using tissue from the subject (autograft) or using tissue from a donor (allograft). These can take the form of a bone–patella tendon–bone or a semi-tendinosus graft. In addition, it is also possible to use an artificial ligament replacement using such materials as dacron and gortex.

Rehabilitation

Rehabilitation is prescribed and carried out by a qualified physiotherapist. Muscle strengthening and proprioceptive training are often used in the rehabilitation process. In addition, the use of a Cybex or Kincom isokinetic dynamometer can be beneficial. Rehabilitation is often sports specific. The use of hydrotherapy and knee bracing can also be a function of this process.

Anterior cruciate ligament rupture in soccer

Knee ligament injuries

The knee is the most complex synovial joint in the human body and the forces transmitted across it during participation in fast athletic activity like soccer are considerable. Hence it is not surprising that when an athlete is tackled and the knee is placed in a vulnerable position that the ligaments of this joint can easily be injured. The most common knee injury in soccer players is rupture of the anterior cruciate ligament (ACL) and/or a combined rupture of the anterior cruciate and medial collateral ligament (MCL). Recent research in this area has shown that female soccer players are particularly at risk and there is a more regular incidence seen in females than in their male counterparts. This section will concentrate on the isolated ACL rupture.

The ACL is one of the main supporting ligaments of the knee and it is responsible for supporting the knee in a movement known as anterior tibial translation, where the tibia is moved anteriorly (forward) with respect to the femur. In addition, the ligament also provides a degree of rotational stability to the joint. The ligaments, together with the muscles, provide joint support and stability and injury to these ligaments of the knee can seriously affect a player's career. *Fig. E6.3* shows the ACL ligament in more detail.

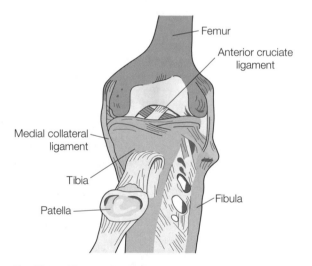

Fig. E6.3.   The anterior cruciate and medial collateral ligament

Mechanism of injury to the ACL

The typical mechanism of injury for the ACL during soccer is where the athlete's leg is in a forced valgus position (often in a contact tackle situation) during which the knee is flexed and there is a degree of internal rotation of the femur on an externally rotated tibia, which is fixed to the ground by the soccer boot. In addition, the ACL can easily be torn when the leg is positioned in severe hyperextension and the force of another player causes the hyperextension to go beyond that normally allowed by the knee joint (i.e., causing excessive anterior translation of the tibia with respect to the femur). Combine these positions with sudden deceleration and any degree of internal or external rotation on a fixed foot (usually because of the studs or bars in the soccer boot) and the ligament is susceptible to partial or complete rupture. The player usually experiences an audible "popping" sound, or a feeling of the knee "giving way" or swelling. *Figs E6.4* and *E6.5* illustrate these positions in more detail.

ACL rupture mechanism in soccer

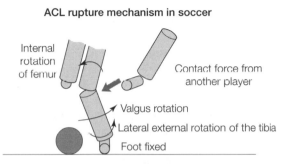

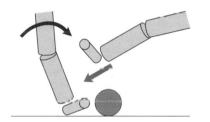

Fig. E6.4.   Valgus and external rotation injury in soccer (ACL injury)

Fig. E6.5.   Knee hyperextension injury in soccer (ACL injury)

Diagnosis
This is usually carried out by an orthopedic surgeon and consists of a clinical test (most often the Lachman test for anterior drawer), a full diagnostic medical history which includes details of other player contact, direction of foot and leg movement, any swelling or audible sounds and any previous history of injury to the knee. Following this assessment the surgeon will usually support his/her diagnoses with one or more of the following techniques: an arthroscopy (cameras are placed inside the knee to view the structures); arthrography (opaque dye is injected in the knee and x-rays are taken); an MRI scan (a magnetic resonance image of the soft tissues structures in the knee); or a KT1000/2000 knee arthrometer test (a device for assessing knee instability). In addition, there are other machines, located in specialized centers, that can also provide an assessment of ligamentous instability. The surgeon may also require the player to have an isokinetic dynamometer assessment to identify the strength of the quadriceps and hamstring muscles and in particular to see if any muscle wasting has occurred. The output from these diagnostic tools provides very important information, which the physiotherapist will use for a successful rehabilitation process.

Non-operative treatment
If the ACL is not considered by the surgeon to be ruptured (either partially or wholly) the surgeon may prescribe non-operative treatment. In this case physiotherapy is used to strengthen the quadriceps and hamstring muscle groups that

support the knee. Other forms of non-operative treatment include the use of knee braces and anti-inflammatory injections. However, there is considerable debate as to the effectiveness of non-operative treatment regimens and in the case of most ACL injuries surgical repair or reconstruction is often required.

Operative treatment

Currently the two most widely used operative procedures for ACL repair include an intra-operative procedure that attempts to reconstruct the ACL as close as possible to the original anatomy of the ligament. Such procedures include either an autograft (harvested from the patient's own tissues) or an allograft (from other human donors) reconstruction/replacement process. Autograft ligament procedures consist of the surgeon using either the bone-patella tendon-bone (BPTB) or the hamstring graft (usually from the tendon of the semi-tendinosus (ST) muscle). Both methods have advantages and disadvantages, and both have currently been shown to be very successful in being able to restore the knee to a stability where athlete is able to return fully functional to sport. Finally, there are a number of artificial ligament replacement methods using such materials as dacron, gortex, and combinations of different fiber composites. However, use of such man-made materials is limited and it is more common nowadays for a surgeon to use either the patient's own biological tissue or that of a donor.

Rehabilitation

This is a critical component of the process of ACL reconstruction and is usually carried out or supervised by a qualified physiotherapist. However, the current strengthening and proprioceptive routines vary significantly. For example, some surgeons require the patients to be moving and weight bearing as soon as possible after the operation, whereas others require a lengthy period of rest and immobilization. Techniques used by physiotherapists include: strength training using isokinetic machines such as a Cybex or Kincom; neuromuscular and proprioceptive exercises involving balance boards and other devices such as wobble boards; plyometric exercises; hydrotherapy; and agility training and specific exercises that prepare the subject for return to their chosen sport. The surgeon may also recommend that a supportive knee brace is worn during this critical rehabilitative procedure.

Search the Internet to see if you can find information on how the two autograft surgical procedures (BPTB or ST (hamstring)) are carried out.

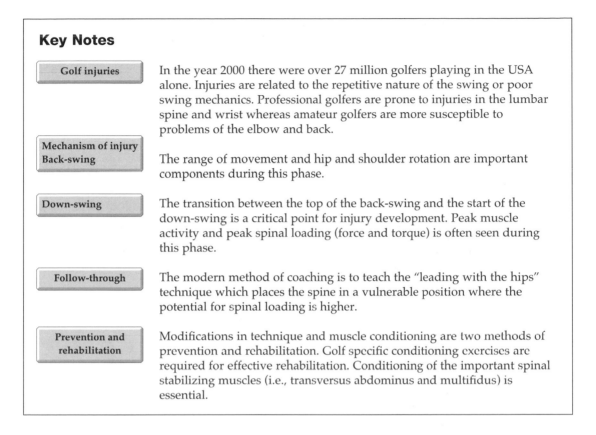

Key Notes

Golf injuries

In the year 2000 there were over 27 million golfers playing in the USA alone. Injuries are related to the repetitive nature of the swing or poor swing mechanics. Professional golfers are prone to injuries in the lumbar spine and wrist whereas amateur golfers are more susceptible to problems of the elbow and back.

Mechanism of injury Back-swing

The range of movement and hip and shoulder rotation are important components during this phase.

Down-swing

The transition between the top of the back-swing and the start of the down-swing is a critical point for injury development. Peak muscle activity and peak spinal loading (force and torque) is often seen during this phase.

Follow-through

The modern method of coaching is to teach the "leading with the hips" technique which places the spine in a vulnerable position where the potential for spinal loading is higher.

Prevention and rehabilitation

Modifications in technique and muscle conditioning are two methods of prevention and rehabilitation. Golf specific conditioning exercises are required for effective rehabilitation. Conditioning of the important spinal stabilizing muscles (i.e., transversus abdominus and multifidus) is essential.

Low back pain in Golf

Golf injuries
In Australia it is estimated that there are over 500 000 playing members of golf clubs around the country and that there are at least four times this amount playing golf at least once every year (2 million golfers). Statistics show that the average amateur golfer playing will have at least one golf related injury per year. In the USA the number of golfers playing in the year 2000 was over 27 million and this number has been increasing ever since. A person can start playing golf from as early as 5 years old and continue to play to well over 60 years of age. Hence, a golfer's playing career can often last as long as 50 years. As a result, injury is complex and can involve a variety of conditions and causes. Professional golfers injure their lumbar spine and wrist more often than amateur golfers who are more prone to problems in the elbow and back. Injuries in golf are increasing as a result of increased participation and they are generally related to either the repetitive nature of the swing or poor swing mechanics.

The phases of the golf swing
The golf swing is generally divided into either three or four phases: the take away or the back-swing, forward-swing or down-swing, early follow-through and late follow-through (although these latter two are often classified simply as the follow-through). In the right-handed golfer, during the take away, it is the left external oblique muscle that is responsible for the initial twisting of the trunk. This activity in the external oblique muscle is proportional to the axial loading on the lumbar spine. From the top of the back-swing to impact the muscles of the

right side of the trunk (primarily right external oblique) are responsible for leading the swing. In this phase the peak muscle activity is linked with the peak loading on the lumbar spine and this is the phase where injury potential is at a maximum. This is particularly true of the point at the very start of the down-swing where there is transition between back-swing and down-swing. During follow-through, after the ball has left the tee, the stroke is primarily governed by the muscles of the shoulder and upper trunk (infraspinatus and supraspinatus, and latissimus dorsi and pectoralis major).

Lower back injury in golf

In the injury-free golfer the right and left para-spinal muscles will fire simultaneously, which is important in their function in stabilizing the lumbar spine. However, in golfers presenting with low back pain this combined action does not take place and there is a non-synchronized pattern of muscle activity. In particular, current research has shown that in a group of male golfers with low back pain there is a delay in the onset of the contraction of the external oblique muscle with regard to the start of the back-swing.

Rehabilitation and prevention

Golfers presenting with low back pain problems are usually subjected to a substantial strength training conditioning program that is both general and golf specific in nature. The muscles of the lumbo-pelvic region (namely the transversus abdominus and the multifidus) are conditioned using golf specific treatment regimens. Once abdominal muscle control has been achieved the subject can then start golf functional rehabilitation. Such rehabilitation for golf would include arm and leg extension exercises in a supine and four-point kneeling position (alternate extensions of the legs and arms in these positions help to condition the transversus abdominus muscle which is an important spinal stabilizing muscle). Next, thoracic and lumbar rotation exercises can be performed in the sitting position using a theraband resistance to standing positions using the same resistive methods. Finally, the subject is put through a series of golf specific conditioning exercises that are designed to establish a degree of functionality in the golf swing (such as a progressive hitting program).

In the rehabilitation process it is important to point out the importance of correct biomechanics of the golf swing. In the modern game of golf, coaching involves developing a technique that uses a considerable amount of hip and shoulder rotation during the swing. For example, modern players are taught to "lead with hips" so they can generate large amounts of rotational torque which is transferred to the club head and consequently to golf ball velocity. However, current research suggests that this technique is one that can potentially lead to lower back problems and it is biomechanically more advantageous (from an injury perspective) to hit the ball with a more squared hip and shoulder position. This is rather like the old method of hitting a golf ball which was used by many early professional golfers. Although this technique may not generate great ball speed and consequently great ball distance, it may serve in the amateur player to prevent future injuries.

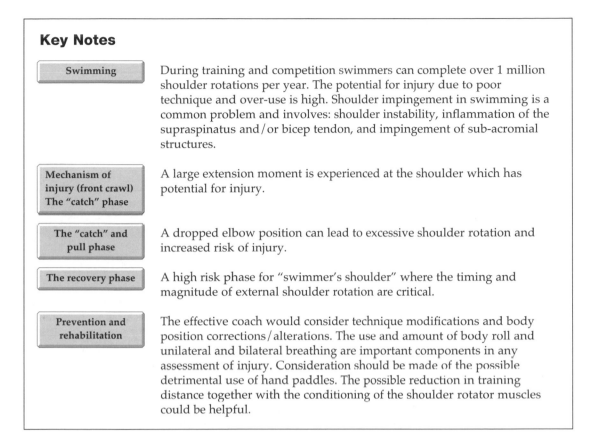

Key Notes

Swimming	During training and competition swimmers can complete over 1 million shoulder rotations per year. The potential for injury due to poor technique and over-use is high. Shoulder impingement in swimming is a common problem and involves: shoulder instability, inflammation of the supraspinatus and/or bicep tendon, and impingement of sub-acromial structures.
Mechanism of injury (front crawl) The "catch" phase	A large extension moment is experienced at the shoulder which has potential for injury.
The "catch" and pull phase	A dropped elbow position can lead to excessive shoulder rotation and increased risk of injury.
The recovery phase	A high risk phase for "swimmer's shoulder" where the timing and magnitude of external shoulder rotation are critical.
Prevention and rehabilitation	The effective coach would consider technique modifications and body position corrections/alterations. The use and amount of body roll and unilateral and bilateral breathing are important components in any assessment of injury. Consideration should be made of the possible detrimental use of hand paddles. The possible reduction in training distance together with the conditioning of the shoulder rotator muscles could be helpful.

Shoulder pain during swimming

Swimming injuries

Swimming velocity can be defined as the speed a swimmer is able to achieve through the water by the movement of the body. This velocity (speed) is dependent upon two factors: stroke rate (stroke frequency) and distance per stroke (stroke length). The stroke rate is determined from the time it takes to complete both the pulling and the recovery phase of the stroke, whereas the propulsive and resistive forces that act on the swimmer govern the distance traveled per stroke. An increase in one component should not, however, be accomplished at the cost or detriment of the other. Stroke frequency can be improved by increasing the number of strokes per pool length (cadence) but one drawback is potential poor technique and possible injury.

Considering the fact that competitive swimmers can easily complete more than 1 million shoulder rotations per year (up to 10 000 m training per day with between 15–25 strokes per 25 m distance) it is inevitable that this will potentially result in injury. Shoulder injury to swimmers is often described by the medical professions as one or more of the following: shoulder joint (glenohumeral joint) instability, inflammation of the supraspinatus tendon and often also the biceps tendon and sub-acromial impingement (impingement of the soft tissue structures lying below the acromion), which are more commonly known or classified as "swimmer's shoulder" (*Fig. E6.6*).

Prior to 1970, propulsive forces in swimming were thought to be generated entirely from action–reaction (Newton's third law) methods, i.e., push backward

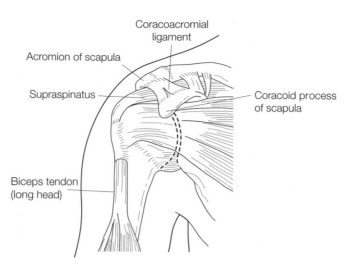

Fig. E6.6.   *Anatomical structures involved in swimmer's shoulder*

and you move forward. This was termed drag propulsion. However, the term lift propulsion was developed in the late 1970s through the research work of James Counsilman (USA) and the modern swimming stroke today has developed into one of a complex combination of both drag and lift propulsion. However, there is still debate as to the exact contribution from both methods of propulsion through water.

The modern arm action in most swimming strokes is a precise sequence that involves the following: first on entry to the water extend the arm, then "catch" the water, next pull the arm through the water in a path that allows elbow bend and inward and outward sweeping movements of the hand, and then finally recovery where the arm is prepared for re-entry into the water. The result has become a modern technique that involves a complex pull and recovery pattern through and over the water in order to generate and utilize both the lift and drag propulsive forces most effectively. *Fig. E6.7* shows an example of this pull sequence in more detail in the different strokes of a modern competitive swimmer.

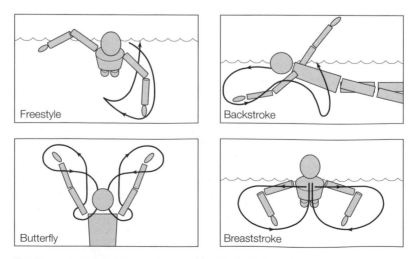

Fig. E6.7.   *The modern competitive swimming stroke (pull phase)*

Mechanism of injury

The "catch" phase (arm entry to the water)
During the initial extension of the arm and "catch" phase of the front crawl swimming stroke the shoulder is required to internally rotate and abduct. The large moment experienced during extension and catch tends to cause the arm to be forcibly elevated.

The "catch" and pull phase
During the pull phase the shoulder is adducted and internally rotated and often the arm follows an inverted question mark pattern under the water. This allows the application of a force for a longer period of time, utilizing the lift principle of propulsion. This action causes the head of the humerus to move under the coracoacromial arch, which can lead to a potential impingement situation. One of the most common technical faults during this phase is the "dropped elbow". The dropping of the elbow during the pull phase causes increased unwanted external shoulder rotation. However, the "high elbow" technique, which provides the muscles with a mechanical advantage, may also present an associated impingement risk. From initial "catch" the hand sweeps down and slightly outwards while ideally maintaining a high elbow position to the deepest point of the stroke. Next, the hand sweeps inwards and upwards until an angle of approximately 90° of the upper arm and forearm is reached. The in-sweep phase may take the hand past the mid-line of the body or under the outside edge of the body. From the mid-stroke position the hand is first sweep outwards and then backwards finishing at the end of the pull past the hips as it exits the water. An increased acceleration of the hand towards the end of the pull and internal rotation and adduction of the shoulder may present a potential impingement problem (*Fig. E6.8* shows both the dropped and high elbow position during the modern front crawl swimming pull phase).

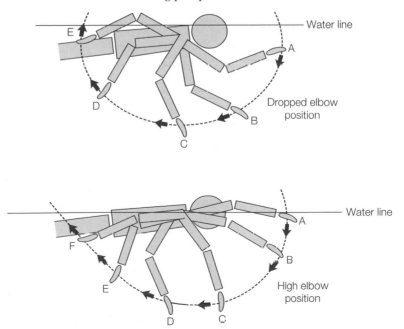

Fig. E6.8.   The dropped and high elbow position during the front crawl pull phase

The recovery phase

Following the pull phase the arm leaves the water usually elbow first (elbow lift) and is required to undergo a recovery phase. During this time the shoulder is required to abduct and externally rotate as the arm is prepared for re-entry in the water. However, the arm at the beginning of this phase is often still internally rotated. The recovery phase is one of the more important phases in terms of risk for possible shoulder impingement. The shoulder is required to rotate externally and abduct to clear the arm over the water and then quickly to prepare for extension and entry to the "catch" phase position. The timing of the external rotation and the magnitude of this action during the recovery is a critical component of risk associated with shoulder impingement and it has been shown that swimmers who execute the recovery phase with a large amount of initial internal rotation of the shoulder are susceptible to potential increased shoulder impingement. External rotation of the shoulder is needed to allow for complete abduction and thus to prevent the greater tuberosity of the humerus making contact with the acromion.

Prevention and rehabilitation

Excessive internal rotation of the shoulder during the pull phase combined with late external rotation during the recovery phase is said to lead to increased risk of shoulder impingement. In addition, reaching across the mid-line on entry, insufficient body rolling, one-sided breathing and asymmetrical muscle balance also contribute to the problem.

Possible solutions to a shoulder impingement problem include the following: during the entry and pull phase of the stroke the swimmer should try and avoid a large elevation angle at entry and rather increase the tilt angle of the arm to achieve the optimum position. Similarly, the swimmer should avoid a fully extended elbow on entry. The swimmer could also help to resist the forcible elevation caused by entry and catch by developing the shoulder extensor muscles: namely latissimus dorsi, pectoralis major, teres major, and triceps brachii. Also, a streamlined hand entry position is advisable. During recovery the swimmer should try to achieve external rotation of the shoulder early in the recovery phase in order to have time to prepare the hand and arm for re-entry to the water.

Other factors which are also recommended for the prevention of shoulder impingement in swimming include not using hand paddles if there is a current problem and possibly changing from distance swimming training to sprint training in order to reduce the number of stroke cycles in a training session. Finally, the swimmer could adopt a bilateral (both sides) breathing technique, assume as horizontal a position as possible in the water (i.e., feet up) and allow some degree of body roll (although there are still current issues of debate as to how much body roll is required for optimum performance and injury prevention).

F1 VIDEO ANALYSIS

Key Notes

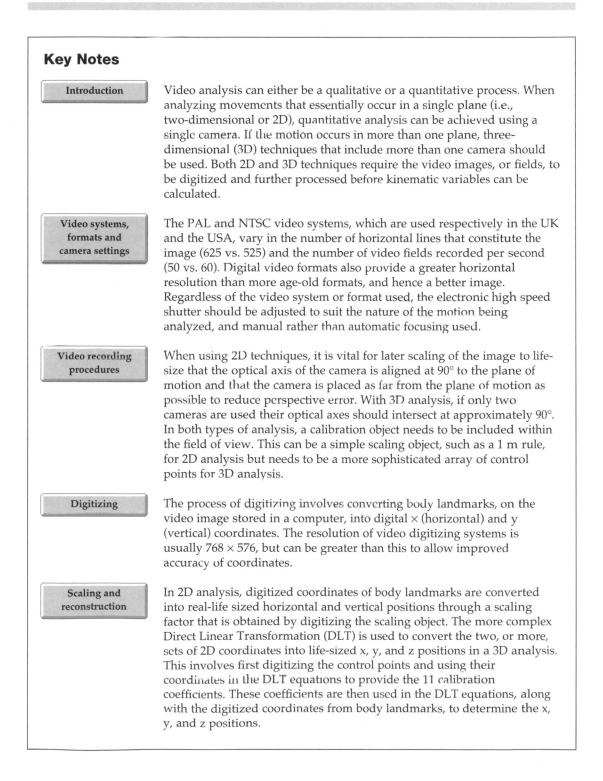

Introduction

Video analysis can either be a qualitative or a quantitative process. When analyzing movements that essentially occur in a single plane (i.e., two-dimensional or 2D), quantitative analysis can be achieved using a single camera. If the motion occurs in more than one plane, three-dimensional (3D) techniques that include more than one camera should be used. Both 2D and 3D techniques require the video images, or fields, to be digitized and further processed before kinematic variables can be calculated.

Video systems, formats and camera settings

The PAL and NTSC video systems, which are used respectively in the UK and the USA, vary in the number of horizontal lines that constitute the image (625 vs. 525) and the number of video fields recorded per second (50 vs. 60). Digital video formats also provide a greater horizontal resolution than more age-old formats, and hence a better image. Regardless of the video system or format used, the electronic high speed shutter should be adjusted to suit the nature of the motion being analyzed, and manual rather than automatic focusing used.

Video recording procedures

When using 2D techniques, it is vital for later scaling of the image to life-size that the optical axis of the camera is aligned at 90° to the plane of motion and that the camera is placed as far from the plane of motion as possible to reduce perspective error. With 3D analysis, if only two cameras are used their optical axes should intersect at approximately 90°. In both types of analysis, a calibration object needs to be included within the field of view. This can be a simple scaling object, such as a 1 m rule, for 2D analysis but needs to be a more sophisticated array of control points for 3D analysis.

Digitizing

The process of digitizing involves converting body landmarks, on the video image stored in a computer, into digital x (horizontal) and y (vertical) coordinates. The resolution of video digitizing systems is usually 768×576, but can be greater than this to allow improved accuracy of coordinates.

Scaling and reconstruction

In 2D analysis, digitized coordinates of body landmarks are converted into real-life sized horizontal and vertical positions through a scaling factor that is obtained by digitizing the scaling object. The more complex Direct Linear Transformation (DLT) is used to convert the two, or more, sets of 2D coordinates into life-sized x, y, and z positions in a 3D analysis. This involves first digitizing the control points and using their coordinates in the DLT equations to provide the 11 calibration coefficients. These coefficients are then used in the DLT equations, along with the digitized coordinates from body landmarks, to determine the x, y, and z positions.

Introduction

Video analysis can either be qualitative and/or quantitative depending on the aims of the investigation. Qualitative analysis involves observation of video and diagnosis of particular aspects of technique that may subsequently be altered for clinical benefit or performance gain. It is a subjective process that seldom requires the camera to be located in a specific or stationary position and often does not require any additional equipment. Quantitative analysis requires kinematic information (i.e., linear position, velocity and acceleration of body segment endpoints and angular position, velocity and acceleration of body segments – see section A) to be obtained from video. Similar to qualitative analysis, following an intervention, such information can be monitored with a view to changing an individual's technique in order to reduce risk of injury or improve performance. Qualitative and quantitative analyses are often combined, particularly by coaches, and a number of software packages are commercially available that allow the user to display a number of images on the same screen, for purposes of comparison, and calculation of simple kinematic variables.

Kinematic information obtained from quantitative analysis can also be used, in combination with body segment parameters (see section C) to calculate center of mass kinematics, segmental energy levels and power (see section D), and joint moments and forces (see sections B and C). Quantitative analysis generally requires the video camera(s) to be stationary and located in a specific position(s), and the images subsequently to be stored in and displayed on a computer. Each image is then digitized to provide horizontal (x) and vertical (y) coordinates of selected points on the body, usually segment end points. These coordinates are then scaled (2D) or reconstructed (3D) to provide real-life coordinates, and smoothed to reduce errors that are inevitably incurred during their collection (see section F). They are also often combined with temporal information to obtain velocities and accelerations. The following text details how video cameras should be used quantitatively to analyze motion that essentially occurs in a single plane (e.g., cycling, running) and multiple planes (e.g., cricket bowling, shot put), as well as digitizing and scaling/reconstruction of coordinates. Smoothing of coordinates will be dealt with in section F3.

Video systems, formats and camera settings

The video system used in the UK is the Phase Alternating Line or PAL System. Each video frame consists of 625 horizontal lines, which is often referred to as the vertical resolution of the system (see *Fig. F1.1a*), although only 576 of these are available for recording the action. During recording of a video frame the odd number lines (i.e., 1, 3, 5, etc.) are scanned from the top to the bottom and from the left to the right of the picture at approximately the same time. The remaining even number lines (i.e., 2, 4, 6, etc.) are scanned 0.02 sec later. Each set of odd and even numbered scanned lines constitute a separate image or video field, separated by 0.02 s, that belongs to the same video frame. During playback, each image can be displayed sequentially to provide 50 fields per sec or 50 Hz, with a time between fields, or temporal resolution, of 0.02 sec. The National Television System Committee (NTSC) system used in the USA has a lower vertical resolution (525 scan lines) but a greater number of fields per sec (almost 60 Hz). High speed video cameras are also commercially available that are able to record more than 1000 images per sec. Such cameras are particularly important for recording detail during rapid movements that would be missed if conventional cameras were used. Irrespective of the video system, the horizontal resolution depends on the video format used and is typically considered as the

(a)   (b)

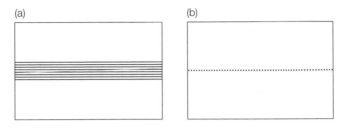

Fig. F1.1.   Depiction of vertical, and horizontal resolution video (see text for details of actual resolution)

number of dots, or pixels, that constitute each of the horizontal lines (see *Fig. F1.1b*). With only 240 dots, VHS should be considered to have too low a resolution to be used for quantitative video analysis, whilst SVHS (up to 400 dots) and digital video (DV – up to 500 dots) formats provide acceptable resolution.

The high speed shutter facility on a video camera should be used to prevent blurred or smeared images of moving bodies. Most cameras have variable electronic shutters which open for 1/120th, 1/250th, 1/500th, and at least 1/1000th of a sec (0.008, 0.004, 0.002, or 0.001 sec respectively), and which allow light to pass into the camera for progressively shorter lengths of time. Rapid movements, such as the motion of a golf club around ball strike, will only appear non-smeared if shutter speeds in excess of 1/1000th of a sec are used; but for slower movements, such as walking, 1/250th of a sec is adequate. The disadvantage of using higher shutter speeds (e.g., 1/500th or 1/1000th of a sec) is that, depending on the light available, the image can appear dark due to the lack of light entering the camera. This is not usually a problem when filming outdoors, but in conditions of poorer light, that typically occur indoors, additional flood lighting may need to be used to improve the quality of the image. High speed shutter settings should not be confused with the temporal resolution of the video system/camera. *Fig. F1.2* demonstrates the relationship between two different high speed shutter settings and temporal resolution for a conventional video camera operating under the PAL System. Much larger high speed shutter settings (i.e., greater fractions of a second) are required for high speed video cameras and are often as high as 1/100000th of a sec (i.e., 0.00001 sec or 0.01 msec or 10 µsec).

The manual focus setting on a video camera should also be used when conducting a quantitative analysis. In automatic focus mode, to which cameras often default, the lens is focused on the object nearest to it, which may not be the subject of interest. Thus, with the camera positioned appropriately (see Video recording procedures that follow) and the camera set to manual focus, the telephoto lens should be used to zoom in as close as possible to the participant standing in the center of the activity area. The focusing ring on the lens is then rotated until the image of the participant is sharp, and the telephoto lens is then used to zoom out to the required image size. This will ensure that the lens remains focused in the plane of motion.

Whilst the high speed shutter and manual focus settings are the most important when using video cameras for quantitative analysis, the quality of the image will also be affected by the white balance setting and other features (e.g., filters for recording in different environments) that are now commonly available on video cameras.

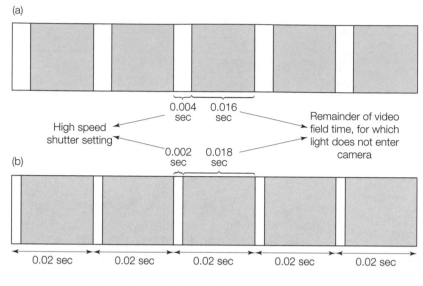

Fig. F1.2.   Depiction of high speed shutter settings of (a) 1/250th sec and (b) 1/500th sec

Video recording procedures

The same video formats and camera settings discussed previously can be used regardless of whether a two-dimensional (2D) or three-dimensional (3D) video analysis is to be undertaken. A 2D analysis typically uses only one camera to record activities that are essentially planar in nature, and in which the plane of motion coincides with the photographic plane (i.e., at 90° to the optical axis of the camera). Alternatively, a 3D analysis uses two or more cameras and should always be used to investigate activities that do not occur in a single plane. The procedures used to record images for 2D and 3D analyses are generally different and are therefore largely dealt with separately.

2D analysis

- The camera should be positioned as far from the plane of motion as possible to reduce the effects of perspective error. The telephoto zoom lens can then be used to bring the image of the participant to the required size in the field of view. This should not be too large, so as to cut some of the activity, or too small so that the individual cannot be digitized accurately. Perspective error occurs when objects or parts of objects that are closer to the lens appear larger than those that are further away. It can be demonstrated by closing your non-dominant eye and looking at your hand, at arm's length, through your dominant eye. If the hand is rotated to a sideways position and moved towards the eye, the thumb appears progressively larger than the little finger, which is further away, even though the two digits are approximately the same length. One of the effects of perspective error (shown in *Fig. F1.3*) is the apparent shortening of body segments when they move out of the plane of motion, which inevitably occurs during even the most planar of activities (e.g., running). In addition, perspective error results in angles between segments becoming more obtuse when they are moved out of the plane (see

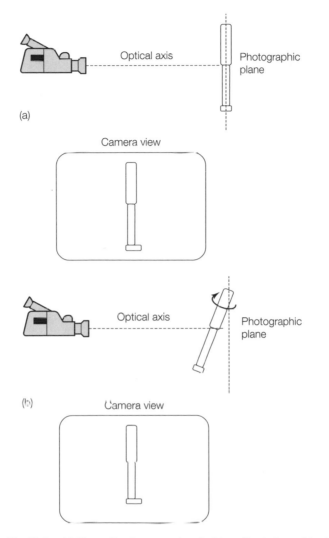

Fig. F1.3.   (a) Plane of body segments coincides with photographic plane. (b) Body segments moved out of photographic plane, showing shorter segment length in camera view

Fig. F1.4). As stated above, 3D analysis should be used when the motion of body segments does not occur in a single plane and information gained from a 2D analysis would largely be inaccurate due to perspective error.

- The optical axis (i.e., an imaginary line passing through the middle of the lens) also needs to be oriented at 90° to the intended plane of motion (see *Fig. F1.5*). Assuming that the plane of motion is vertical (e.g., during running), this can be partly achieved by placing a spirit level on top of the camera and positioned both parallel and perpendicular to its optical axis. This is, of course, assuming that the top surface of the camera is both horizontal and level. If not, the height of the center of the lens can be measured and a marker placed in the plane of motion at the same height. The telephoto lens can then be used to zoom in on the marker, which should remain in the mid-line of the image. When the required field of view size is established, the bottom of the

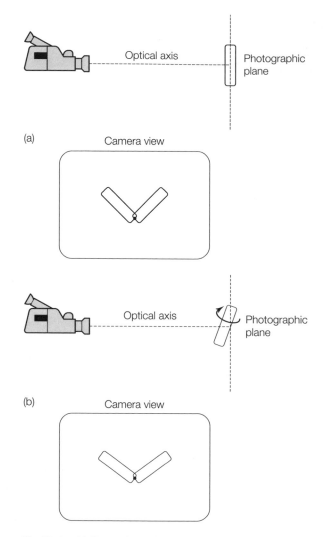

Fig. F1.4. (a) Plane of body segments coincides with photographic plane. (b) Body segments moved out of photographic plane, showing more obtuse segment angle in camera view

image should also be parallel with a line that is known to be horizontal (e.g., the ground). A 3–4–5 triangle (or multiples thereof) can be used to ensure that the optical axis is aligned at 90° to the plane of motion, in the horizontal plane (see *Fig. F1.5b*). A plumb-line should be used to ensure that the apex of the triangle, or a line extending from this point, is positioned directly below the center of the lens (i.e., the optical axis).

- Vertical and horizontal scaling objects (e.g., a 1 m rule) need to be placed in the photographic plane/plane of motion and included in the field of view during recording. A vertical reference (e.g., a plumb-line) should also be recorded.
- If the action occurs over a relatively long path (e.g., long jump or bowler's run-up) use of a single camera will result in a field of view where the individual is too small to be digitized accurately. In such conditions a number of

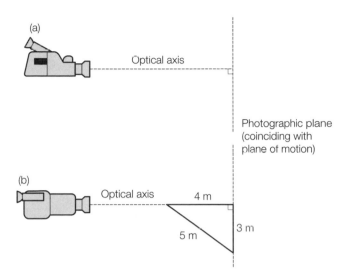

Fig. F1.5. Optical axis of the camera aligned at 90° to the photographic plane (a) when viewed from the side, and (b) when viewed from above

synchronized cameras can be used, with their fields of view overlapping slightly.

- In conditions of poor lighting (e.g., indoors with a high speed shutter setting in excess of 1/250th of a sec) the action should be illuminated with flood lights positioned at approximately 30° to the plane of performance.

2D and 3D analysis

- The camera should be mounted on a rigid tripod and, normally, once in the required position should not be moved during recording.
- An information board should be included in the field of view of all cameras including the date, time, participant code, trial number, and so on.
- Following recording, body segment end points (often joint centers) are digitized, as detailed later. To facilitate this process, individuals should ideally wear minimal and tight fitting clothing that contrasts with the color of the background during recording. The background should also be un-cluttered and non-reflective. Joint centers can also be marked on the skin, either directly with a soluble pen or using stickers that contrast with the color of the skin. While such markers are useful in identifying joint centers, they should not be relied upon accurately to represent the underlying segment end points; particularly when segments rotate out of the photographic plane.

3D analysis

- Two or more cameras should be used to film the activity. Ideally their optical axes should intersect at approximately 90°, but this angle can range between 60–120° (see Fig. F1.6).
- Ideally, the cameras should be gen-locked so that their shutters open at exactly the same time, enabling video fields from separate cameras to be synchronized. If this is not possible, due to the cabling required (see Fig. F1.6) between the cameras, a timing device should be included within the field of view of all cameras.

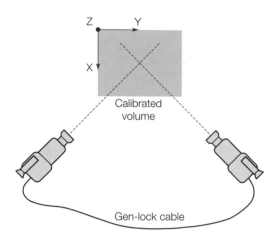

Fig. F1.6.   Plan view of typical camera position used in a 3D video analysis

- In replacement of the scaling object used in 2D analysis, at least six control points should be distributed throughout the volume in which the activity takes place. Use of more than six points provides more accurate reconstruction of real-life positions of body markers from digitized coordinates. For activities that occur in a relatively small volume the control points are usually contained within a calibration frame that can be dismantled after use (see *Fig. F1.7*). The exact location of each control point must be known, and is usually expressed in relation to one of the points on the frame that forms the origin of three orthogonal (X, Y, and Z) axes (see *Figs F1.6* and *F1.7*). Where the activity takes place in a larger volume, the frame can be repositioned throughout it, or alternatively a series of poles containing control points can be used. Each control point must be visible by all cameras, and the structure containing them obviously must be removed prior to the activity being recorded.
- Similar to 2D analysis, in conditions of poor lighting (e.g., indoors with a high speed shutter setting in excess of 1/250th of a sec) the action should be illuminated with flood lights positioned beside each camera.

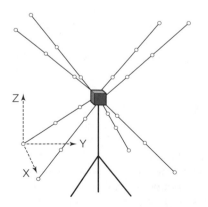

Fig. F1.7.   Example of a calibration frame

Digitizing

Once video images have been stored in the computer's memory in digital format, in the PAL System they are made up of 768 horizontal pixels and 576 vertical pixels. The process of digitizing essentially places an imaginary grid over the video field, or image, with the same number of coordinates as pixels. The bottom left-hand corner of the grid coincides with that of the image and shows the x and y coordinates as 0,0 (see *Fig. F1.8*). From this point each of the grid's horizontal lines or pixels represents a new vertical or y coordinate, which increases from the bottom to the top of the image. Similarly, each vertical line or pixel constitutes a different horizontal or x coordinate that increases from left to right. The process of digitizing involves using the computer mouse to move a cursor over the image to locate points of interest, which are usually body segment end points. Clicking a mouse button then records the x and y co-ordinates of the point (see *Fig. F1.9*). If the kinematics of the whole body center of mass are required then 18 points on the body are typically digitized, although this depends on the anthropometric model used (see section C4). Digitized points are often joined together by the computer to form a stick figure (see *Figs F1.8* and *Fig F1.9*) or more humanoid figure.

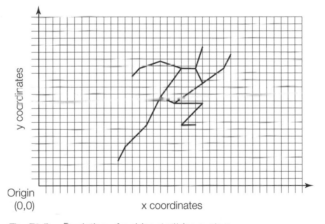

Fig. F1.8.   Depiction of a video digitizing system

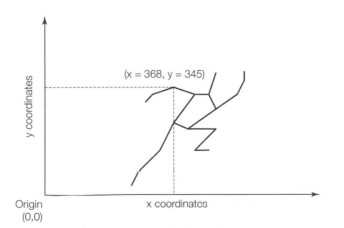

Fig. F1.9.   Depiction of digitizing of the elbow joint

The number of x and y coordinates that form the digitizing grid is known as the resolution of the digitizing system, which represents the smallest change in position that it can detect. Digitizing systems with a resolution of 768 × 576 are generally considered to yield less accurate coordinates than those that can be obtained from systems used to digitize 16 mm cine film. However, recent software developments that enable the image to be zoomed and allow multiple coordinates to be obtained from single pixels have improved the resolution of video systems. The advantage of improved digitizer resolution (see *Fig. F1.10*) improves the accuracy of the digitized coordinates.

(a)   (b)

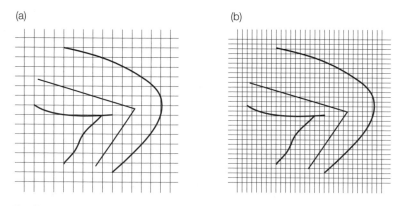

Fig. F1.10.   Effects of improved resolution of the digitizing system from (a) to (b)

Scaling and reconstruction

Digitized coordinates need to be converted into real-life horizontal and vertical positions of body landmarks. This is commonly referred to as "scaling" in 2D video analysis and, assuming that the plane of motion coincides with the photographic plane, is achieved by firstly digitizing the horizontal and vertical scaling objects that were included in the field of view. The length of the scaling object (in meters) is then divided by the number of coordinates that it represents to produce a scaling factor, ideally for both horizontal and vertical directions. Coordinates of body landmarks are then multiplied by the respective scaling factor to give their true position (in meters) in relation to the origin, which usually coincides with the bottom left corner of the image.

In a 3D analysis, following digitizing, 2D coordinates of body landmarks are obtained from the images recorded by each camera. Before these sets of x–y coordinates can be reconstructed into real-life horizontal (X and Y) and vertical (Z) positions, it is imperative that the images from each camera are synchronized. If the cameras are not gen-locked, then the timing device included in the fields of view can be used for synchronization. This can be done by selecting one of the cameras to be the time base. A curve consisting of a series of third degree polynomials, known as a cubic spline, is then fitted to the coordinate time data from the other camera(s) and the data interpolated to the same time base.

Now that two (or more) sets of synchronized x-y coordinates have been obtained they can be reconstructed into a single set of X, Y, Z real-life coordinates. This is most commonly done using the Direct Linear Transformation (DLT) that was first introduced by Abdel-Aziz and Karara (1971) and is represented in the equations following.

$$x = \frac{L_1 X + L_2 Y + L_3 Z + L_4}{L_9 X + L_{10} Y + L_{11} Z + 1}$$

$$y = \frac{L_5 X + L_6 Y + L_7 Z + L_8}{L_9 X + L_{10} Y + L_{11} Z + 1}$$

The 11 calibration coefficients: or parameters $(L_1\text{–}L_{11})$ included in the DLT equations above represent the location and orientation of the camera and the characteristics of the digitizing system, and replace the scaling factor used in 2D analysis. In order to calculate these parameters, first the six (or more) control points that were included within the field of view of each camera need to be digitized. Assuming that only six control points were used, this results in 12 new equations for each camera; one for each digitized x and y coordinate from each control point. As the X, Y, and Z coordinates of each control point are known, the 12 equations can be solved using a least squares approach to obtain the 11 DLT parameters.

With the 11 DLT parameters known, the DLT equations can now be used to find the X, Y, and Z coordinates of the digitized body landmarks. Again assuming that only two cameras were used, each body landmark will have two pairs of digitized x–y coordinates. These are inserted into the DLT equations to form four new equations, which can be re-arranged and solved to find the X, Y, and Z real-life coordinates of the body land marks.

The scaled, life-sized coordinates, whether 2D or 3D, need to be smoothed (see section F3) to reduce errors incurred in the digitizing process, prior to any linear or angular kinematic variables being calculated.

Reference

Abdel-Aziz, Y.I. and Karara, H.M. (1971) Direct linear transformation from comparator coordinates into object space coordinates in close-range photogrammetry. In: *ASP Symposium on Close Range Photogrammetry*. American Society of Photogrammetry, Falls, Church, pp. 1–18.

F2 OPTOELECTRONIC MOTION ANALYSIS

Key Notes

Optoelectronic motion analysis	Optoelectronic motion analysis uses a series of cameras which project infra-red light onto reflective spheres called targets. The reflected light is optically registered by the cameras and electronically converted to information registering the location of the targets in space. This optoelectronic process can automatically register the location of the targets in space thus making the process of motion analysis more simple and less time consuming than traditional manual methods. Further, by using a sufficient number of cameras it is possible to obtain three-dimensional (3D) data. The major advantage of optoelectronic motion analysis is ease of data collection. The major disadvantage is cost of the cameras and the need for specialized software.
Optoelectronic cameras	The cameras used are based on video technology. Around the lens is a series of infra-red light emitting diodes. Infra-red light cannot be seen by the human eye so using this type of light does not affect the performer. When these diodes flash, infra-red light is reflected back from the targets and recorded by the camera. This makes an "image" for that flash. This image is then transmitted back to the host computer as digital information. The cameras can repeat this operation quickly with sample rates of 240 Hz being common, and sample rates of up to 1000 Hz possible.
Optoelectronic targets	The passive targets are usually made from polystyrene balls covered in reflective tape. The targets can be of any diameter depending on application but typically need to cover about 1/200th of the field of view. Thus, for a field of view of 3 m (3000 mm) the target diameter needs to be around 15 mm. Target diameters available are as small as 3 mm to as large as 30 mm.
Calibration	The space within which the performer operates and in which the cameras are able to detect targets must be calibrated before use. Each manufacturer has developed their own system of calibration but a common method is to place a calibration object on the floor in the movement volume. A wand with two or more markers of known separation is moved around the whole of the movement volume to calibrate the volume. This process is known as "dynamic calibration". Using the manufacturer's recommended procedure it is possible to obtain reconstruction accuracies of less than 1 mm, and reconstruction precisions of around 0.2 mm.
Target sets and biomechanical models	Different target sets can be used for different applications. A 16 target set for use in general whole body human movement analysis consists of targets are placed on the 2nd metatarsal–phalangeal joint (2, left and right), ankle joint (2), knee joint (2), hip joint (2), shoulder joint (2,

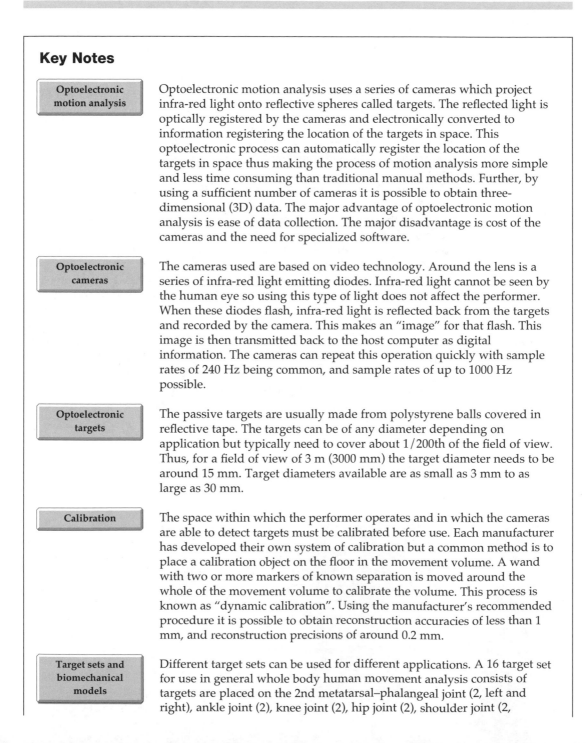

superior surface of the acromion), elbow joint (2), wrist joint (2), C7, and vertex. These markers define a common 12 segment biomechanical model consisting of foot (2), shank (2), thigh (2), upper arm (2), lower arm and hand (2), torso, and head. One of the limitations of this target set is that targets which are intended to represent the joint centers are placed on the outside of the joint center. Some software "correction" is required to account for this. A second limitation is that the full 3D motion of the segments cannot be obtained unless more markers are placed on each segment. Other target sets are available for detailed study of the lower body and which overcomes these limitations.

Introduction

Motion analysis is one of the major data collection tools in biomechanics. Its main purpose is to **collect data** on objects (usually humans) as they move around performing a task or activity. Data on their motion is obtained from a **recording of the motion** (for example, frames from a video recording) and then a registration of points (usually joint centers) from the video frame, a process known as **digitizing**. This process based on video frames can be quite **lengthy** and **time consuming** particularly if the digitizing is carried out manually.

Optoelectronic motion analysis tries to reduce the complexity of data collection in motion analysis and speed up the process. This optoelectronic process can automatically register the location of the targets in space thus making the process of motion analysis more simple and less time consuming. Further, by using a sufficient number of cameras it is possible to obtain **three-dimensional** (3D) data. A "passive" system does this by using a series of cameras which project infra-red light onto reflective spheres called **targets**. The reflected light is **optically** registered by the cameras and **electronically** converted to information registering the location of the targets in space. An "active" system uses cameras to receive signals produced by energized targets. To energize the targets a power source is needed which usually adds extra weight and complexity to the target set up, although this does have some advantages when identifying targets. Most optoelectronic systems used in biomechanics are passive systems so only these will be considered in this section.

The major **advantage** of optoelectronic motion analysis is ease of data collection. The major **disadvantage** is cost of the cameras and the need for specialized software.

Optoelectronic system and data collection

A passive **optoelectronic system** is based on a number of cameras. Usually for a 3D system 6–8 cameras are required which are spread around the volume in which measurements are to be made. A typical set up is illustrated in *Fig. F2.1*, which shows the location of eight cameras around a measurement volume of approximately 27 m³ ($3 \times 3 \times 3$ m) . At the center of the volume is a series of points which represent the location of each target attached to the body. *Fig. F2.1* also shows a close up of one of the cameras.

A typical human body marker placement for use in the analysis of a vertical jump is illustrated in *Fig. F2.2*. (Note that the targets are bright because they reflect the visible light generated by the camera flash when taking the picture).

The **targets** are usually made from polystyrene balls covered in reflective tape. The targets can be of any diameter depending on application but typically need to cover about 1/200th of the field of view. Thus, for a field of view of 3 m

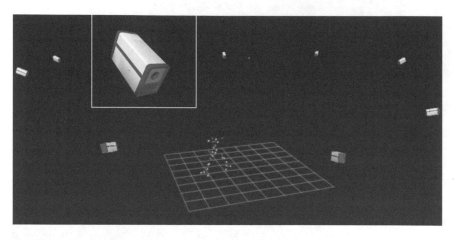

Fig. F2.1.   A typical optoelectronic camera configuration

Fig. F2.2.   Typical target placements for a whole body movement

(3000 mm) the target diameter needs to be around 15 mm. Target diameters available are as small as 3 mm to as large as 30 mm.

The **camera** used is based on video technology. Around the lens is a series of infra-red light emitting diodes. Infra-red light cannot be seen by the human eye so using this type of light does not affect the performer. When these diodes flash, infra-red light is reflected back from the targets (in the same way as seen in *Fig. F2.2* for visible light) and recorded by the infra-red light sensitive chip within the camera. The light is focused onto this chip by the camera lens in exactly the same way as a normal camera. This makes an "image" for that flash. A typical image seen by one camera is given in *Fig. F2.3*. This image is then transmitted back to the host computer as digital information. The cameras can repeat this operation quickly with sample rates of 240 Hz being common, and sample rates of up to 1000 Hz possible.

The **data** representing the location coordinates of each target are produced by the manufacturer's **software**. The host computer takes the images from all the cameras and "reconstructs" the data to provide the coordinates for each target. With the camera set up as in *Fig. F2.1*, 3D coordinates can be obtained. An

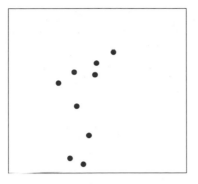

Fig. F2.3. A typical image of the targets seen by one camera

important condition is that each target must be seen by a minimum of two cameras in order to do this. Thus one limitation of the optoelectronic system is that targets must be "seen" by the cameras so cannot be placed in locations which can be obscured, for example by loose fitting clothing or long hair. The software also enables the markers to be identified and tracked. This is at best a semi-automatic part of the system as human intervention is required to solve data reconstruction difficulties that originate form target mis identification, cross-over of targets, and target drop out. For a well set-up system with appropriate activity and target placement these interventions should be minimal

Operational procedures

Calibration

As with any motion analysis system, the space within which the performer operates must be calibrated before use. Each manufacturer has developed their own system of calibration but a common method is to place a calibration object on the floor in the movement volume. A wand with two or more markers of known separation is moved around the whole of the movement volume to calibrate the volume (*Fig. F2.4*). This process is known as **dynamic calibration**. Using the manufacturer's recommended procedure it is possible to obtain reconstruction accuracies of less than 1 mm, and reconstruction precisions of around 0.2 mm.

Target sets and biomechanical models

Different target sets can be used for different applications. A target set is depicted in *Fig. F2.2* for use in **general whole body** human movement analysis. This is a **16 point target set** in which targets are placed on the second metatarsal–phalangeal joint (2, left and right), ankle joint (2), knee joint (2), hip joint (2), shoulder joint (2, superior surface of the acromion), elbow joint (2), wrist joint (2), C7, and vertex. These markers define a common **12 segment biomechanical model** consisting of foot (2), shank (2), thigh (2), upper arm (2), lower arm and hand (2), torso, and head. One of the limitations of this target set is that targets which are intended to represent the joint centers are placed on the outside of the joint center. Some software "correction" is required to account for this. A second limitation is that the full 3D motion of the segments cannot be obtained unless more markers are placed on each segment. Another target set is depicted in *Fig. F2.5* which is for a detailed study of the lower body and which overcomes these limitations.

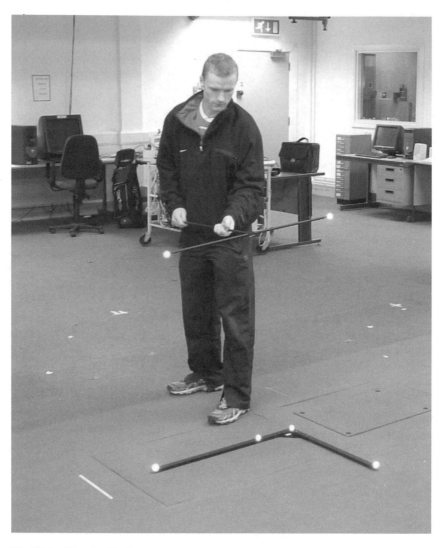

Fig. F2.4. Wand calibration

Calculation of kinematic variables

The main purpose of the optoelectronic system is to obtain the 3D (x, y, z) coordinates of each target as a function of time. These data are usually output by the system to a data file. This data file can be used as input to software (either from the manufacturer or a third party) to compute a range of kinematic data. It is also possible to access this data and to display it – or even perform calculations – using commonly available spreadsheet programs.

Other applications

Optoelectronic systems collect kinematic data but other instruments can be integrated into the data collection system. A common addition is the inclusion of a force platform. Any other data system needs to be **synchronized** with the kinematic data collected by the cameras but is usually provided for by the manu-

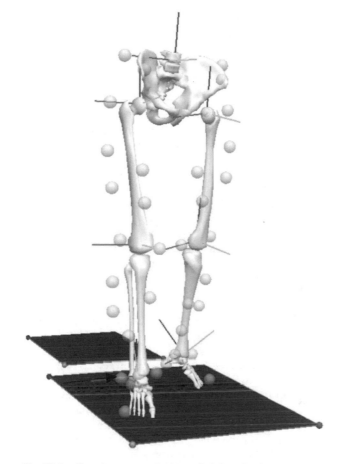

Fig. F2.5.   Twenty-seven point lower limb target set

facturer. If kinetic data is simultaneously collected then the software provided by the manufacturer (or third party) enables complex biomechanical data, such as joint moments and powers, to be computed. These data form the basis of **gait analysis**, and integrated optoelectronic systems have a major role to play in this area.

F3 DATA SMOOTHING

Key Notes

Data smoothing	Data smoothing is required to reduce the effect of errors in the data that occur in the process of data collection. These errors are particularly troublesome when first and second derivatives (i.e., velocity and acceleration) are computed. Several data smoothing algorithms are available for use in computer analysis software to provide data smoothing, a common one being the Hanning algorithm.
Errors in motion analysis data	Motion analysis requires the digitization of points representing, for example, the joints of the human body. The process of digitization introduces errors in the data and these come from three main sources: 1) image recording (foreshortening, depth and obliquity); 2) point location and resolution; and 3) event timing.
Hanning algorithm	The Hanning algorithm is one method that can be used to reduce the effect of errors by "smoothing" the data. The Hanning algorithm gets rid of the "jaggedness" associated with raw data and improves the estimates of velocity and acceleration which may be computed through the process of numerical differentiation. The Hanning algorithm is sometimes referred to as a "moving average" algorithm as it is applied over the first three points of the data (i.e., points 1, 2, and 3) and then moves on one point to the next set of three (i.e., points 2, 3, and 4). This is repeated over the whole data set. The Hanning algorithm has some limitations and other algorithms are available, such as the Butterworth 4th order algorithm, which allow more flexibility in the smoothing required.

Errors in experimental data

Experimentally collected data in sport and exercise biomechanics always has some error associated with it. This error is introduced due to the process used to collect data. It is most noticeable in motion analysis data but it exists in all other forms of data. The error in motion data can come from a number of sources categorized as 1) **image recording errors**, 2) **digitization errors**, and 3) **timing errors**.

Image recording errors
These occur in two-dimensional (2D) analysis due to: 1) **foreshortening** error, which is when a length is oriented towards or away from the camera and appears to be smaller than it really is; 2) **depth** error, which is when a length closer to the camera appears larger than when it is further away; 3) **obliquity** error, which is the increased error in measurement at the edges of the image. These errors can be minimized by filming perpendicular to the **plane of action**, by making sure the movement to be analyzed is **planar** in the plane of action, and by restricting the action to the central area of the film (i.e., avoid making measurements at the edges of the image).

Digitization errors
These occur due to 1) **point location error**, which is due to the difficulty of identifying reference points and joint centers; 2) **resolution errors**, which are due to the resolving ability of the digitizing system, the size of the image and the actual size of the field of view.

Timing errors
These occur due to 1) **timing mechanisms** used, whether it be a clock or electronic oscillator but this is usually very small, or 2) **event timing** error, which is usually obtained to ±1 sample (for example, heel strike in running can only be judged to ±1 frame). Note: in video analysis both "frames" and "fields" can be used where two fields make up one frame. In the context of this section the term frame is used in the general sense and refers to successive images regardless of how they are composed.

It is sensible when collecting data to try to reduce the errors as much as possible by the use of appropriate procedures. It is impossible to remove all the errors at source so various error reduction methods have been developed.

The effect of errors in data

Errors in data make the data look "ragged" but the main problem is the inexactness with which any single point can be estimated. For example, consider the path of the center of gravity in the long jump take-off. The original or "raw" data obtained from a motion analysis based on video are given in *Fig. F3.1*. On this figure is marked the frame at which touch-down and take-off occur. Due to the raggedness of the curve it is not possible to be really sure about the height of the center of gravity at touch-down or take-off.

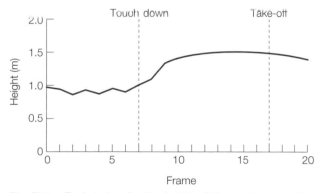

Fig. F3.1. Typical data for the height of the center of gravity during the touch-down to take-off phase in a long jump

A further problem is the effect that these errors have on the "processed" data. In sport and exercise biomechanics displacement data are collected (as, for example, in motion analysis) but other kinematic quantities are then required, such as velocity and acceleration. Velocity is the rate of change of displacement and acceleration is the rate of change of velocity and these are calculated according to the equations:

velocity (v) = change in displacement (d_2-d_1)/change in time (t) (F3.1)

acceleration (a) = change in velocity (v_2-v_1)/change in time (t) (F3.2)

When these are implemented on data that contains errors the effect of the error is magnified in the velocity calculation and magnified even further in the acceleration calculation. This is illustrated in *Fig. F3.2* for an object that is moving at constant velocity. The left-hand panel shows displacement data and the right-hand panel shows the corresponding velocity as calculated from equation F3.1. When the data has no errors (top row), the result is a velocity value that is constant, reflecting the constant velocity condition that is being analyzed. When the data has one error (middle row – the error introduced by the digitizing process) the effect is to overestimate one of the velocity calculations, but underestimate the next velocity calculation. This gives a spike in the velocity data so, rather than a flat line, the velocity data now gives incorrect values for some of the velocity points. This problem gets worse if the data has two errors in it (bottom row) and it can be seen that the spike in the velocity data becomes worse.

In the general case when all of the data have some error, it can be difficult to obtain a value for velocity that has any practical value. As noted above the calculation of acceleration using equation F3.2 becomes more difficult as it is based on the already affected velocity data.

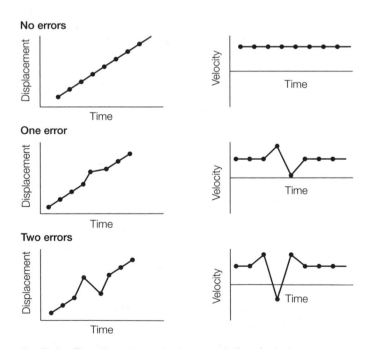

Fig. F3.2.   The effect of errors in the computation of velocity

Smoothing and the reduction of the effect of errors in data

The errors noted above can be reduced by applying an algorithm to **smooth** (or **filter**) the data. The algorithm has a similar effect to that achieved by hand when you draw a smooth curve through some ragged data.

A common smoothing algorithm is **Hanning**. This is described as a "moving average" algorithm and has the form of:

$$y_i = 0.25\ x_{i-1} + 0.5\ x_i + 0.25\ x_{i+1} \text{ for } i = 1 \text{ to } (N-1) \tag{F3.3}$$

where the variable x is the original data, the variable y is the newly computed "smooth" value, and N is the number of frames. This formula is applied to all of the data as a moving average. It can be applied more than once for increased smoothing. To illustrate its application, consider the data from *Fig. F3.1*, which is tabulated in *Table F3.1*.

Table F3.1   Typical raw data for a long jump take-off with the smoothed data based on a Hanning algorithm.

Frame	Raw data	Smoothed data
1	0.970	0.955
2	0.940	0.928
3	0.860	0.898
4	0.930	0.898
5	0.870	0.904
6	0.945	0.915
7	0.901	0.943
8	1.025	1.012
9	1.095	1.137
10	1.333	1.296
11	1.423	1.403
12	1.433	1.448
13	1.501	1.487
14	1.512	1.497
15	1.463	1.406
16	1.505	1.501
17	1.532	1.503
18	1.443	1.473
19	1.473	1.463
20	1.463	1.448
21	1.393	1.428

The smoothed data (y) for frame 2 is calculated as:

$$Y2 = 0.25^* (1) + 0.5^* (2) + 0.25^* (3)$$
$$= 0.25^*(0.970) + 0.5^*(0.940) + 0.25^*(0.860)$$
$$= \underline{\mathbf{0.928}}$$

Similarly, for frame 3

$$Y3 = 0.25^*(0.940) + 0.5^*(0.860) + 0.25^*(0.930) = \underline{\mathbf{0.898}}$$

This is a time consuming process so it is best done by computer, either in a spreadsheet or a specially written computer program.

It should be noted that the moving average algorithm cannot calculate data for the first or last data points in the array as it needs to have a data point to represent the (i–1) or (i+1) data which does not exist for the first and last points respectively. To overcome this, end point routines are used. For the Hanning algorithm these are:

$$y_1 = 0.5 ^*(x_1 + x_2) \tag{F3.4}$$

$$y_N = 0.5 ^*(x_{N-1} + x_N) \tag{F3.5}$$

The results of this are presented in *Fig. F3.3*.

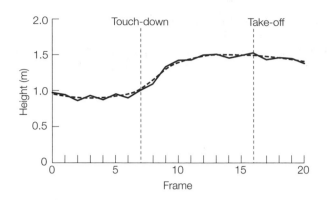

Fig. F3.3.   Raw and smoothed data plotted from Table F3.1

Other smoothing algorithms

The **Hanning** algorithm has the advantage that it can easily be programmed on a spreadsheet or customized computer program. Although it does successfully smooth the data, it is **not very flexible** for this purpose. Frequently in sport and exercise biomechanics it is necessary to have **greater smoothing** than the Hanning algorithm can supply. For this reason other algorithms have been developed.

The **Butterworth second order** smoothing algorithm is an important algorithm which uses similar principles to the Hanning, but is more flexible. It is frequently referred to in the biomechanical literature as a **Butterworth fourth order**, as this algorithm is often applied twice. This is because one of its side effects is to cause a temporal distortion of the data (where the data are shifted forward in time) which is corrected if the algorithm is applied for a second time in the reverse direction (i.e., starting at the last data point and working forwards to the first data point). The Butterworth fourth order algorithm is highly versatile and very effective for smoothing a wide variety of biomechanical data. Normally this algorithm will be available in the motion analysis software used for producing kinematic data.

Splines are another method to produce effective smoothing of biomechanical data. These come in two forms (cubic splines or quintic splines) with the latter providing more flexibility for complex data structures. The essential principle of this method is that the algorithm tries to fit a smoothed curve over three (cubic) or five (quintic) adjacent data points and moves onto the next group of data and repeats the data fitting until the end of the data structure is reached. This localized smoothing is also very effective in smoothing out errors in the data. As this is also a complex algorithm then it will also be available in the motion analysis software used for producing kinematic data.

F4 ACCELEROMETERS AND OTHER MOTION MEASURING DEVICES

Key Notes

Acceleration	Acceleration is defined as the time rate change in velocity. It is calculated from the change in velocity divided by time and is the second derivative of position–time data. Acceleration is a vector quantity, and has both magnitude and direction.

Acceleration calculated from the double differentiation of displacement–time data is often contaminated with high levels of noise (errors) and is often unsuitable for analysis.

Accelerometer	Accelerometers are devices which can be used to directly measure the acceleration of a body.

Newton's 2nd Law	Newton's second law states that if a mass, m, experiences an acceleration, a, then there must be a net force F acting on the mass such that:

$$F = m.a.$$

Hooke's law	Hooke's law states that if a spring of stiffness k is stretched from its equilibrium position, then there must be a net force acting on the spring. The force F is related to the stiffness and the degree of deformation such that:

$$F = k.x$$

where x is the change in spring length.

Goniometry	Goniometry is the direct measurement of joint angles. The term goniometer comes from the Greek word for angle which is gōniā. An electro-goniometer is a device that responds to changes in angular position by producing a detectible change in its electrical characteristics.

Acceleration

Motion is described by displacement, velocity and acceleration. The displacement and velocity can be measured with reasonable accuracy using kinematic methods such as video analysis. Since these systems are based on the measurement of position data some form of differentiation must be used to determine the velocity and acceleration. Each time the original data set is differentiated the effect of any small measurement error in the data is multiplied. To determine acceleration from position data requires the calculation of the second differential (change in displacement/time is the first differential; change in velocity/time is the second differential) and consequentially acceleration data are often contaminated with a large amount of error.

Suppose the velocity and acceleration of an athlete during an activity are to be studied. By filming the athlete side-on with a video camera and then measuring the displacement frame-by-frame (or field-by-field if a 50 Hz analysis is required), the information shown in *Table F4.1* can be obtained. By calculating the changes in displacement between consecutive frames and dividing by the time (differentiating) it is possible to determine the velocity. The same process can be followed to find the frame-to-frame changes in velocity to thus allow the acceleration to be found.

In *Table F4.1* the actual displacement data with the subsequent calculation of velocity and acceleration can be seen. In *Table F4.2* a small amount of error (plus or minus 2 cm) has been added to the data. These errors occur as a consequence of the measurement process. It can be seen than even very small amounts of error in the displacement data lead to large errors in the acceleration making it necessary to find an alternative approach to determining acceleration during movement.

Table F4.1. Velocities and accelerations calculated from "error free" coordinate data

Frame	Position/m	Change in displacement/m	Change in time/s	Velocity/ms^{-1}	Change in velocity/ms^{-1}	Acceleration/ms^{-2}
1	50.00					
2	50.40	0.40	0.04	10.00	0.50	12.50
3	50.82	0.42	0.04	10.50	0	0
4	51.24	0.42	0.04	10.50	1.00	25.0
5	51.70	0.46	0.04	11.50	1.00	25.0
6	52.20	0.50	0.04	12.50		

Table F4.2. Velocities and accelerations calculated from coordinate data containing small errors

Frame	Position/m	Change in displacement/m	Change in time/s	Velocity/ms^{-1}	Change in velocity/ms^{-1}	Acceleration/ms^{-2}
1	**50.02**					
2	**50.38**	0.36	0.04	9.00	2.50	62.5
3	**50.84**	0.46	0.04	11.50	2.00	−50.0
4	**51.22**	0.38	0.04	9.50	2.50	62.5
5	51.70	0.48	0.04	12.00	0	0
6	**52.18**	0.48	0.04	12.00		

An alternative method for determining acceleration would be to measure the forces acting upon a body and to use Newton's second law ($\Sigma F = m \times a$) to calculate the resultant acceleration. However, this method is only possible when it is practical to measure the contact forces acting upon the body of interest. There are many applications where this is not possible either because the body of interest is not in contact with any surfaces or the movement of interest occurs in a situation where contact forces can not be easily measured.

However, obtaining accurate and reliable acceleration data is essential to many areas of biomechanics. For example, good acceleration data are necessary for the calculation of joint reaction forces (the internal forces acting across joints in the human body obtained through the process of mathematical modeling). There are also many applications of the use of acceleration data to drive control devices in the automotive and aeronautical industries. To solve the difficulties associated with obtaining acceleration indirectly using displacement or force data an alternative method is to measure the acceleration directly. This involves the use of an **accelerometer.**

In applications that involve flight, such as aircraft and satellites, accelerometers are very often based on the properties of rotating masses. However, the most common design in human movement is based on a combination of Newton's law of mass acceleration and Hooke's law of spring action.

Newton's 2nd law and Hooke's law

Newton's second law states that if a mass, m, experiences an acceleration, a, then there must be a net force F acting on the mass and this is given by $F = m.a$. Hooke's law states that if a spring of stiffness k is stretched from its equilibrium position, then there must be a net force acting on the spring given by $F = k.x$ (where x is the change in spring length). If these two equations are combined it reveals that the displacement of the spring will be proportional to the acceleration such that:

$$F = m.a = k.x$$

Therefore:

$$a = k.x \: / \: m$$

The figure below (*Fig. F4.1*) shows an accelerometer constructed of a small mass attached to a spring. When there is no acceleration the spring rests at its natural length (x_1) and there is no force acting upon the mass. If the system is accelerated to the right the spring must exert a force on the mass to bring about its acceleration. This requires the spring to lengthen. As it lengthens force is developed until the mass is experiencing an acceleration equal to that of the remainder of the system. If the displacement of the mass is measured it is possible to calculate how great the acceleration was using the equation $a = k.(x_2 - x_1)/m$ from above.

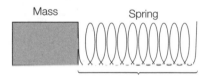

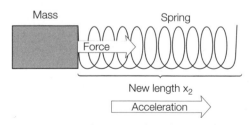

Fig. F4.1. A spring mass based accelerometer demonstrating the link between the deformation of the spring ($x_2 - x_1$) and the acceleration ($a = k (x_2 - x_1)$/mass). Note that the acceleration of the accelerometer to the right requires the spring to lengthen until such time as the mass is experiencing a spring force sufficient to make it accelerate

The **spring–mass principle** is used in many accelerometer designs. Accelerometers differ in how they measure the displacement of the mass. Common types of accelerometer include:

Sensor category	Key technologies
Capacitive	Metal beam or similar produces capacitance; change in capacitance related to acceleration
Piezoelectric	Piezoelectric crystal mounted to mass–voltage output converted to acceleration
Piezoresistive	Beam or similar whose resistance changes with acceleration
Hall Effect	Motion converted to electrical signal by sensing of changing magnetic fields
Magnetoresistive	Material resistivity changes in presence of magnetic field
Heat Transfer	Location of heated mass tracked during acceleration by sensing temperature

The most common accelerometers used in sport and exercise biomechanics are either capacitive or piezoelectric. Piezoelectric accelerometers are generally more expensive than those using capacitive technology.

It is important to remember that acceleration is a vector quantity and thus has both magnitude and direction. In the example above, only the magnitude of the acceleration in the direction that the spring is being stretched can be measured. This means that the accelerometer is able to measure in one dimension only and that the acceleration calculated is only representative of acceleration in that direction. To gain a complete picture of the acceleration of a body it is necessary to have three accelerometers, one aligned with each of the planes of motion. Some accelerometers are produced containing separate sensors in each plane to allow 3D measurements with a single device.

It is important to be aware of the effect that changing the orientation of the accelerometer has on the output. Consider the spring-mass system described previously; if the spring is oriented, as shown in *Fig. F4.1*, such that it is aligned with the horizontal plane, then at rest there will be no force acting between the spring and the mass. If the system is rotated 90°, so the mass hangs down below the spring (*Fig. F4.2*), then the spring will exert a force on the mass equal to its weight. This force is the result of the acceleration due to gravity (F = m.g). In this situation the baseline acceleration is said to be equal to 1g – where g represents the acceleration due to gravity (9.81 ms^{-2}).

If the system were oriented the opposite way around, the mass would compress the spring and thus a negative displacement would be recorded. Here an acceleration of minus (–) 1g would be recorded. It is thus important that careful thought is given to the orientation of the accelerometer. Acceleration is reported in either metres per second per second (m.s^{-2}) or relative to the acceleration due to gravity (g). Typical values for acceleration are given below.

Earth's gravity	1 g
Passenger car in corner	2 g
Bobsled rider in corner	5 g
Human unconsciousness	7 g

When using accelerometers one of the most important considerations is the mounting of the accelerometer on the body. For the accelerometer output to give an accurate representation of the acceleration of the body it is essential that the motion of the accelerometer is the same as that of the body being measured. This requires a firm mounting between the transducer and the body. In the case of

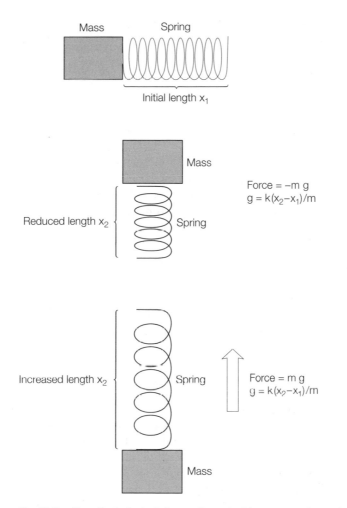

Fig. F4.2.   *The effect of orientation on the output from an accelerometer*

human movement it is important to realize that not all parts of the body experience the same motion, and that the movement of the soft tissues (skin, fat, muscle) will be different to that of the skeletal system.

When measuring the accelerations associated with human movement it is normal to see that the accelerometers are attached to the body at sites with minimal soft tissue between the skin and the underlying skeleton. The malleoi, head of fibula, greater trochanter, and acromion process are all common examples of locations where accelerometers are fixed on the body. Accelerometers have also been mounted on bite-bars held between the teeth. Even at these bony sites a firm mounting is necessary and combinations of bees wax and other mounting glues as well as tape have been used to ensure good fixation. Some experiments have even mounted accelerometers on pins driven into the bone to provide a close approximation between the accelerometer and bony movement.

When looking at the shock absorbing qualities of the body accelerations at different parts of the body can be compared to see how these change as the force is absorbed by the body tissues. The normal effect of the body systems is to gradually dissipate the force and lower accelerations are found higher up the body than at the foot (see table below).

Activity	Location	Acceleration
Walking barefoot	Tibia	~2.5 g
Running barefoot	Tibia	~9 g
Running in shoes	Tibia	~8 g
Running in shoes	Head	~3 g

Goniometry

If the range and rate of joint motion, rather than the position or orientation of the body in space are of most interest, then an alternative to video motion analysis is to use **goniometry**. **Goniometry** is the direct measurement of joint angles. The term goniometer comes from the Greek word for angle, which is gōniā.

In the simplest form a goniometer can be thought of as a protractor with extending arms (*Fig. F4.3*). To use this type of goniometer:

1. Align the fulcrum of the device with the fulcrum or the joint to be measured
2. Align the stationary arm of the device with the limb being measured
3. Hold the arms of the goniometer in place while the joint is moved through its range of motion
4. The angle between the endpoints represents the entire range-of-motion.

Whilst this type of goniometer may provide a cheap and simple method for measuring the range of motion at a single joint, under controlled conditions, it is not suitable for measuring how joints move during dynamic activities. Here the facility to sample the joint angle at regular intervals throughout the movement is required. This is achieved through the use of **electro-goniometers.**

An **electro-goniometer** is a device that responds to changes in angular position by producing a detectible change in its electrical characteristics. An example would be an angular potentiometer. As the joint angle changes so the position of the contacts on the potentiometer change and the resultant change in resistance can be measured using a simple electric circuit. The size of the resistance change would be proportional to the angular displacement. It is thus possible to gain a record of joint motion without the need for laborious digitisation of video.

Fig. F4.3.   A long-arm goniometer showing the alignment of the two measurement arms along the axes of the limbs and the central protractor

Despite their relative simplicity, goniometers have never formed a major role in the analysis of human movement for a number of reasons. Initially goniometers were only able to detect changes in angle about a single axis, thus requiring multiple devices and separate mounting to detect movement about other axes. For example, to measure plantar-flexion/dorsi-flexion and inversion/eversion at the ankle would require two separate goniometers to be attached in the sagittal and frontal planes respectively. This is difficult at the ankle, especially if the measurements were to be performed whilst the participant was wearing any kind of footwear. This difficulty has been overcome to some extent with the development of tri-axial goniometers where a single device is able to measure angular displacement about three separate planes.

A further issue with goniometery is the difficulty of aligning the device with the joint axis of rotation, especially in those joints where the axis is not stationary. For example at both the knee and shoulder joints there is significant gliding and rolling of the joint axis such that the axis of rotation changes depending upon the joint's absolute position. In addition, the data from the goniometer only provides information of the relative orientation of the two adjoining limbs and does not provide information about the absolute position of the body in space, something that is often required.

To help address the problem of obtaining joint displacement data without the need for the manual digitization of film or video various **opto-electronic** devices have been developed. These devises use automated procedures to track markers in space and to plot their coordinates. In essence the procedure is the same as manual digitisation of film or video, however the identification of the points is done automatically by computer.

The use for opto-electronic systems began as early at the 1960s but only became really viable with the advancement of computer technology in the 1980s. To work, opto-electric systems need to be able to identify the points of interest in the body and hence require clear contrast between the background and the desired object to be tracked. This is usually achieved by using reflective markers attached to known body landmarks. The markers are illuminated by infra-red light and tracked by infra-red sensitive cameras (e.g. Qualysis, Vicon, Elite systems). Thus only the motion of the markers are detected and can be tracked and plotted to provide motion data. An alternative approach has been to use markers which light-up in sequence (e.g., CODA).

Opto-electronic systems have been widely used in sport and exercise, their most common application being to the measurement of gait. However, they are generally limited to laboratory based analysis and are not suitable to measurement of competitive performances or field measures (see section F2).

F5 FORCE PLATE

Key Notes

The force plate	Force platforms measure the ground reaction force (GRF) which, in accordance with Newton's third law of motion, is equal in magnitude and opposite in direction to the action force that is applied to the plate. In the UK the vertical component of GRF is normally denoted as Fz, and the two horizontal components are denoted as Fy and Fx. Force plate transducers are usually piezoelectric or strain gauge types, which display high linearity, low hysteresis and minimal cross-talk between axes.
Interpreting GRF–time curves	In accordance with Newton's second law of motion, the sum of all of the forces acting on a body in a particular direction is proportional to the acceleration experienced by the body in the same direction. For example, during running the magnitude of Fz minus the athlete's body weight determines the magnitude of the vertical acceleration of their center of mass (C of M). Similarly, the magnitude of Fy minus the force of air resistance determines the runner's horizontal acceleration in the direction of running.
GRF related variables	In addition to peak forces, impulse and loading rate, a number of other variables can also be obtained from force platform information. The center of pressure (C of P) is the position of the resultant GRF vector in a plane that is parallel to the surface of the plate. Two-dimensional coordinates (Ay and Ax) are used to locate the C of P in relation to the center, or origin, of the plate. The free moment (Tz or Mz') is the turning force or moment around a vertical axis through the C of P.

The force plate
Newton's third law of motion dictates that for every (action) force that is applied by one body to another body, a (reaction) force is exerted by the second body on the first that is equal in magnitude and opposite in direction, as depicted in *Fig. F5.1*. In sport and exercise biomechanics the reaction force exerted by the ground on an individual is often studied; and is termed the ground reaction force (GRF). The force plate, or platform, embedded into the ground in a variety of settings (e.g., a laboratory or athletics track) is used to measure the GRF. Force data can be combined with the velocity of the C of M to obtain the power of the whole body (see section D1), and with kinematic and anthropometric data to determine joint reaction forces (see section C9).

Platforms measure force using transducers. When a force is applied to the plate each transducer experiences a deformation that is proportional to the magnitude of the force. A voltage, measured from the transducer, also alters in proportion to the amount which the transducer has deformed. Thus, the change in voltage measured by the transducer is proportional to the magnitude of force that it experiences. Force plates used in sport and exercise biomechanics either

Fig. F5.1. *Depiction of Newton's third law of motion*

use strain gauge or piezoelectric transducers. Piezoelectric platforms are more sensitive to rapid changes in force, but suffer from a change in output voltage with no change in applied force (i.e., drift). As such, they are more suited to measure forces from relatively short-lived, dynamic activities such as walking, running and jumping. Strain gauge plates are less susceptible to drift and are not as sensitive as piezoelectric models, so are preferred for recording forces from longer, less dynamic activities such as archery or shooting.

Regardless of the type of transducer used, there should ideally be a linear relationship between the force applied to the platform and the measured voltage (see *Fig. F5.2a*). Assuming linearity, the gradient of the relationship is effectively the calibration coefficient, which is used to convert volts into Newtons. In situations where the relationship is non-linear (see *Fig. F5.2a*), a higher order polynomial (e.g., quadratic) can be fitted to the data points to provide the calibration coefficient. Force plates should also display minimal hysteresis (see *Fig. F5.2b*), so that the relationship between force and voltage that is observed when the plate is loaded is the same as when it is unloaded. Transducers are arranged in force plates so that they measure three components of GRF that are parallel to the plate's three orthogonal axes (see *Fig. F5.3*). There should be minimal cross-talk, which is the detection of force by the transducers in one

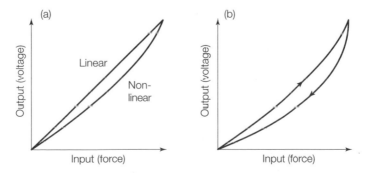

Fig. F5.2. *Depiction of (a) linearity and (b) hysteresis of a force platform*

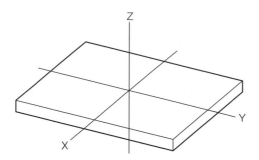

Fig. F5.3. Orthogonal force plate axes

direction (e.g., parallel to the z axis) and also by those in one or both of the other directions (i.e., parallel to the x and or y axes). In addition, the force plate should have a natural frequency that is much higher than the highest frequency of the signal being measured, and ideally higher than 800 Hz. This is so that the force being applied to the plate does not cause it to vibrate, which would affect the magnitude of the force being detected.

The voltage output from the transducers must be amplified before being recorded and stored, usually on a computer. Sampling of the signal into a computer should also use an analog-to-digital-converter (ADC) that has at least 12 bits (ideally 16 bits) to ensure that as small a change in force as possible can be detected. To satisfy the Nyquist theorem, the signal should normally be sampled at a minimum of 500 Hz, particularly if forces are recorded during impacts.

Two conventions exist to identify the three components of GRF that force plates measure. The convention shown in *Fig. F5.4*, which is commonly used in the UK, labels positive Fz in the vertical upwards direction, normal to the surface of the plate. Positive Fy acts along the forward horizontal direction, parallel to the long axis of the plate, and positive Fx occurs in the positive right lateral direction. It therefore follows that negative Fz, Fy, and Fx act downward, backward and in the right medial direction. The convention adopted by the International Society of Biomechanics (ISB) replaces Fz with Fy, Fy with Fx, and Fx with Fz.

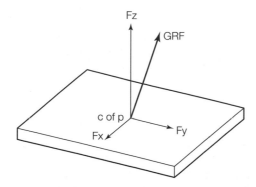

Fig. F5.4. Horizontal (Fx and Fy) and vertical (Fz) components of ground reaction force (GRF)

Interpreting GRF–time curves

Knowledge of Newton's laws of motion is imperative in understanding and interpreting GRF–time curves. Forces acting on a runner during the stance phase include the weight of the athlete (Fw), air resistance (Fa), and the components of GRF (see *Fig F5.5*). In accordance with Newton's second law of motion (see section B2), the sum of all the forces acting in each (x, y, z) direction are proportional to the acceleration experienced by the athlete in that direction (i.e., $\sum F = m \cdot a$); as shown in equations F5.1–3. Dividing both sides of each equation by the mass of the runner would yield the acceleration of the runner's c of m. Assuming that both Fw and Fa are constant, the shape of the resulting acceleration–time curves would be identical to that of the force–time curves.

$$F_z - F_w = m \cdot a_z \tag{F5.1}$$

$$F_y - F_a = m \cdot a_y \tag{F5.2}$$

$$F_x = m \cdot a_x \tag{F5.3}$$

Hypothetical Fz– and Fy–time curves are shown in *Fig. F5.6*, together with free body diagrams that coincide approximately with three points during the stance phase. The Fx–time curve has been omitted from *Fig. F5.6* as the magnitude of this component is much smaller than the other two and therefore

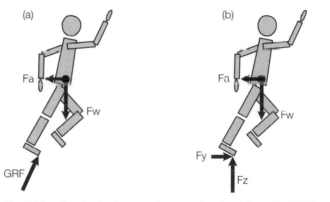

Fig. F5.5. Free body diagram of runner showing (a) resultant GRF, and (b) Fy and Fz components of GRF

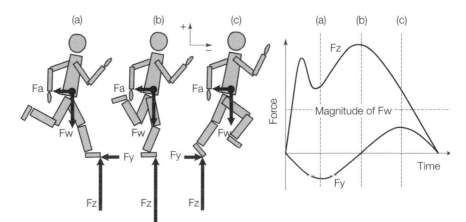

Fig. F5.6. Free body diagram and Fz and Fy–time curves during (a) the braking phase, (b) transition between braking and propulsion, and (c) the propulsive phase of running

has less effect on the acceleration of the body, and generally shows far greater inter-individual variability. In the vertical direction, as the athlete's foot first contacts the ground the magnitude of the Fz component of GRF is smaller than that of the body weight (i.e., Fz < Fw). In accordance with equation F5.1, this results in a negative (i.e., downward) force and, therefore, acceleration acting on the runner. As the runner is already moving downwards at this time, acceleration in the same direction results in an increase in the downward velocity of the c of m. This pattern is quickly reversed as Fz exceeds Fw, resulting in a positive acceleration that acts in an upwards direction. Initially, this decreases the downwards velocity of the runner until it reaches zero and their downwards motion is arrested. This occurs approximately halfway through the stance phase, after which the positive acceleration causes the runner's c of m to move upward, with increasing velocity, until just before toe-off. Here, the magnitude of Fz again drops below that of Fw, causing the acceleration to act in a downward direction. However, unlike at the start of the stance phase the runner is moving upward at this time so the negative acceleration causes the velocity to decrease in this direction immediately before toe-off.

Interpretation of the Fy–time curve from the stance phase of running is generally simpler than the Fz curve as motion only occurs in one direction (i.e., forward along a line parallel to the y axis). Assuming air resistance to be so small as to be negligible, as the runner's foot contacts the plate in front of their c of m, a braking force (i.e., negative Fy, prior to point b in *Fig. F5.6*) is experienced that acts in a backward direction. Again, in accordance with Newton's second law (see equation F5.2), this force acts to decelerate the forward motion of runner. This situation continues until the runner's c of m passes over the point of support (i.e., point b in *Fig. F5.6*) and the reaction force changes from negative to positive (i.e., forward). Positive Fy (i.e., after point b in *Fig. F5.6*) causes a positive horizontal acceleration that increases the forward horizontal velocity of the runner. Thus, when running at an approximately constant velocity, a braking force (negative Fy) acts during the first half of the stance phase that causes the horizontal velocity of the runner to decrease. Through the second part of the stance phase a propulsive force (positive Fy) dominates, which causes the runner's c of m to accelerate in the direction of motion.

Precisely how much the velocity of the runner's c of m changes in any direction during the stance phase can be determined using the impulse–momentum relationship (see section B3). Graphical integration of the force–time data, using Simpson's or the Trapezium rule, would yield the area bounded by the curves and hence the impulse (see *Fig. F5.7*). In accordance with the impulse–momentum relationship, the change in velocity of the runner's c of m can be obtained by dividing the net impulse by his/her mass. With regard to the forces acting in the direction of motion (i.e., Fy), if the braking impulse is greater than the propulsive impulse (see *Fig. F5.8a*) the runner will lose velocity during the stance phase. Conversely, if the braking impulse is less than the propulsive impulse the runner will gain velocity as he/she passes over the plate (see *Fig. 5.8b*); and if the two impulses are equal (i.e., zero net impulse), then the runner will complete the stance phase with the same velocity with which they started it (see *Fig. F5.8c*). In the unlikely situation of a runner experiencing zero net impulse over successive strides, their overall velocity would decrease due to effect of air resistance during swing phases of each stride. Thus, in order to maintain a constant running velocity, the propulsive impulse should be slightly greater than the braking impulse during each stance phase.

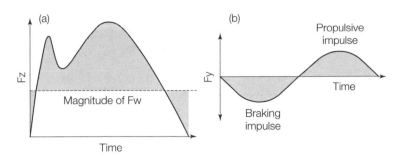

Fig. F5.7.   *Shaded areas depict (a) vertical and (b) horizontal impulse during the stance phase of running*

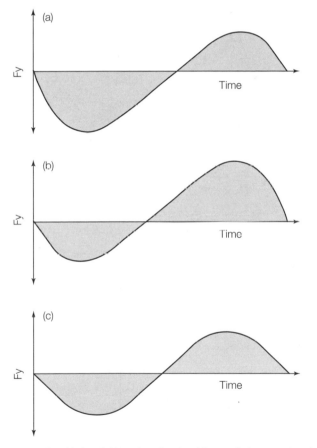

Fig. F5.8.   *Horizontal impulse showing (a) overall decrease in velocity, (b) overall increase in velocity, and (c) no change in velocity of center of mass during stance phase of running*

GRF related variables

The rate at which the vertical component of the GRF is applied to the body is often measured during running, landing, and so on, together with the peak force, as an indication of the risk of chronic injury due to such activities. The instantaneous loading rate can be obtained by measuring the gradient of a tangent that is drawn at the point on the Fz–time curve where the rate of change is considered to be greatest (see *Fig. F5.9a*). Alternatively, an average loading rate can be obtained by measuring the rate at which Fz rises by a force equal to the participant's body weight after an initial brief period whilst the body is loaded with, for example, 50 N, as recommended by Miller (1990; see *Fig. F5.9b*). This method obviously masks the peak loading rate provided by the instantaneous method, but produces a more reliable and objective measure due to the systematic way in which it is calculated.

In addition to the three components of GRF and their impulses, the center of pressure (c of p) and the free moment are often calculated from force plate data. The c of p is the position of the GRF vector in relation to a plane parallel to and just below the surface of the plate (see *Fig. F5.10*). Two coordinates (Ax and Ay)

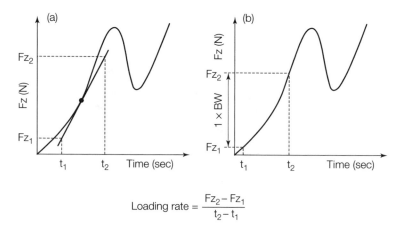

$$\text{Loading rate} = \frac{Fz_2 - Fz_1}{t_2 - t_1}$$

Fig. F5.9.   Depiction of (a) instantaneous, and (b) average loading rate from the initial phase of Fz–time curve during running

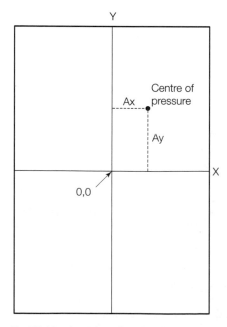

Fig. F5.10.   Location of centre of pressure, at one point in time, using coordinates (Ax and Ay) in relation to the center of the plate (0,0)

give its location in relation to the origin, or center, of the plate, as shown in *Fig. F5.10*. Ax and Ay can be either positive (one side of the respective axis) or negative (the other side of the axis), which designates the quadrant of the plate in which the c of p lies. If an individual runs across the platform or stands on it, the c of p will lie somewhere beneath their foot (see *Fig. F5.11a*). Alternatively, if a two footed stance is adopted then the c of p will lie roughly midway between the two points of contact (see *Fig. F5.11b*). A common application has been to examine the pattern of movement of the c of p beneath the foot during the stance phase of running. Motion of the c of p also mirrors that of the c of m during standing or during activities that require the body to be as stationary as possible (e.g., archery or shooting), so it has also been used as a measure of stability during such activities.

The free moment is the moment or torque about the vertical axis through the c of p, which coincides with the Fz vector (see *Fig. F5.12*), and is commonly referred to as either Mz' or Tz. It must also be remembered that, like the components of force, Mz' is equal in magnitude but opposite in direction to the

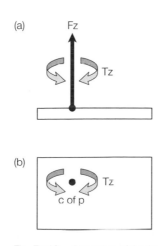

Fig. F5.11.   Location of center of pressure during (a) one-legged, and (b) two-legged standing

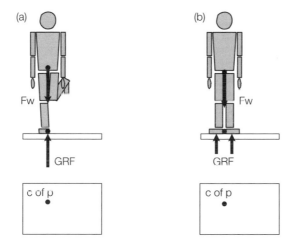

Fig. F5.12.   Depiction of the free moment looking (a) along the plates' x axis, and (b) along the z axis

moment that is applied to the plate about the vertical axis. Mz' is therefore often used to measure the reaction to the moment exerted by individuals during activities that involve twisting around a vertical axis.

Reference   Miller, D.I. (1990) Ground reaction forces in distance running. In: *Biomechanics of Distance Running* (P.R. Cavanagh ed.). Champaign, IL.: Human Kinetics, pp. 203–224.

F6 PRESSURE MEASUREMENT

Key Notes

Pressure	Pressure is defined as force per unit area such that: Force (N) / Area (m²) = Pressure (N/m²)
Units of pressure	There are a number of common units used to report pressure, all are derived from the basic units of Newtons per meter squared (N/m²). The **Pascal (Pa)** is the pressure created by a force of 1 N acting on an area of 1 m². Since 1 Pa represents a relatively low force spread over a large area it is more common to see **Kilopascals (kPa)** reported, where 1 kPa is equal to 1000 Pascals. **Atmospheric** or **barometric pressure** is the pressure caused by the weight of the air in the Earth's atmosphere. **Standard pressure** is a pressure of one normal (standard) atmosphere defined as: 1 Atmosphere = 101325 Pa or 101.325 kPa If measured using a mercury barometer pressure may be reported as the height in mm of the column of mercury (Hg). 1 Atmosphere = 760 mmHg at 0°C (32°F)

Pressure

Pressure is a mechanical parameter that is relevant in many applications. Pressure is defined as force per unit area and thus has units derived from this:

Force (N) / Area (m²) = Pressure (N/m²)

Another common unit used to describe pressure is the **Pascal (Pa)**. One Pascal is the pressure created by a force of 1 N acting on an area of 1 m². Since 1 Pa represents a relatively low force spread over a large area and in many applications would yield very high values it is not uncommon to see **Kilopascals (kPa)** reported, where 1 kPa is equal to 1000 Pascals.

An alternative to the use of Pascals for the reporting of pressure is to compare the measured pressure with the ambient pressure due to the Earth's atmosphere. **Atmospheric** or **barometric pressure** is the pressure caused by the weight of the air in the Earth's atmosphere. Imagine a column of one square meter cross-section extending from the Earth's surface to the edge of the atmosphere. This column will contain a certain number of air particles that will collectively create a force due to their weight. The force will depend upon the air density and the distance between the Earth's surface and the atmosphere.

Standard pressure is a pressure of one normal (standard) atmosphere defined as:

1 Atmosphere = 101325 Pa or 101.325 kPa

or 1 Atmosphere = 760 mmHg at 0°C (32°F) (this definition will be explained later)

The pressure experienced in any situation will be dependent upon both the magnitude of the applied force and area over which it acts. For example, if a person of body weight 750 N were to stand on one foot, and the area under the foot was 0.01 m^2 then the pressure would be:

$$\text{Force } / \text{ Area}$$
$$750 \text{ N } / \ 0.01 \text{ m}^2 = 75{,}000 \text{ N}/\text{m}^2 = 75 \text{ kPa}$$

or

$$75 \text{ kPa } / \ 101.325 \text{ kPa} = 0.74 \text{ Atmospheres}$$

If the same person were to put on a pair of shoes with a pointed heel (e.g., stilettos) the area in contact with the ground would decrease. If the new contact area was 0.002 m^2 the pressure in this situation would be:

$$750 \text{ N } / \ 0.002 \text{ m}^2 = 375{,}000 \text{ N}/\text{m}^2 = 375 \text{ kPa}$$

or

$$375 \text{ kPa } / \ 101.325 \text{ kPa} = 3.70 \text{ Atmospheres}$$

In both the above situations the force was the same but the pressure differed significantly. This is important in the study of human movement as looking at pressure gives an indication of the distribution of the load. From the injury perspective, the body is more likely to suffer damage and pain from a force concentrated in a small area (and thus a high pressure) than if the same load was distributed more widely. For example, it is more painful to have your foot trodden on by someone wearing stilettos than someone wearing flat shoes.

In many situations it is desirable to reduce the pressure by increasing the contact area, this is seen in the design of protective equipment such as helmets, shin pads, and so on that serve to distribute the load over a larger area and thus reduce the pressure exerted on the underlying tissues and so lower the potential for injury. When moving on a soft or fragile surface, for example, snow or ice, it is advantageous to spread the load to prevent the surface collapsing; this is seen in the design of snow shoes, skis, and so on.

By contrast there are situations where it is desirable to maximize the pressure and to have as small an area of contact as possible. Many cutting or piercing tools have an obvious point to focus the load into a small area to allow penetration without the necessity for a large force.

To measure pressure it is necessary to have an indication of both the area and the force applied. There are many different approaches to measuring pressure; a number of the common methods are addressed below.

The simplest method for measuring pressure is a **manometer.** A manometer generally consists of two connected columns of fluid; when both columns experience an equal pressure the fluid in each will rest at the same level (*Fig. F6.1*). If a greater pressure is experienced on one side than the other, the level of fluid on that side will go down and the level on the other side will rise (*Fig. F6.2*). The difference in height of the two columns of fluid will be related to the applied pressure, the cross-sectional area of the column and the density of the fluid, such that:

Weight of fluid column = Volume of fluid $\times$ Density (ρ) $\times$ gravity (g)

Volume = height (h) $\times$ cross-sectional area (a)

Weight of fluid column = $\rho \times h \times a \times g$

Pressure = Force / Area

Pressure = $\rho \times h \times a \times g \ / \ a = \rho \times h \times g$

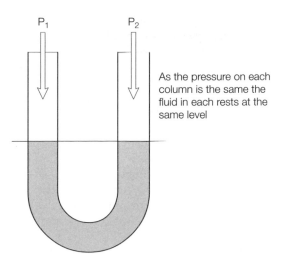

Fig. F6.1.   A U tube manometer under equilibrium conditions

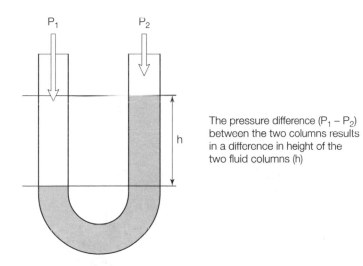

Fig. F6.2.   A U tube manometer under conditions of uneven pressure

If the reference end of the manometer is sealed in a vacuum then the height of the fluid will represent the absolute pressure. This is the case in the design of the common mercury barometer. Pressure is sometimes reported in the units mmHg which represents the height in millimeters of a column of mercury (Hg is the chemical symbol for mercury) in a manometer. At a pressure of 1 Atmosphere (101.325 kPa) the column height would be 760 mmHg. The common reference to blood pressure as 180/60 (or similar) is recorded in relation to the height of a mercury column and thus should properly read as 180 mmHg/60 mmHg.

Thus far, pressure measurements under steady-state conditions have been considered. These are called **static pressure measurements**, or **steady-state measurements**. This technique is useful in applications in which equilibrium or only very slow changing conditions are experienced. If the rate of change or pattern of change of pressure over short intervals of time is required these are called **dynamic pressure measurements**.

To measure dynamic pressure changes usually requires the use of an **electro-mechanical pressure sensor.** Electromechanical pressure sensors, or pressure transducers, convert motion generated by a pressure sensitive device into an electrical signal. The electrical output is proportional to the applied pressure. The most common pressure transducer types are strain gauge, variable capacitance, and piezoelectric.

A common application of dynamic pressure measurement is in the analysis of the pressure distribution beneath the foot during standing or gait. Pressure sensitive insoles are made up from a thin layer of material containing a large number of pressure sensors distributed throughout the sole. These allow detection of areas of high and low pressure beneath the foot at any moment during movement. The pressure profile created is usually presented as a series of colors to represent the different pressures (*Fig. F6.3*) or using a 3D graph with bars of different heights to indicate the magnitude of the pressure. Foot pressure analysis has been widely used to investigate the effect of different types of footwear and the link between pressure patterns and different injury patterns. Generally, higher pressures are indicative of a greater risk of injury.

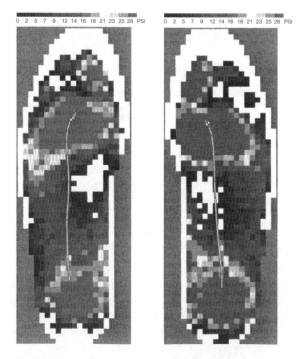

Fig. F6.3.   Pressure profile indicating the peak pressures beneath the foot during walking

F7 ELECTROMYOGRAPHY

Key notes

The electromyogram (EMG)

Electromyograms (EMGs) are recordings of electromyographical signals that emanate from muscle fibers prior to their contraction. Surface electromyography in sport and exercise biomechanics invariably involves recording signals from muscle fibers belonging to many different motor units.

Electromyography equipment

EMGs are usually detected using a pair of recording electrodes that are connected to a differential amplifier, which amplifies the difference between the signals detected from the two electrodes. Amplifiers can be connected directly to a PC for storage of EMGs (hard-wired system). Alternatively, amplified signals can be sent via radio waves to a receiver connected to a PC (telemetry system), or stored in a memory card before later being transferred to a PC (data logger system).

Recording EMGs

Surface electrodes should ideally be located between a motor point and a tendon, parallel to the direction of the underlying muscle fibers. Prior to this the skin should be shaved, washed, and ideally rubbed with an alcohol wipe to reduce skin–electrode impedance. Unwanted signals emanating from other adjacent muscles (cross-talk) should be minimized prior to recording EMGs.

Time domain processing

To quantify the magnitude of muscle activity over time, the raw EMG can be processed using the Average Rectified Value (ARV), Root Mean Square (RMS) or Linear Envelope. Both the ARV and the RMS are typically calculated over time windows that have a width of between 10 and 200 msec. The Linear Envelope typically constitutes a second order Butterworth filter with a cut-off frequency between 3 and 80 Hz.

Normalizing EMGs

EMGs that have been processed in the time domain can only be compared with those from the same muscle at different times without removal of electrodes. To compare EMGs between different muscles and individuals, processed EMGs should be normalized by dividing them by the EMG, processed in exactly the same way, from a reference contraction. This can be an isometric submaximal or maximal voluntary contraction. Alternatively, if the aim is to improve the homogeneity of EMGs from a group of individuals, then each processed EMG should be normalized by dividing it by the mean or peak processed EMG from the same task.

Frequency domain processing

The frequency content of the raw EMG can be revealed by a Fast Fourier Transform (FFT). Typically the FFT is calculated over intervals of 0.5–1 s, and the median frequency (MDF) is obtained from the resulting Power Density Spectrum. Changes in the MDF over time have traditionally been used as a measure of the fatigue state of the muscle. More recently, concerns over the use of the FFT on non-stationary signals has led to the development of more sophisticated joint time–frequency domain analysis techniques (e.g., wavelet analysis).

The electromyogram (EMG)

The fundamental unit of the neuromuscular system is the motor unit, which consists of the cell body and dendrites of a motor neuron, the multiple branches of its axon, and the muscle fibers that it innervates. Prior to tension being developed within a muscle, an action potential is generated by the motor neuron which propagates along the axon and then the muscle fiber. At rest, muscle fibers have a potential difference of –60 to –90 mV with respect to the outside of the muscle. Propagation of the action potential along the muscle fiber reduces the potential difference (depolarization) until it becomes positive (i.e., hyper-polarization) before it returns to the resting level (repolarization) after the action potential has passed. During a sustained muscle contraction repeated cycles of depolarization and repolarization, also known as the firing rate, often occur in excess of 20 times per sec.

Changes in the electrical potential of muscle fibers can be detected using electrodes placed either inside the muscle (fine-wire electrodes) or on the surface of the skin overlying the muscle (surface electrodes). The majority of sport and exercise science applications use surface electrodes which, depending on their size, can detect the signal from thousands of muscle fibers belonging to many (e.g., 20–50) different motor units. More recently, arrays containing many tiny surface electrodes have been developed that have the potential to be able to detect signals from fibers belonging to individual motor units. Regardless of the type of electrodes used, once the detected signal has been amplified and recorded, it is known as the electromyogram (EMG). A typical raw EMG recorded using surface electrodes is shown in *Fig. F7.1*.

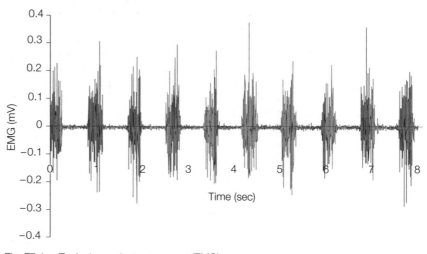

Fig. F7.1.   Typical raw electromyogram (EMG)

Electromyography equipment

It is generally accepted that the peak amplitude of the raw EMG recorded using surface electromyography does not exceed 5 mV and that its frequency spectrum is between 0–1000 Hz; with most of the usable energy limited to below 500 Hz and the dominant energy between 50–150 Hz. When detecting and recording EMGs a major concern should be that the fidelity of the signal is maximized. This is partly achieved by maximizing the signal-to-noise ratio (i.e., the ratio of the energy in the electromyographical signal to that in the noise). Noise can be considered as any signals that are not part of the electromyographical signal and can include movement artifacts, detection of the electrocardiogram, ambient noise from other machinery, and inherent noise in the recording equipment. Maximizing the fidelity of the EMG is also achieved by minimizing the distortion (i.e., alteration of the frequency components of the signal) that it receives during detection and recording. Both the equipment and procedures used to detect and record EMGs have a major influence on their fidelity, and should be given careful consideration.

Most commercially available electromyographical systems can be classified as either hard-wired, telemetry, or data logger systems. A data logger or telemetry system is necessary if data are to be collected away from the main recording apparatus; however data loggers typically do not allow on-line viewing of EMGs as they are being recorded and telemetry systems can be prone to ambient noise and cannot be used in areas with radiated electrical activity. Hardwired systems do not suffer from these limitations, but obviously preclude data collection outside of the vicinity of the recording apparatus. The fidelity of the recorded EMG is dependent on the characteristics of the (differential) amplifier that is connected to the electrodes, which are listed below together with recommended minimum specifications.

- Input Impedance (>100 MΩ)
- Common Mode Rejection Ratio (CMRR) (>80 dB [10,000])
- Input Referred Noise (<1–2 μV rms)
- Bandwidth (20–500 Hz)
- Gain (variable between 100 and 10,000)

Whilst the requirements of amplifiers are generally agreed on by electromyo-graphers, the configuration of electrodes and the material from which they are made are not. Some prefer pre-gelled silver/silver chloride (Ag/AgCl) electrodes that are circular with a diameter of 10 mm and a center-to-center distance of 20 mm. Others recommend silver bar electrodes that are 10 mm long, 1 mm wide, have a distance of 10 mm between them and are attached without the use of a gel. *Fig. F7.2* shows a schematic diagram of the equipment needed to detect and record EMGs.

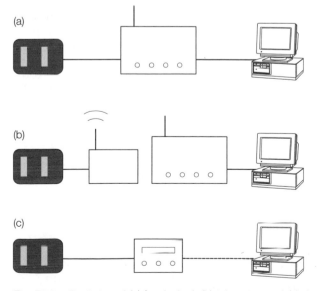

Fig. F7.2. Depiction of (a) hard-wired, (b) telemetry, and (c) data logger electromyography systems

Recording EMGs

In order to maximize the amplitude of the EMG the recording electrodes should be located between a motor point and a tendon. If the motor point cannot be located using, for example, a stimulator, electrodes can be placed in the center of the belly of the muscle whilst under contraction; although it should be recognized that this location could coincide with a motor point. Differential amplifiers subtract the signal detected by one recording electrode from that detected by the other. Thus, locating electrodes either side of a motor point will lead to the cancellation of symmetrical action potentials that are traveling in opposite directions from the neuromuscular junction and that reach the electrodes at approximately the same time. However, if both electrodes are placed to one side of a motor point the signal is not canceled to the same extent, as one electrode detects the signal slightly earlier than the other (see *Fig. F7.3*). Following the location of an appropriate site, the electrodes should be oriented along a line that is parallel to the direction of the underlying muscle fibers.

An improvement in the input impedance that is offered by many of today's amplifiers has diminished the need to reduce the skin–electrode impedance to, for example, below 10 kΩ. Skin preparation techniques that involve abrasion with fine sandpaper or scratching with a sterile lancet are, therefore, now largely redundant. Some preparation of the skin (to below 50 kΩ) is, however, still necessary in order to obtain a better electrode–skin contact and to improve the fidelity of the recorded signal. Typically, this involves cleansing the skin with soap and water and dry shaving it with a disposable razor. Additional rubbing with an alcohol soaked pad and then allowing the alcohol to vaporize can be used to reduce further impedance in individuals with less sensitive skin. In addition to the recording electrodes, differential amplifiers require the use of a reference electrode that must be attached to electrically neutral tissue (e.g., a bony landmark). The degree of skin preparation given to the reference electrode site should be the same as that afforded to the muscle site. Most electromyographers also advise using an electrode gel or paste to facilitate detection of the underlying electromyographical signal. This can be accomplished either through the use of pre-gelled electrodes or by applying a gel or paste to the skin or electrode prior to attachment. Use of gel or paste is not always necessary when using so-called "active electrodes" (i.e., those that are mounted onto a pre-amplifier). Here, the electrolytic medium is provided by the

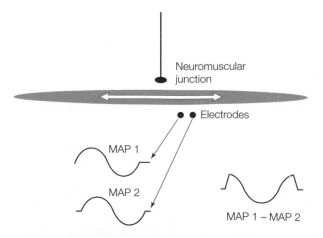

Fig. F7.3. Depiction of the recording of a single muscle fibre action potential by a differential amplifier

small amount of sweating that takes place when dry electrodes are applied to the skin.

Even if surface electrodes are placed close to the belly of the muscle it is possible that the detected signal may contain energy that emanates from other, more distant muscles. This is particularly pertinent when recording EMGs from muscles that are covered by thicker than normal amounts of subcutaneous fat, such as the gluteals and abdominals. The presence of cross-talk has traditionally been detected using functional tests that involve getting the participant to contract muscles that are adjacent to the one under investigation, without activating the one of interest. The detection of a signal from electrodes overlying the muscle of interest is, therefore, an indication of cross-talk. If possible, decreasing the size of the electrodes and/or the spacing between them reduces the chances of recording cross-talk. However, the most effective way of reducing it to almost negligible levels is to use a double differential, rather than a (single) differential amplifier. This has three, rather than two, detecting electrodes that are equally spaced apart, which calculates the difference between the signals detected by electrodes 1 and 2, and electrodes 2 and 3. These two (single differentiated) signals are then further differentiated (double differentiation) by the amplifier. This procedure works by significantly decreasing the detection volume of the three electrodes, and thereby filtering out signals from further away.

The Nyquist theorem dictates that electromyographical signals, which are detected using surface electrodes, should be sampled at a minimum of 1000 Hz (ideally 2000 Hz) to avoid aliasing (i.e., loss of information from the signal). Sampling of the signal into a PC should also use an analog-to-digital converter (ADC) that has at least 12 bits (ideally 16 bits) to ensure that as small a change in muscle activity as possible is able to be detected by the system.

Time domain processing

Raw EMGs have been processed in numerous ways, particularly since the advent of computers. Today, if the electromyographer wishes to quantify the activity of a muscle or investigate how this activity changes over time, raw EMGs are processed in what is known as the time domain. This is achieved using either the Average Rectified EMG, Root Mean Square EMG or Linear Envelope, all of which provide an estimate of the amplitude of the raw EMG in μV or mV.

Calculation of the Average Rectified EMG value (ARV) involves first either reversing all of the negative phases of the raw EMG (full-wave rectification). The integral of the rectified EMG is then calculated over a specific time period, or window (T), and the resulting **integrated EMG** is finally divided by T to form the ARV (see equation F7.1).

$$ARV = \frac{1}{T}\int_0^T |X(t)| dt \qquad \text{(F7.1)}$$

where $X(t)$ is the EMG signal
         T is the time over which the ARV is calculated

The Root Mean Square (RMS) EMG is the square root of the average power of the raw EMG calculated over a specific time period, or window (T) (see equation F7.2).

$$RMS = \sqrt{\frac{1}{T}\int_0^T X^2(t) dt} \qquad \text{(F7.2)}$$

Both the ARV and RMS are recognized as appropriate processing methods and are commonly used by electromyographers, although the RMS yields a larger amplitude than the ARV (see *Fig. F7.4*).

As well as using a single calculation of the RMS or ARV, the raw EMG is often processed by making successive calculations throughout its duration; with the resulting series of values forming a type of moving average. For this, the duration (or width) of successive time windows (T) can vary between 10 to 200 msec, depending on the duration and nature of the raw EMG. Selection of short duration window widths (e.g., 10–50 msec) may allow the detection of rapid alterations in activity, but the resulting curve will still resemble the rectified EMG (see *Fig. F7.5*). Thus, peak amplitudes from repetitions of the same task will remain highly variable. Adoption of longer widths (e.g., 100–200 msec) will reduce the variability of peak amplitudes, but the resulting curve will lose the trend of the underlying EMG (see *Fig. F7.6*). As such, rapid changes in muscle activity may go undetected. A possible solution is to use a moving average (either RMS or ARV) in which the time windows overlap instead of including discrete sections of the EMG. Overlapping the windows by a progressively greater amount results in a curve that increasingly follows the trend of the underlying rectified EMG, but without the variable peaks that are evident in the rectified EMG (see *Fig. F7.7*).

The Linear Envelope is also a popular processing method for use on EMGs from dynamic contractions. Similar to the moving average, this involves smoothing the rectified EMG with a low pass filter (see section F3), and also results in a curve that follows the trend of the EMG. When using the Linear

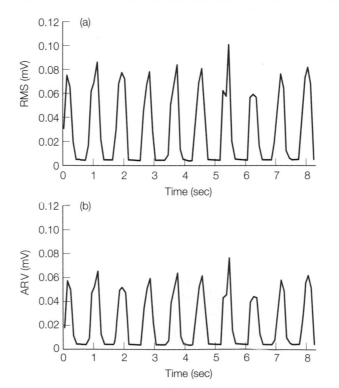

Fig. F7.4.   *The raw EMG in Fig. F7.1, processed using (a) the Root Mean Square (RMS), and (b) the Average Rectified Value methods with a time window width of 100 msec*

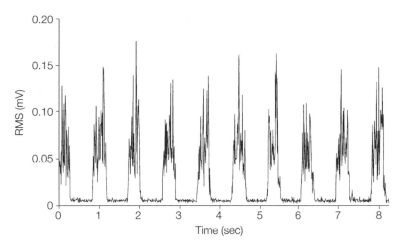

Fig. F7.5.   *The raw EMG in Fig. F7.1, processed using the RMS method with a time window width of 10 msec*

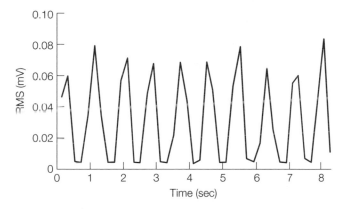

Fig. F7.6.   *The raw EMG in Fig. F7.1, processed using the RMS method with a time window width of 200 msec*

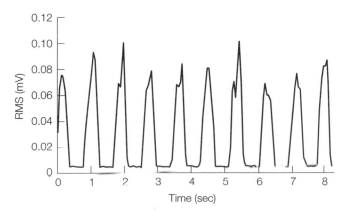

Fig. F7.7.   *The raw EMG in Fig. F7.1, processed using the RMS method with a time window width of 100 msec overlapped at 50 msec*

Envelope the type, order and cut-off frequency need to be selected. Traditionally, a second order Butterworth filter has been applied with a cut-off frequency between 3 and 80 Hz. Deciding on the cut-off frequency is similar to choosing the width and amount of overlap of the time window when using a moving average. A low frequency will result in a very smooth curve, which will be unable to detect rapid changes in activation. Conversely, a higher frequency will closely follow rapid changes in activity, but will still bear the peaks that characterize the rectified EMG.

Following processing the EMG is often used to estimate when a muscle is active (i.e., on) or inactive (i.e., off). Typically, in order to determine the amplitude threshold at which the muscle is considered to be active, the baseline EMG (or noise) is treated as a stochastic (or random) variable. The mean of this baseline is, for example, calculated over 50 msec and the muscle is deemed to be active when the EMG amplitude exceeds 2 standard deviations above the mean baseline activity for 20 msec or more.

Normalizing EMGs   EMGs processed in the time domain can only be compared with those recorded from the same muscle at another time without the removal of electrodes (i.e., during the same testing session). Re-location of electrodes over the same muscle on subsequent occasions will invariably result in the detection of signals from different motor units. The skin–electrode impedance will also differ between sessions, regardless of how well skin preparation techniques are adhered to, which will affect the shape of the underlying signal. These and other factors will, therefore, affect the amplitude of the processed EMG. The amplitude of EMGs recorded from the same muscle on different occasions, as well as from different muscles and different individuals, cannot therefore be compared directly, even if they have been processed using the same method. This problem can be solved by normalizing EMGs after they have been processed; which involves expressing each data point of the processed EMG from the specific task as a proportion or a percentage of the peak EMG from a reference contraction that has been processed in the same way. The reference contraction is usually an isometric sub-maximal or maximal voluntary contraction (MVC) of the same muscle. Use of the EMG from an isometric MVC has the added potential of revealing the activity of a muscle, in relation to its maximal activation capacity, during a specific task (see *Fig. F7.8*). However, in order to achieve reliable and valid EMGs from isometric MVCs indi-

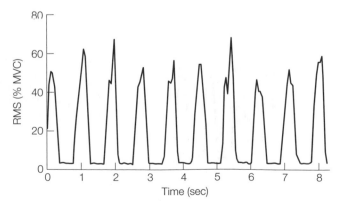

Fig. F7.8.   *The processed EMG in Fig. F7.7 normalized to a maximal voluntary contraction (MVC) of the same muscle*

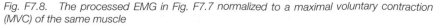

viduals must practice them extensively. Previously unrehearsed MVCs will result in torque or force, and hence muscle activity, that is far from maximal.

Normalization can also been used to reduce inter-individual variability of EMGs recorded from the same task. It is now well established that dividing each data point within the task EMG by either the mean or the peak EMG from the same task is the most effective way of improving group homogeneity. However, due to the nature of the denominator used in their normalization equation, normalizing EMGs in this manner cannot be used to compare the amplitude of EMGs between different muscles and individuals.

Frequency domain processing

Raw EMGs are processed in the frequency domain primarily to investigate changes in the signal that accompany muscular fatigue. It is now well established that fatigue is associated with a compression of the frequency spectrum towards the lower frequencies (see *Fig. F7.9*), that occurs largely due to a decrease in the conduction velocity of action potentials.

Transformation of a raw EMG from the time domain to the frequency domain is typically achieved using a Fast Fourier Transform (FFT), which is usually performed between 0.5 and 1 sec. The output of the FFT is typically represented as the power spectrum density (PSD), which shows the relative magnitudes of the range of frequencies present in the raw signal (see *Fig. F7.9*). One of two parameters is commonly obtained from the PSD in order to quantify it. The median frequency (MDF) is defined as the frequency that divides the PSD into equal halves, and the mean frequency (MNF) is calculated as the sum of the product of the individual frequencies and their own power divided by the total power. The MDF is less sensitive to noise and more sensitive to spectral compression than the MNF and, as such, is more commonly used.

Regardless of which parameter is chosen, it is typically obtained from consecutive time windows, to enable changes in the signal that occur as a consequence of fatigue to be monitored. Successive values from the contraction period are then analyzed using (linear) regression; with the intercept of the regression line being the initial frequency and the gradient representing the fatigue rate. In addition to fatigue, the frequency spectrum of the raw EMG is affected by a host of other factors. Similar to analysis in the time domain, specific frequencies (e.g., MDF) cannot, therefore, be compared directly when they are calculated from EMGs recorded from different muscles or individuals, or from the same muscle when the electrodes have been re-applied. However, comparisons can be made between the gradient of the regression line in order to investigate differences in fatigue rates between different muscles, occasions or individuals.

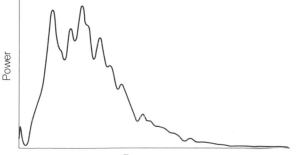

Fig. F7.9. Power Spectrum Density obtained from a section of the raw EMG in Fig. F7.1

The FFT should only be used on EMGs that display high stability; typically those recorded from sustained force isometric contractions between 20% and 80% MVC. EMGs recorded from dynamic contractions typically reduce the stability of the signals, largely as a consequence of recruitment and de-recruitment of different motor units. As such, the FFT should only be used in such circumstances when signal stability is reasonably high and parameters (i.e., MDF or MNF) should only be calculated at the same phase of repetitive cyclic events. The problem of obtaining spectral parameters from non-stationary signals has largely been overcome by using the joint time–frequency domain approach which estimates the change in frequency as a function of time. The simplest method that conforms to this approach is the short-time Fourier transform, which splits the EMG into small continuous or overlapping time windows, applies a FFT to each and calculates the MDF or MNF as above. Recently, more sophisticated methods of time–frequency domain analysis have been applied to EMGs. These include the Wigner–Ville transform from which the instantaneous MDF or MNF is calculated, the Hilbert transform from which the averaged instantaneous frequency is obtained, and wavelet analysis which produces intensity spectra.

F8 ISOKINETIC DYNAMOMETRY

Key Notes

Isokinetic dynamometer	This is a device that is used either to assess or exercise the agonist and antagonist muscles that are located about the joints in the human body. The device can be used to test almost all the joints of the human body. The machine provides a constant pre-determined angular velocity of movement. The limb of the body moves the lever arm of the machine at this pre-determined angular velocity. Throughout the full range of movement the limb will experience an equal and opposite resistance to the force it is applying to the lever arm in order to move it. There are various types of isokinetic dynamometers available and these are usually presented in the form of rehabilitation devices that are located in hospitals or universities. The device can examine the effectiveness of a strength training program that is prescribed following surgery or injury to a joint. Modern isokinetic dynamometers have the ability to test the limb at various speeds and in either an isokinetic, isometric, or isotonic mode of assessment.
Isokinetic	Involves a fixed speed with a variable resistance. This occurs throughout the full range of movement of the limb/lever system.
Isometric	This is usually when the joint or position of the limb is held in a fixed angular position. The muscle develops tension but there is no change in the length of the muscle.
Isotonic	This involves a situation of an equal tension developed in the muscle throughout the exercise. The muscle develops equal tension while the muscle length changes. This is technically a very difficult situation to achieve accurately in practice.
Application	The device can be used in sport to exercise muscles to an optimum strength, or in medicine where it can be used to assess the degree of muscle wasting following injury or surgery to a limb/joint. The device can develop strength in a muscle throughout a full range of movement. The various modes of testing and various pre-determined angular velocities available allow a complex combination of both assessment and exercise. These devices are expensive and are rarely seen in gymnasiums that are solely used for exercise purposes.

The term **isokinetic** is a word used to describe muscle contraction when the rate of movement (velocity) is held constant. The term **dynamometer** is a word used to describe an apparatus for measuring force or power especially during muscular effort. An **isokinetic dynamometer** is a device, which is usually electro-mechanical (both electrical and mechanical) in operation, that assesses isokinetic torque curves of muscles during different movement patterns.

Isokinetic dynamometers

Isokinetic dynamometers are used extensively within many forms of human movement, for example within sports as exercise devices that develop specific muscles and muscle groups, and within rehabilitation and medicine to condition muscles following injury or muscle wasting.

There are many different commercial companies that produce isokinetic dynamometers and they can be used to test almost any joint of the human body. Some examples of the more common commercial isokinetic dynamometer testing machines include: KINCOM; ARIEL; CYBEX; BIODEX and AKRON. All these machines will incorporate a device that will control the rate of movement of the arm-crank of a machine to which the limb or lever of the body is attached. *Fig. F8.1* helps to illustrate this in more detail.

Isokinetic devices can be set up to examine almost any joint within the human body. *Fig. F8.1* shows an application on the shoulder during a flexion and extension movement. The machine in this case would assess the **agonist** and **antagonist** shoulder muscle function. The **agonist muscle** is defined as the muscle that contracts while another muscle resists or counteracts its motion (i.e., the antagonist). The **antagonist muscle** is defined as the muscle that offers a resistance during the action of the agonist muscle. This muscle contraction can take the form of both concentric and eccentric types. **Concentric contraction** is defined as when muscle tension is developed and the muscle shortens. In this case the muscle contracts concentrically and the fibers of the muscle **shorten** (i.e., origin and insertion are drawn together). **Eccentric contraction** is when muscle tension is developed and the muscle lengthens. As the muscle contracts eccentrically its fibers **lengthen** and origin and insertion are drawn apart. During the shoulder movement portrayed in *Fig. F8.1* the machine would assess the torque/strength possessed by both the flexor (pectoralis major and deltoid) and the extensor (latissimus dorsi and teres major) muscles of the shoulder joint.

Muscular contribution to joint stability is invaluable and it is helpful in the prevention of injuries. The knee joint is a typical example where muscular contribution to stability is essential. At the knee joint, muscular stability is provided by the quadriceps (extensors) and the hamstrings (flexors). Following both injury and surgery to the knee joint (such as in the case of a ligament replacement) the muscles usually undergo severe muscle wasting. The isokinetic dynamometer is a device that can provide a measure of the amount of muscle wasting and hence a measure of the rehabilitation that is needed to regain the muscle balance (strength).

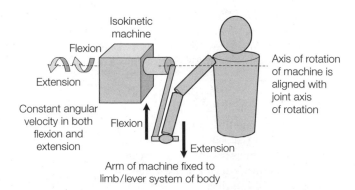

Fig. F8.1. Isokinetic dynamometry (shoulder flexion/extension application)

The isokinetic device usually has the ability to subject the limb to one of three testing or exercise modes: **isokinetic, isometric,** and **isotonic.**

Isokinetic

Isokinetic assessment involves a fixed speed with a variable resistance that accommodates the muscle's ability to generate force. It is characterized by a constant velocity at pre-selected rate. The resistance offered by the machine varies to match the exact torque applied (or created) by the muscle (i.e., a force applied at a distance from an axis of rotation). This also occurs throughout the full range of movement (ROM). The unique application of this form of testing is that the joint is tested throughout its full range of movement (i.e., a situation that is similar to the actual condition within sport or exercise).

Isometric

Isometric or static testing is a situation when the muscle develops tension and there is no muscle length change (muscle contraction against resistance in which the length of the muscle remains the same). Since in these cases the joint is usually held in a fixed angular position (i.e., an isometric exercise where isometric in this context means equality in dimension) the resistance automatically varies to match the force applied. An example of an isometric muscle contraction would be in the case of pushing against a wall.

Isotonic

Isotonic form of testing involves situations where there is equal tension developed in the muscle (i.e., constant force). This type of exercise is technically difficult to achieve correctly and it involves muscle contraction in which the muscle remains under constant tension while the length of the muscle changes. An example of this type of muscle contraction/exercise could possibly be seen in the case of performing free weights (i.e., the movement of the weights and bar in an arm curl exercise). The weights (fixed amount) provide the constant tension and the athlete moves the bar through flexion and extension at the elbow joint (depending on how exactly you exercise with the weights). However, to be strictly defined as isotonic the velocity of the movement would need to be controled (i.e., no acceleration).

Isokinetic dynamometers are used in medicine when the limb requires regular exercising after surgery to restore muscle power and prevent any seizing or limitation of movement. They can also be used to monitor the effectiveness of physiotherapy strength training programs. In sport the device can be used to exercise and develop the power (strength) of certain muscle groups to an optimum level. Maximal exercise can take place throughout a full range of movement. The machine can be adjusted to simulate the exact movement used in the specific sport and even monitor the progress of an athlete in training and/or rehabilitation.

Within the modern gymnasium or training area there are often many machines that use or are labeled with the term isokinetic. However, it is important to identify that these machines are not the same devices as the sophisticated rehabilitation and training models produced by companies such as KINCOM and CYBEX. The machines that are seen in the gymnasium are often only a modified exercise device that uses the principle of Cams (where they are able to change resistance at specific joint angles or position) to create different exercise effects. It would be unusual to see a true isokinetic dynamometer in this environment.

Operation of the isokinetic dynamometer

The isokinetic dynamometer incorporates an electro-mechanical device which keeps the limb at a constant pre-determined angular velocity during the movement. Any effort applied encounters an equal and opposite resistance force. The resistance developed is in proportion to the amount of force exerted. A maximal effort can be experienced as if a maximal load were being applied at all the points throughout the arc of motion. The anatomical axis of rotation of the joint (where the torque is created) is aligned with the machines axis of rotation (where the torque is transmitted). Various arms and levers of adjustable length are available so the device can accommodate most of the joints of the human body and varying sizes of subjects. The results are presented as a measure of torque against angular displacement. Torque values can be plotted against the position of the limb at any instant. *Fig. F8.2* depicts a torque against angular displacement trace for a knee flexion–extension assessment using an isokinetic dynamometer.

Considering *Fig. F8.2* it is possible to see that the trace begins at 90° of knee flexion. As the quadriceps muscle extends the leg to full extension (180°) the device registers the torque generated by the quadriceps. Resistance is offered from the machine at every single point (angular position or displacement) throughout the full range of movement from 90° to full extension (i.e., the 180° position). This resistance matches the torque generated by the muscles. During testing, the limb is set to move at a constant pre-determined angular velocity. This angular velocity can be from as low as 30°/s to in excess of 240°/s. Angular velocities that are low (30°/s) are said to measure the endurance capacity of the muscles. For example, it would be difficult for a muscle to maintain maximum force (torque) against a lever arm that is only moving very slowly (i.e., the contraction and force application would be required for a longer period of time). Conversely angular velocities that are high (> 240°/s) are said to examine the maximum strength capacity of the muscles. For example, at the 240°/s speed the muscle will reach its maximum torque very quickly and it will not have to main-tain this for too long. The torque traces produced by the muscles at different speeds (angular velocities) will be different.

Fig. F8.2 illustrates a trace measured at 120°/s and it is possible to see the maximum torque (force) generated by the quadriceps (left-hand trace) occurs at approximately 110° of knee flexion (or after only 20° of extension movement from the initial 90° flexion position). Considering the torque created by the hamstrings (right-hand side of the figure) it is possible to see that the maximum torque generated is less than that generated by the quadriceps and this occurs in a distinctly different pattern. For example, the quadriceps produce a high acceleration and high peak torque whereas the hamstrings produce less acceleration (i.e., it takes longer for them to accelerate the lever arm to the pre-determined velocity) and a lower peak torque that is maintained for a longer

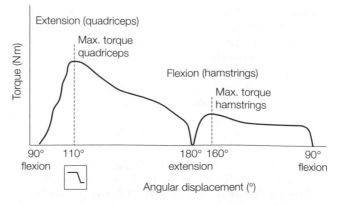

Fig. F8.2.   Torque vs. angular displacement (knee flex/ext) (measured at a test speed of 120°/s)

period of time. This indicates that the device is clearly able to show the differences between the function of the two muscle groups that support the knee joint. In this context it may be important to clarify how acceleration is produced on a machine that provides a constant angular velocity of movement. At the beginning of the movement the limb is stationary and the athlete is told to begin the exercise. The lever arm of the machine must hence be accelerated in order for it to reach this pre-determined angular velocity (i.e., going from a stationary position to one of constant velocity). It is during this phase that the muscles cause an acceleration of the lever arm.

Application

Isokinetic dynamometers are able to measure several variables in relation to muscle function during both exercise and rehabilitation. Such variables as **peak torque, angle of peak torque, time to peak torque, agonist to antagonist muscle strength ratios, work done, power** and **torque decay rate** are found on most modern machines. In addition it is also possible to have different test speeds for different functions of the muscle. For example, it is possible to assess the quadriceps at a test speed of 120°/s during their extension movement and yet at the same time test the hamstrings at 30°/s during their flexion movement. Nevertheless, the machine does have a limitation in this respect and the maximum angular velocity provided by these devices rarely exceeds 300°/s. This is considerably less than the angular velocity encountered during dynamic human movement activities such as in kicking a soccer ball.

However, within isokinetic exercise and testing two problems can occur if they are not accounted for. These are identified as follows:

1. During the test the subject must exert a maximum effort throughout the full range of movement. For example, for the machine to maintain the constant pre-determined test speed (angular velocity) the subject must maintain a maximum effort to the lever arm. In older type isokinetic machines it was possible for the subject to exert an effort that was less than maximum and thus move the lever arm at a test speed that was less than that of the pre-selected rate.
2. In the context of testing using an isokinetic dynamometer it is possible to observe that sometimes the movement under test will be acting against gravity (as in the case of knee extension) and sometimes the movement will be acting with gravity (as in the case of knee flexion). In this case it would be necessary to have to correct the torque generated by the muscles for the effects of resistance or assistance due to gravitational acceleration.

Most modern isokinetic devices have these correction factors built into the software that comes with the machines. It is now possible to see if the pre-set angular velocity has in fact been achieved by the subject at all the points throughout the movement. In this case it would be easy either to correct the subject's movement pattern by encouragement or at least be aware of the errors associated with not reaching actual test speed values. Similarly in the case of gravity correction most modern machines will allow the tester to weigh the subject's limb before testing begins. This information is then input into a formula within the software, which is used to gravity correct all the torque values that are produced.

Isokinetic devices are a significant part of modern exercise testing and prescription and they are becoming essential for rehabilitation following injury. Unfortunately, however, the significant cost of these devices limits their availability to only those technologically advanced and financially viable universities, hospitals, or private industry laboratories.

F9 ANTHROPOMETRY, BIOMECHANICS, AND SPORTS EQUIPMENT DESIGN

Key Notes

Anthropometry	Anthropometry is the measurement and study of the human body, its parts, and capacities. Athletes body shapes are changing, and it is clear that they are becoming stronger, fitter, and faster. These changes occur as a result of the need to improve and the need to accommodate changes in equipment design.
Sports equipment design	The technical demands of sports events are increasing and sports equipment design can be used either to enhance performance or to impede it.
Cycling	Technological advances in the sport of cycling have developed from the sport of triathlon. Both athlete posture and body orientation on the bike are critical components to effective performance. It is clear that both the rider and the bike can significantly affect performance. Current rules and restrictions cause riders to have to modify riding positions in order to achieve success.
Javelin	The "new rules" javelin was introduced for safety reasons in 1986. This introduction caused athletes to have to change their technique to accommodate the new device. As a result the javelin event became more technically demanding. Today, shorter more technical athletes may have the advantage over the taller stronger athletes.
Tennis	Differences in athlete anthropometry can create different requirements in both the skill and the equipment used in tennis. It is evident that taller, stronger players may have an advantage with respect to the service action. Latest equipment is developing at a rapid rate which may allow athletes of different anthropometries to potentially become world champions.

Anthropometry

Anthropometry can be broadly defined as the measurement and study of the human body and its parts and capacities. Biomechanics is one area of study, which combines anthropometry, and the design of sporting equipment particularly well and although the terms biomechanics and anthropometry are relatively new their applications have been used extensively to study, aid, and enhance human movement for over 500 years.

As an example (and continuing from section E5 – Propulsion through a fluid), the new body suit for swimmers has become one of the latest scientific applications in sports equipment design, which is attempting to reduce world records. Back in 1875, while swimming across the English Channel, Mathew

Webb wore a swimsuit that would have weighed around 10 lbs (4.55 kg). In 2004, at the Athens Olympics the Speedo one piece ("Fastskin") costume weighed only a few ounces (0.09 kg), a saving of over 98% in weight since the original costume of 1875. While wearing these suits, it is speculated that there is an 8% lower drag resistance and they are even better than swimming with no costume on at all. This lower drag resistance is achieved through a series of resin stripes or ridges printed on the fabric, which cause tiny vortices of water to form around the suit allowing the body to cut and glide through the water with minimum friction (rather like the skin of a shark). In swimming 90% of the drag resistance is caused by the shape of the swimmer and only 10% is attributed to the friction caused between the skin, the costume and the water. Hence, not only is the suit of critical importance but also is the swimmer's anthropometric body shape. As an example the Australian men's 4 × 200 m freestyle relay team for Athens 2004 had an average height of over 191 cm (over 6 feet 3 inches) and an average weight (mass) of over 83 kg.

Considering these statistical facts it is clear that the sports person's body shape and size is continually changing and rapidly developing. As a result, appropriate modifications are required in sporting equipment. This section examines some recent biomechanical applications towards changes and needs in anthropometry.

Sports equipment design

There is no doubt that sporting equipment can significantly affect the performance of athletes either detrimentally through injury or spectacularly through world record performances. The future will see significant changes in sports surface construction; sport environment development and sport equipment design that will be needed to both prevent records being continually broken and yet sometimes assist in their achievement. Further, there is also no doubt that athletes are getting physically fitter, much stronger and significantly faster and that their specific anthropometry is rapidly changing. As a result, the technical demands of the event and the equipment will dramatically change and it will be interesting to see how the future will respond to these changes.

All this technology must inevitably pose the important social and scientific questions of how far will we go to enhance performance and when does the point arrive that we are measuring the equipment and surroundings rather than the individual athlete? Alternatively, has this day already arrived and it is the "fast" pools like Sydney or Athens and the "sprint" track in Tokyo that will decide how medals are won and lost?

This section concentrates on how sports equipment has changed to accommodate changes in body shape and structure and how body shape has changed to accommodate new sports equipment. Again, for clarity the section will specifically look at selected sports. These will be examples from cycling, javelin, and tennis.

Cycling

Technological advances in cycling have developed primarily from the growth of the sport of triathlon (such as aerobars, steep seat angled frames, and forward riding positions). This has caused the traditional racing time trial posture to be subjected to much experimentation and change. The posture adopted on the bike is a direct function of the cyclist's body configuration and shape (hip, knee, and ankle angles, body position (cyclist relative to pedals) and body orientation (trunk angle with the horizontal)).

This specific change and importance of posture and anthropometry probably first emerged as a significant factor in 1989 when Greg Lemond cycled to a 57 s victory in the final time trial of the Tour de France. Lemond attributed this success and significant average speed of 54.545 km/h to the new aerodynamic riding position and posture.

In 1992 at the Barcelona Olympic Games, Chris Boardman (GBR) shocked the cycling world with an astounding win in the 4000 m individual pursuit event. This win with its new "cycling position" and "technological bike" from Lotus created a revolution in bike design characteristics. The position adopted by Boardman allowed the rider to rest the arms on the tri-handlebar arrangement, which allowed an almost perfect time trailing combination of both athlete and bike (*Fig. F9.1*).

This advance into bike design and body position continued to develop rapidly after this historic ride by Chris Boardman in 1992, and many new cycling positions and innovative bikes were suddenly seen evolving. However, in 1993, while riding a homemade bike put together from spare mechanical parts, one rider, Graeme Obree, established a riding position that was to stay and change the shape of time trialing yet again. The "Obree" position allowed the rider to rest the torso on the arms, which were tucked away underneath, thus totally eliminating them from the aerodynamic equation and reducing the drag by 15%. This theoretically would give a speed gain of more than 2 km/h at speeds of 50 km/h. In addition, as a result of the particularly narrow bottom bracket he managed to reduce the drag even further by riding almost totally "knock-kneed".

In 1993, Graeme Obree set the 4000 m single pursuit time to a new world record of 4 min 20.9 s beating the record previously held by Chris Boardman. The following year saw the "Obree" position adopted by many other riders together with the same and also often modified bike design (chest pad added to aid comfort on the longer rides). Despite this interest and acceptance by many riders, this new bike design and anthropometric racing position could still not conquer the 1 h speed record set previously by Chris Boardman on the Lotus bike. Nevertheless, on 27 April 1994 Graeme Obree returned using the new "Obree" position, to set the 1 h speed record to an astounding 52.513 km/h.

Unfortunately, this success was to be limited, as in May 1994 the UCI (*Union Cycliste Internationale*) decided that the position would be banned from the forthcoming world championships. The position was still used in the triathlete circuit where the UCI rules were not valid and many riders were seen taking between 3 and 5 s off each of their kilometer times.

Following the UCI ban Graeme Obree returned to the cycling circuit in 1995 with a new modified position and bike. This time, Obree used conventional equipment: a normal bike with a particularly long stem, with aerobars extended to their limits. The arms were now in an outstretched position with the hands

Fig. F9.1. The cycling position first shown in 1992

about 30 cm in front of the front wheel hub. This position was now adopted and termed what is known today as the "superman position".

During the Atlanta Olympic Games in 1996, many other riders used the "superman position" and the position achieved great success in both the men and women's 4000 m individual single pursuit titles. The world record for 4000 m had now been improved to a time of 4 min 19 s.

In the same year at the world championships in Manchester, England, Chris Boardman was seen with an addition to his Lotus bike of 1992, which included a custom-built handle bar that allowed him also to adopt the new "superman position". This combination of machine, athlete, and aerodynamics was too much and the cycling world saw the 4000 m individual time trial record plummet to 4 min 11.114 s. Then to add to this spectacular performance and only 1 week after the world championships, Chris Boardman then set the 1 h speed record to an outstanding 56.375 km/h.

However, to add more controversy and confusion, the UCI then decided also to ban this "superman position" and they created rules that stated the handlebar must not exceed the front wheel hub by more than 15 cm. In addition, they also added that the distance between the front wheel hub and the bottom bracket (i.e., at the pedals) could now be 75 cm. This value was exceptionally long considering that most conventional bikes had only 60 cm distances. This new value of 75 cm appeared to put the specifications at a ridiculous level and all that was needed for the new specifications to adopt the banned "superman position" was to lengthen the front part of the frame. However, it was still particularly difficult to achieve the exact same "superman position" previously adopted by Graeme Obree, Chris Boardman, and now many other riders.

The years 1997 to 1999 saw a considerable number of changes to the rules presented by the UCI and in 1999; they changed the specifications yet again. This time the maximum distance permitted between the front wheel hub and the bottom bracket (i.e., at the pedals) was to be 65 cm. This change had a dramatic effect and it meant that now it was only possible for smaller riders to have any chance of reaching the previously successful "superman position".

In the year 2008 at the Beijing Olympic Games we will see bike specification and athlete position acquire more constraints, and perhaps even preventing athletes from achieving what is considered to be any "normal" aerodynamic position. Handlebar extensions may now not project more than 10 cm past a vertical line which passes through the front wheel spindle (i.e., the front wheel hub), which is a new regulation that means taller riders have no chance of adopting any comfortable aerodynamic posture and the previous record breaking "superman position" is, for these riders, now totally impossible to achieve.

Such significant changes in the specifications by the UCI and the adoption of only "standard" frame design will mean that the anthropometry of world class riders will need to change. As a result record breakers will not be tall athletes but will require the athlete to have a stature that suits the bike specifications and the need to achieve some degree of aerodynamic control within this new restriction. Perhaps it may even be the case that the old records set with the "superman position" will become like many other records set in conditions that can now not be repeated. Thus, becoming a part of sport history when athletes, coaches and spectators will only talk of records that were achieved at a specific track, with a particular bike and with a certain type of athlete. Perhaps this day has already arrived and it is the equipment and conditions that are being judged rather than the athlete who is taking part?

Search the Internet to see if you can find the new specifications presented by the *Union Cycliste Internationale* for the Beijing Olympic Games in 2008.

Javelin

The men's javelin event and the method adopted in throwing it is a subject that has a particularly stringent set of specifications, rules, and conditions. Indeed, optimization of training techniques has resulted in significant improvements in the physical skills of the athlete. In addition, the evolution of the javelin from a wooden implement to a device made of light alloys, that appears to float in the air, has also contributed to this dramatic development. *Fig. F9.2* identifies the men's javelin world record performance between the years 1912 and 1996.

In the years previous to 1984, the world record for the men's javelin event was set by Tom Petranoff (USA) at a distance of 99.72 m. At this time, the IAAF (International Amateur Athletics Federation) expressed serious concern over the distances that the male javelin throwers were achieving. The javelin would often "float" through the air and then hit the ground and even slide onto the running track, making the event dangerous for other competitors. In some cases, such as with a strong cross-wind, the javelin would often land directly on the running track. Then in 1984, an athlete from the German Democratic Republic, Uwe Hohn threw the implement a massive distance of 104.80 m. This performance finally convinced the IAAF to change the javelin specifications in order to make the event safer.

The new specifications javelin was introduced in April 1986 after much controversy and confusion. The new device was to possess the same weight (mass) as the old javelin but the center of gravity would be moved 40 mm towards the metallic head of the spear. These factors in aerodynamics terms meant that center of pressure in different angles of attack would be behind the center of gravity, thus preventing it from "floating" as in the old model. Furthermore, it was speculated that based on a given set of release parameters, the new javelin could only achieve distances that were 10% less than those acquired with the old model. In addition, the new javelin would always land point first, making it easier to record and also stopping it from sliding

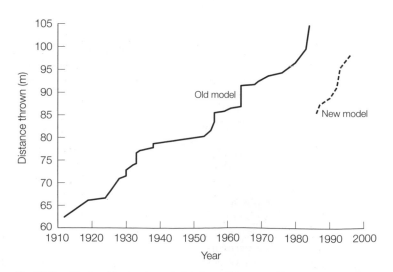

Fig. F9.2.   The world record men's javelin performance 1912 to 1996

dangerously into the running track. Finally, it was also speculated that the device would not be directly affected by cross winds and should therefore travel in a straight line.

Around the time of this change many athletes expressed both genuine concern and a degree of confusion as to the effects of the new model on the sport. Many athletes thought that it would stop the event being technical and it would now be dominated by the taller, heavier and stronger men. It was often stated that it was now an unfair event as the smaller, more technical throwers would not have any chance of winning the competitions using the new rules javelin.

The ideal javelin athlete should have a combination of the components of speed, strength, coordination, flexibility, and a good throwing arm with a kinesthetic "feel" or "sense". Furthermore, several mechanical factors are also critical and these can be briefly summarized as: release speed; release angle; release angle of attack; release angle of attitude; front foot to foul line distance; angular velocity components about the longitudinal axis (spin); perpendicular horizontal axis (pitch); and an axis mutually perpendicular to these two (yaw).

Considering these factors it is important to point out that almost all of them can be affected by the anthropometric composition of the athlete. According to some researchers the single most important component for success in javelin is release velocity, however there are also many researchers who disagree with this statement and claim that there are a number of critical components to throwing success.

One clear anthropometric factor that will be affected by physical stature and condition of the athlete is termed height of release. *Table F9.1* identifies the comparison of the height of release variable using both the old and the new specification javelin.

Considering *Table F9.1*, it is interesting to note that the height of release of the javelin using the new specification model has actually been reduced in order to throw the implement with any degree of success. It is also interesting to note that Petranoff is the only athlete to appear on both lists, perhaps indicating that he was one of the few athletes who could make the transition from the old model to using the new rules javelin.

Table F9.1   Height of release variable for "old" and "new" model javelin

Athlete	Old type javelin Height of release (m)	Athlete	New "rules" javelin Height of release (m)
Nemeth	2.05	Raty	1.81
Megla	2.21	Zelezny	1.64
Ershov	1.97	Petranoff	1.72
Olson	1.82	Yevsyukov	1.71
Colson	2.01	Hill	1.69
Lusis	1.86	Mizoguchi	1.57
Luke	1.91	Wennlund	1.69
Zirnis	1.68	Shatilo	1.81
Petranoff	2.09		
Mean	1.96		1.71
SD	0.16		0.01

Using the old specification javelin, it has been suggested by researchers that there was a correlation between speed of release and the distance thrown. This correlation was said to be 0.93. Whereas the correlation between speed of release and distance thrown with the new implement may be significantly lower, with values being reported to be in the range 0.80–0.87. Further, research on the topic went on to suggest that the technical requirements of the new javelin may actually be greater than those of the old model, which was in fact opposite to what most athletes and coaches believed.

In 1987 the IAAF produced a technical report from the world championships, which used the new rules javelin. Raty (Finland) won the event with a release velocity of 29.6 m/s, Zelezny (USSR) with the highest release velocity came third and Hill (GBR) with the next highest release velocity came 7th.

In summary, it would appear that the new rules javelin event was indeed not dominated by the bigger, taller and stronger athletes and that the event had actually become more technical. The difference in distance between the world number 1 and the 50th place in 1985 with the old javelin was 12 m. In 1986, this gap was reduced to only 8 m, making the competition much closer and perhaps even reflecting greater accuracy in recording the distances by the officials. However, what is clear is that some throwers managed to adapt to the new javelin better than others. Two examples of this were seen in the athletes Yevsyukov (USSR) and Gampke (GDR) who following the introduction of the new implement became world-class performers. Perhaps this clear adaptation was due to the anthropometric composition of these athletes, which were better suited to the technical demands of the new device.

Search the Internet to see if you can find out what is the current world record for both the men and women's javelin event.

Tennis

The tennis serve is probably the most important stroke in the game. The action involves a smooth, coordinated movement of different body parts delivered at an optimum height. The body adopts a link system of movement, initiated from the legs, which produces increases in velocities from one segment to the next. Finally, this velocity is transferred to the hand and racket resulting in maximum desired power of service.

According to empirical research the tennis serve makes up to 20–30% of all shots in both singles and doubles and accounts for 12% of winning shots on clay and 23% of winning shots on grass. Furthermore, it is the only stroke that the opponent cannot directly affect and therefore it places the server at a distinct advantage.

The inter-relationships between different body positions, different types of service delivery (flat, topspin, and slice), different racket trajectories, segment velocities and spin have all been analyzed previously. However, it is clear that there is limited published research that directly measures changes in these parameters due to contrasting anthropometry.

For an effective execution of the flat serve it is necessary for the ball to be contacted as high above the ground as possible. This allows the ball to be struck at a flat or downward trajectory. According to researchers this requires a contact height of approximately 3 m to hit the service area with an adequate margin for error. However, few players can achieve this height, so the optimum serve becomes a compromise of compensation by varying projection angles by between 4° and 7°. *Fig. F9.3* illustrates the service action and identifies this

Fig. F9.3.   The modern elite tennis service action

height of contact position requirement and this flat or downward trajectory in more detail.

In order to achieve maximum contact height during the service, almost maximum extension of the joints such as the knee, hip, and elbow (180°) is required. Research has suggested that contact heights in effective serves are approximately 150% of the standing height, with joint extensions for the elbow and knee at approximately 173° and 165° respectively. Further research work in the area has identified that a number of elite players are actually off the ground at the moment of ball and racket contact during the service action. This results from the rigorous leg drive towards the ball, which increases contact height.

It is suggested by some that the taller players will serve harder, with a higher velocity and with a much more controlled "flatter" trajectory. The smaller players will therefore have to assume the "up" and "out" service method and hitting the ball at a much reduced velocity. Furthermore, it is proposed that because of this clear anthropometric difference, the smaller players will need to use the "foot up" technique of serving in order to try and increase the impact height relative to their standing posture. The taller players could therefore use the "foot back" technique and hence assure faster progression towards the net, as, for example, in the case of Pete Sampras from the USA.

As a result of the need for effective height during the serve action different players, depending upon their anthropometry and technique, will be naturally better at serving than others. *Tables F9.2* and *F9.3* indicate the basic anthropometry (height and weight (mass)) of some of the world's top tennis players and the basic anthropometry of the fastest servers on the tennis circuit in the year 1999.

It is interesting to note that the average height of the male and female top five seeds in the 1999 US Open ATA (Association of Tennis Professionals) rankings are 6 feet 1 inch for the men and 5 feet 11 inches for the women (indicating little difference between the sexes).

Again, it is interesting to point out that of the men's fastest servers at this time none of them were in the top five seeds of the 1999 US Open ATA rankings, whereas in the women's data both Venus Williams and Monica Seles were both

Table F9.2   Basic anthropometry of the 1999 US Open top five seeded players

Athlete (seeded)	Height (ft in)	Weight (mass) – (lbs)	Age (yrs)
Male			
Pete Sampras (1)	6' 1"	170	27
Andre Agassi (2)	5' 11"	165	29
Yevgeny Kafelnikov (3)	6' 3"	179	25
Patrick Rafter (4)	6' 1"	175	26
Gustavo Kuerten (5)	6' 3"	167	22
Female			
Martina Hingis (1)	5' 7"	130	18
Lindsay Davenport (2)	6' 2"	175	22
Venus Williams (3)	6' 1"	168	19
Monica Seles (4)	5' 10"	155	25
Mary Pierce (5)	5' 10"	150	24

Table F9.3   Basic anthropometry of some of the fastest servers in the world in 1999

Athlete	Height (ft in)	Weight (mass) (lbs)	Service speed (mph)
Male			
Greg Rusedski	6' 4"	190	143
Mark Philippoussis	6' 4"	202	142
Julian Alonso	6' 1"	180	140
Richard Krajicek	6' 5"	190	139
Female			
Venus Williams	6' 1"	168	124
Brenda Schultz–McCarthy	6' 2"	170	123
Jana Novotna	5' 9"	139	116
Kristie Boogert	5' 10"	142	111
Monica Seles	5' 10"	155	109

seeded. However, it is often argued that the great success achieved by Pete Sampras (USA) was attributed to the powerful and accurate serve and volley technique, which allows him clearly to dictate the pace of the game. Similarly, Lindsay Davenport was also said to possess strokes which needed less power in her racket. Both Sampras and Davenport have fast swing speeds and long "loopy" type strokes, which is only usually observed in around 15% of top tennis players today.

Consequently, because of these different anthropometric components (such as the ability to possess long fast powerful strokes) within tennis players, each player will adopt a technique and indeed a racket that suits their own individual style and anthropometry in order to achieve optimum performance. For example, players like Lleyton Hewitt (AUS), at a height of 5 feet 9 inches, are significantly smaller than the average height of 6 feet 1 inch for the US Open top five seeds in 1999, yet he is still a very successful athlete and he has in more recent years regularly been rated as the world number one player. Hence he must have adopted a technique and indeed equipment that benefits his particular game.

The tennis racket has evolved dramatically since the wooden rackets of the 1970s, when the most popular choice was Canadian Ashwood, which was cut into long strips and then steam glued and pressed together. In addition, metal rackets molded from aluminum (chosen for the high strength to weight ratio) were also a popular choice among players, for example, Jimmy Connors of the USA in the late 1970s. The typical racket of 1970 possessed a string area of 70 square inches (450 cm²); it had a weight (mass) of 12.5 ounces and a racket frame of approximately 18 mm deep. Today, tennis rackets are complex highly engineered components that are subjected to much research and development. However, the question that should be asked is: can these technological advances in racket design really make up for the significant differences in performance created from different anthropometry and in particular differences in serving power and speed?

In 1976, the Prince racket emerged and was to create a revolution in tennis racket design and construction. The aluminum Prince racket head was almost double the original size with a string area of 130 square inches (839 cm²). The resulting years saw many copies of the Prince racket develop, some of which had an even larger surface area. This continued until 1980 when the ITF (International Tennis Federation) limited the string area to a maximum of 15.5 inches in length (approx 40 cm) and 11.5 inches in width (approx 30 cm). This was hence to set a theoretical maximum available area of 178 square inches (1148 cm²).

This standardization had a dramatic effect on the industry and the resulting years saw manufacturers molding rackets of graphite and glass fiber, which produced a racket that was to be far more powerful than the older wooden rackets, yet at the same time was also significantly lighter.

At this time research into racket design at Pennsylvania University in the USA showed that increasing the mass of the racket head by 33% produced only a 5% rise in the speed of the ball once it had been struck. However, a 33% increase in racket head speed increased the ball speed by 31%. This clearly showed the important link between the athlete and the racket (as the athlete is required to accelerate the racket to the ball) and the critical importance of lighter rackets (i.e., so the athlete can use the same force but move the racket much quicker). In 1984, racket development then observed another significant change when an inventor from Southern Germany (Siegfried Kuebler) developed the wide-bodied frame. The theory was that the racket could be made even stiffer and lighter by increasing the frame depth or side cross-section, while narrowing the front profile. Basically, the stiffer the racket the more energy is returned to the ball. When ball and racket connect, there is only a short time of contact and any flex in the racket is wasted energy.

Stiffer rackets generate more power and also have significantly larger "sweet spots", which help with directional control. The "sweet spot" is considered to be the most responsive part of the racket's frame. If a ball and racket make contact at the "sweet spot" it produces the most powerful shot with no, or very little, vibration or shock. Current rackets will have a very large "sweet spot", which is nowadays located at the top part of the racket where most top players in the modern game will hit the ball.

As with the Prince racket design many manufacturers rushed into producing racket head frames up to 39 mm deep and only 10 mm wide. Materials such as kevlar and complex thermo-plastics were used in construction which helped allow the strings of the racket to act almost independently of the frame, leading to a reduced ball contact time, less lost energy and consequently more power.

In the year 1992, the modern racket had a surface area of 115 square inches ($742\ cm^2$); it was 39mm thick with an aerodynamic profile and all with a weight (mass) of only 10 ounces. Compared with the wooden racket of the 1970s, it was 64% bigger, 116% thicker and yet 20% lighter. The racket of the future will have an even larger "sweet spot" and it will probably contain multiple "sweet spots", which will account for shots that are even miss-hit. This racket will provide good ball speed without effort, will be easy to maneuver, be aerodynamically efficient, and will not cause unnecessary fatigue. It will be constructed of titanium or hyper carbon and finally will not be responsible for any injury to the athlete. To add to all this it will interact with the anthropometry of the player providing performance or rebound efficiency, movability, precision, power and at the same time comfort.

Today's elite tennis players will select a racket that is carefully balanced to their particular needs and indeed to some extent individual anthropometry. For example, if they are a powerful serve and volley player it is likely that this player will use a different racket to a player whose strength is to play from the baseline. This customization and precise selectivity will continue to play a critical role in the development and progression of the sport of tennis into and beyond the 21st century.

References

Bartlett, R. M. and Best, R. J. (1988) The Biomechanics of Javelin Throwing: a review. *Journal of Sports Sciences* **6**, 1–38.

Elliot, B. (1996) The Super Servers: Pete Sampras and Goran Ivanisevic have two of the fastest and most feared serves in men's tennis. *Australian Tennis Magazine* **21** (6), 46–47.

Faria, I. E. (1992) Energy expenditure, aerodynamics and medical problems in cycling. *Sports Medicine* **14** (1), 43–63.

APPENDIX 1 FREE BODY DIAGRAMS

Free body diagrams are pictures (diagrams) of forces acting on a body. They allow us to be able to analyze the effect of all the external forces acting on a body more easily (i.e., the effect of the net force). As we have seen within human movement there are a number of different types of forces that can act on a body: gravitational force (weight); frictional force; normal reaction forces; applied contact forces; tensile, shear and compressive forces; muscle and joint forces; and centripetal, tangential and centrifugal forces. In human movement it is often the case that several forces will act on the body simultaneously. As we have seen earlier, force is a vector quantity and thus a force can be expressed or represented by lines with both magnitude and direction. The net effect of these forces (the resultant) acting on a body can be determined through representing all the forces acting on a body using a **free body diagram**.

In drawing free body diagrams there are a number of steps that we should go through in order to assist us in the accurate representation of all the forces acting on a system or body. These can be outlined as follows:

First: isolate the body from its surroundings. Then draw the body upon which the forces are considered to act. For example, if we are interested in the forces on the lower arm (the ulna and radius) then only draw the ulna and radius. Do not draw any other body that the body of interest may or may not be in contact with.

Second: take time to identify all the external forces that are acting on the body or system. This is usually the most difficult part. As a guide, it is useful to systematically go through the different forces that could be acting. For example, if the body has mass then there will be a weight force acting and this will be through the center of mass of the body; if the body is in contact with any other body there must be a normal force (acting perpendicular to the surfaces in contact) and a frictional force (acting along/parallel to the surface of contact) acting between the two bodies; when there is no rotational component of force (torque) the position and location of our force vectors is not so important so long as we maintain consistent lines and directions of force (orientation) application; when expressing forces on a free body diagram the line of action of the force is located through the point of application; if two forces are equal and opposite and they lie on the same line of action the resultant effect of the forces will be zero and we could represent these forces anywhere along the line of action; often it is useful to break the forces down into their component parts (i.e., horizontal and vertical components of a resultant force).

Third: once all the components in the second stage have been classified then the next stage is to draw all these forces on the diagram and include the magnitude and direction (angles and orientations) of each known force. In addition represent all the unknown forces on the diagram.

Finally: select a coordinate system of conventional representation in order to identify the positive and negative components of the force.

Note: if we are concerned with torques and moments the positioning of these components on the free body diagram is critical.

Example   A mass (i.e., a body) is being pushed up an inclined plane by a horizontally directed force (i.e., parallel to the inclined plane). Draw the free body diagram of the external forces acting on the body.

Free body diagram

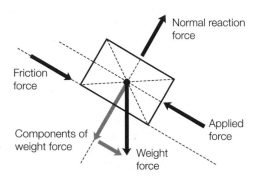

During the analysis of movement it is often important to collect data at more than one single moment. Indeed, in most biomechanical analyses it is important to be able to record changes in the key variables over a period of time. Most measurement devices work by sampling and recording data at regular intervals during the measurement period. The number of samples in any given period is known as the **sample frequency** and is usually recorded in **hertz (Hz)**, that is, the number of measures per second. For example most video recorders operate at 25 frames per second (25 Hz) which can allow for analysis on a field by field basis at 50 Hz, while it is common to sample force platform data at around 1000 Hz.

Consider the following simple example to illustrate the effect of sample rate on the data recorded. If the changes in the knee angle and the vertical ground reaction force during a drop-jump take-off are recorded, the input signal being recorded is the knee angle and this is changing in a continuous fashion, in other words there is always a knee angle present that could be recorded. When data are sampled a series of "snap shots" of these data are recorded creating a discontinuous record of the magnitude of the knee angle (see *Fig. App2.1*).

By looking at these two graphs it is possible to see that whilst it may be acceptable to sample the knee angle data at 25 Hz it would not be acceptable for the force data as key moments in the input signal are missed. This is because the knee angle and forces change at different rates and are said to have a different **frequency content**.

The frequency content describes the make-up of the signal and reflects the rate at which changes in the magnitude of the variable happen. In the case of the knee angle data it is clear that the knee angle undergoes a gradual change from flexion to extension with only one major direction change in the movement. On the other hand the force data are more complex with more rapid changes in the magnitude.

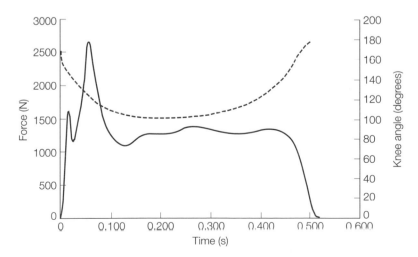

Fig. App2.1   Knee angle and vertical ground reaction forces during a drop-jump. Original data sampled at 50 Hz for the knee angle and 1000 Hz for the force. Note the knee angle would be largely unaffected by the change in sample rate whilst the force data would produce a very different result particularly for the early part of the movement

In order to record all such changes it is important to sample at a sufficiently fast rate.

To understand better frequency content it is often helpful to consider how a signal may be composed. *Fig. App2.2* shows how three separate signals can be added together to form a composite, by adding together the three sets of data. If the three input signals are sampled separately these would create the three data

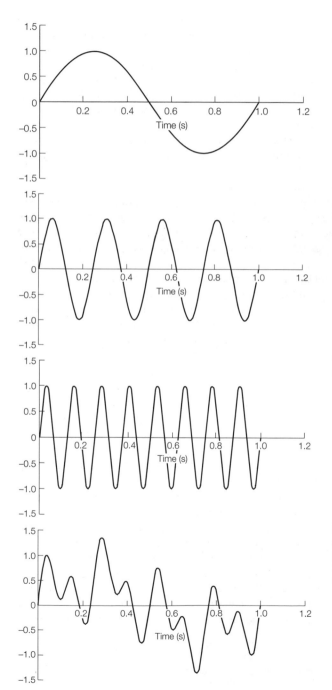

Wave 1. This wave has a frequency of 1 Hz as it undergoes a single cycle in 1s

Wave 2. This wave has a frequency of 4 Hz as it experiences four cycles in the 1 s period

Wave 3. This wave has a frequency of 8 Hz experiencing eight cycles in the same 1 s period

Wave 4. This is the composite of the three waves above. By adding the data from the first three waves this more complex wave form is produced. It contains 1 Hz, 4 Hz and 8 Hz components and has a bandwidth of 7 Hz (8 – 1)

sets shown in the figure; each has a different frequency of oscillation. If these three signals were each part of a single composite signal then the data set produced would be the summation of the three independent signals at each moment in time (i.e., at each sample moment the three input signals would be added together). The final wave form would look different from each of the three initial graphs. The frequency content of the final waveform would have a lower value of 1 Hz and an upper value of 8 Hz.

Before starting an experiment it is therefore important that some thought is given to selecting the most appropriate sample frequency. This can be done by applying the **sampling theorem.** The sampling theorem is sometimes called the **Nyquist–Shannon sampling theorem** or **Whittaker–Nyquist–Kotelnikov–Shannon sampling theorem** after the scientists credited with its development. The theorem was first formulated by Harry Nyquist in 1928 but was only formally proved by Claude E. Shannon in 1949. The theorem states that:

The sampling frequency must be greater than twice the bandwidth of the input signal in order to be able to reconstruct the original perfectly from the sampled version.

It is important to note that the theorem refers to the **bandwidth** of the signal and not simply to the greatest frequency within the signal. The bandwidth is considered to be the range captured between the highest and lowest frequencies in the signal. In Fig. App2.2 this would refer to the range between 1 Hz and 8 Hz. Thus the bandwidth is 7 Hz, whereas the upper frequency would give 8 Hz. Whilst in human movement the law is often simplified by only considering the upper frequency value, there are other sampling conditions where consideration of the actual bandwidth is critical to ensure that the appropriate sample frequency is used.

If the sampling condition is not satisfied, then frequencies will overlap and the nature of the recorded signal will be different from the input signal. This overlap is called **aliasing.** To prevent aliasing, either 1) increase the sampling frequency or 2) introduce an anti-aliasing filter or make the anti-aliasing filter more stringent. The anti-aliasing filter is used to restrict the bandwidth of the signal to satisfy the sampling condition. This holds in theory, but cannot be satisfied in practice as there may be some elements of the real signal that fall outside of the sampled range and thus the recorded signal will not include all of the real signal. However, in most situations the amount of information lost may be small enough that the aliasing effects are negligible.

If the sampling frequency is exactly twice the highest frequency of the input signal, then phase mismatches between the sampler and the signal will distort the signal. For example, sampling $\cos(\pi * t)$ at $t = 0,1,2...$ will give a discrete signal $\cos(\pi * n)$, as desired. However, sampling the same signal at $t = 0.5,1.5,2.5...$ will generate a constant zero signal, because the cosine of 90°, 270°, 450° (0.5 π, 1.5 π, 2.5 π) and so on, will be zero. These two sets of samples, which differ only in phase and not in frequency, give dramatically different results because they sample at exactly the critical frequency. It is thus important that the selected sample frequency is **more** than twice the signal bandwidth and not exactly twice this value.

Appendix III MATHS REVISION: ALGEBRAIC MANIPULATION

The revision examples within this section are those that are commonly used within biomechanics and are representative of some of those that have been used within this text.

Algebra

Algebra refers to the branch of mathematics that generalizes arithmetic by using variables for numbers (i.e., $x + y = y + x$).

The Rules of Signs

+	×	+	=	+	PLUS
+	×	−	=	−	MINUS
−	×	+	=	−	MINUS
−	×	−	=	+	PLUS

Any number multiplied by zero (0) equals zero (0)

Example of multiplication of different signs

$$-8 \quad \times \quad 3 \quad \times \quad -6 \quad =$$

Carry out parts of the calculation first and introduce brackets.

(−8	×	3)	×	−6	=
(−8	×	3)	=	−24	(Part 1)
−24	×	−6	=	+144	(Part 2)

(Note: it is not normally necessary to put the plus sign before a number)

Summary of rules for division of positive and negative integers
An integer is classified as a number that may be expressed as the sum or difference of two natural numbers. A natural number is any positive integer (i.e., 1, 2, 3, 4 etc.).

+a	÷	+b	=	+	(a/b)
+a	÷	−b	=	−	(a/b)
−a	÷	+b	=	−	(a/b)
−a	÷	−b	=	+	(a/b)

Example

64	÷	8	÷	−2	÷	2	=	
(64/8)	÷	−2	÷	2	=			
(8/−2)	÷	2	=					
−4	÷	2	=	−2				

Solution of problems involving two or more arithmetic operations

Rules of precedence:

1. Evaluate terms in brackets
2. Multiplication and division
3. Addition and subtraction

Order of working can be remembered by using the BODMAS rule

B	O	D	M	A	S
( )	of	÷	x	+	−
First					**Last**

BODMAS tells you the order in which to perform calculations if you have a choice. Brackets first, then of (such as square root **of** 4 or 3 to the power **of** 5 (i.e., operations)), then divide, multiply, add and subtract.

Example

3	(2	+	5)	÷	6	(7	−	4)	=	
3	(7)			+	6	(3)			=	
21				+		18			=	39

Percentages

32% of 69

$$= \frac{32 \times 69}{100}$$

$$= \underline{\mathbf{22.08}}$$

What % of 79 is 37?

$$\frac{37}{79} \times 100$$

$$= \underline{\mathbf{46.84\%}}$$

Decimals

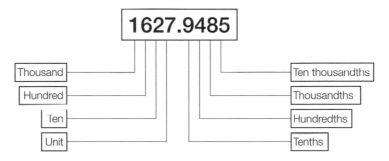

Stating to the correct number of specified decimal places

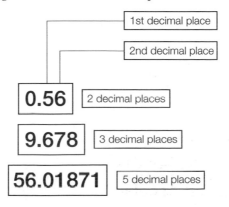

Rules

The last decimal place that is specified is **unchanged** if the digit that follows it is 4 or less. The last decimal place specified is **increased** by 1 if the digit that follows it is 5 or more.

Example

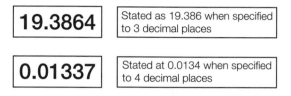

19.3864   Stated as 19.386 when specified to 3 decimal places

0.01337   Stated at 0.0134 when specified to 4 decimal places

Powers

4×4 Raised to the power of 2 or squared
$4 \times 4 \times 4$ Raised to the power of 3 or cubed
$6 \times 6 \times 6 \times 6$ Raised to the power of 4

Written expression

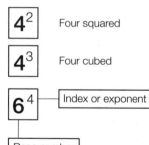

4^2 Four squared

4^3 Four cubed

6^4 ── Index or exponent

── Base number

Rules of indices

Multiplication rule

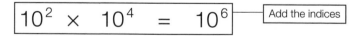

$$10^2 \times 10^4 = 10^6$$ Add the indices

This works when the base numbers are the same

$$2^2 \times 2^4 = 2^6$$
$$4 \times 16 = 64$$

However when the base numbers are not the same

$$3^4 \times 5^5 = 81 \times 3125$$

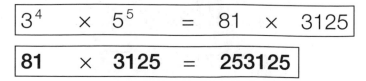

$$81 \times 3125 = 253125$$

Any number expressed to the power of zero equals 1.

$$6.65^0 = 1$$
$$3^0 = 1$$

Division rule

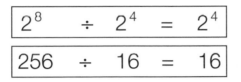

$$10^5 \div 10^2 = 10^3$$ Subtract the indices

This works when the base numbers are the same

$$2^8 \div 2^4 = 2^4$$

$$256 \div 16 = 16$$

However when the base numbers are not the same

$$8^4 \div 3^6 = 4096 \div 729$$

$$4096 \div 729 = 5.62$$ (2 decimal places)

Raising indices to a power rule

Multiply indices

$$(10^3)^2 = 10^{3 \times 2} = 10^6$$

$$(8^4)^3 = 8^{4 \times 3} = 8^{12}$$

Summary

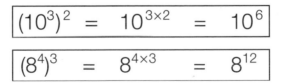

MULTIPLICATION $a^m \times a^n = a^{m+n}$ add indices

DIVISION $a^m \div a^n = a^{m-n}$ subtract indices

Applies when base numbers are the same

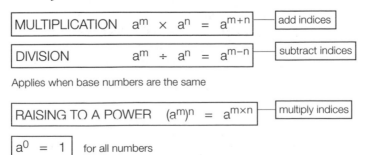

RAISING TO A POWER $(a^m)^n = a^{m \times n}$ multiply indices

$$a^0 = 1$$ for all numbers

Transposition of formula

$$6x - 10 = \frac{3x}{3}$$

Solve for x

Multiply both sides by 3 to remove the fractional component

$$3(6x - 10) = 3 \; \frac{3x}{3}$$

Cancel out

$$3(6x - 10) = \cancel{3} \; \frac{3x}{\cancel{3}}$$

$$3(6x - 10) = 3x$$

$$
\begin{aligned}
3(6x - 10) &= 3x \\
18x - 30 &= 3x \\
18x - 3x - 30 &= 0 \\
18x - 3x &= 30 \\
15x &= 30 \\
x &= \frac{30}{15} \\
x &= 2
\end{aligned}
$$

$$
\begin{aligned}
3x + 4 &= 2 \\
3x &= 2 - 4 \\
x &= \frac{2 - 4}{3} \\
x &= -\frac{2}{3}
\end{aligned}
$$

Example

$$
\begin{aligned}
9x + 8 - 4 &= 6x \\
9x - 6x + 8 - 4 &= 0 \\
9x - 6x &= -8 + 4 \\
3x &= -4 \\
x &= -\frac{4}{3} \\
x &= -1 \cdot \dot{3}\dot{3}
\end{aligned}
$$

General rules regarding transposition of formula

Negative quantity on one side of the equation becomes a positive quantity when it is transferred (transposed) to the other side of the equation. Similarly a divisor on one side of the equation becomes a multiplier when it is transferred to the other side of the equation (i.e., either side of the equal sign).

Solve for x

$8(x + 2) = 3(x - 3) + 45$

Multiply out the brackets

$8x + 16 = 3x - 9 + 45$

Transpose formula

$8x - 3x = -9 + 45 - 16$

$5x = 20$

$x = \dfrac{20}{5}$

$x = 4$

Solve for x when

$$\dfrac{4x + 9}{3} + \dfrac{3x + 7}{4} = 11$$

Determine the lowest common denominator $= 3 \times 4$

Reduce the fractions by multiplying both sides 3×4

$$3 \times 4 \left(\dfrac{4x + 9}{3} + \dfrac{3x + 7}{4}\right) = 3 \times 4 \times 11$$

Multiply out brackets

$$3 \times 4 \times \left(\dfrac{4x + 9}{3}\right) + 3 \times 4 \left(\dfrac{3x + 7}{4}\right) = 3 \times 4 \times 11$$

Cancel out where possible

$$\cancel{3} \times 4 \times \left(\dfrac{4x + 9}{\cancel{3}}\right) + 3 \times \cancel{4} \left(\dfrac{3x + 7}{\cancel{4}}\right) = 3 \times 4 \times 11$$

Multiply out brackets

$4(4x + 9) + 3(3x + 7) = 3 \times 4 \times 11$

$16x + 36 + 9x + 21 = 3 \times 4 \times 11$

$16x + 9x = (12 \times 11) - 36 - 21$

$25x \quad - 132 - 57$

$25x = 75$

$x = \dfrac{75}{25}$

$x = 3$

Powers

ay^2 means $a \times y \times y$

$2ay$ means $2 \times a \times y$

Solution of an Expression

Example 1

Find the value of

$xy + 2yz + 3zx$

When $x = 3$, $y = 2$ and $z = 1$

$= (3 \times 2) + (2 \times 2 \times 1) + (3 \times 1 \times 3)$

$= 6 + 4 + 9$

$= 19$

Example 2

Find the value of

$\dfrac{2a + 4ab + 3c}{a + 2b + 3c}$

When $a = 6$, $b = 3$ and $c = 2$

$= \dfrac{(2 \times 6) + (4 \times 6 \times 3) + (3 \times 2)}{6 + (2 \times 3) + (3 \times 2)}$

$= \dfrac{12 + 72 + 6}{6 + 6 + 6}$

$= \dfrac{90}{18}$

$= 5$

Multiplication and Division of algebraic functions

$(-a) \times (-b) = +ab$

$(+a) \times (+b) = +ab$

$(-c) \times (+d) = -cd$

$(+c) \times (-d) = -cd$

To multiply two or more expressions the rule is, find the product of the coefficients and prefix this product by the sign obtained from applying the rule of signs

Examples

$4b \times 2b = 8b^2$

$-3a \times 4a = -12a^2$

$-4x \times -6x = 24x^2$

$+5y \times y = 5y^2$

Power (indices)

$4b^3 \times 3b^2 = 12b^5$

$6bc^2 \times 5b^4c^3 = 30b^5c^5$

$2xy^5 \times 8x = 16x^2y^5$

$3y^2 \times 4x^3y^4 = 12x^3y^6$

Expressions containing two or more terms

$(2a + 2)(4a + 3)$

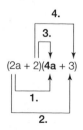

Order of operation

$= 8a^2 + 6a + 8a + 6$

$= \mathbf{8a^2 + 14a + 6}$

Example 1

$(3x^2 + 6)(4 - 2x)$

$12x^2 - 6x^3 + 24 - 12x$

Example 2

$(2x^2 + 4y - 2)(2 + 3x - 4y)$

$4x^2 + 6x^3 - 8x^2y + 8y + 12xy - 16y^2 - 4 - 6x + 8y$

Combine like terms

$4x^2 + 6x^3 - 8x^2y + 16y + 12xy - 16y^2 - 4 - 6x$

APPENDIX IV MATHS REVISION: TRIGONOMETRY

Throughout biomechanics there is a considerable use of trigonometry and it is important to have a good understanding of the more common relationships. Essentially this will be a revision of the trigonometry used within mathematics studied at school level, however it is important in the context of this text.

Many of the principles used within biomechanics are based on the right-angled triangle which is shown in the following examples.

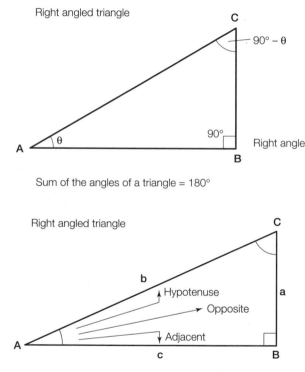

Right angled triangle

$90° - \theta$

θ

$90°$

Right angle

A

B

C

Sum of the angles of a triangle = 180°

Right angled triangle

b

Hypotenuse

Opposite

Adjacent

c

a

A

B

C

Pythagoras Rule
$b^2 = a^2 + c^2$
$b = \sqrt{a^2 + c^2}$

a = opposite
b = hypotenuse
c = adjacent

Lengths of the sides of the triangle

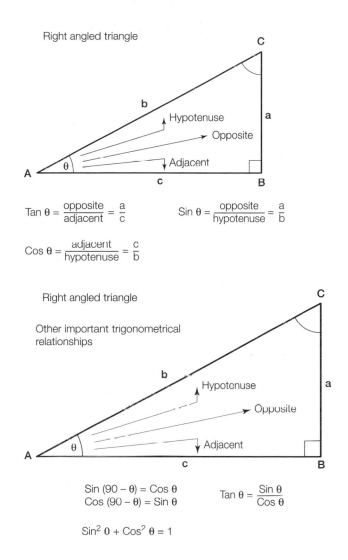

Right angled triangle

$$\text{Tan } \theta = \frac{\text{opposite}}{\text{adjacent}} = \frac{a}{c} \qquad \text{Sin } \theta = \frac{\text{opposite}}{\text{hypotenuse}} = \frac{a}{b}$$

$$\text{Cos } \theta = \frac{\text{adjacent}}{\text{hypotenuse}} = \frac{c}{b}$$

Right angled triangle

Other important trigonometrical
relationships

$$\begin{array}{l} \text{Sin } (90 - \theta) = \text{Cos } \theta \\ \text{Cos } (90 - \theta) = \text{Sin } \theta \end{array} \qquad \text{Tan } \theta = \frac{\text{Sin } \theta}{\text{Cos } \theta}$$

$$\text{Sin}^2 \, 0 + \text{Cos}^2 \, \theta = 1$$

Within biomechanics there are also many examples where the application of trigonometry is required in triangles that are not right-angled. The following formulae are useful in this context.

Application of Trigonometry – all triangles

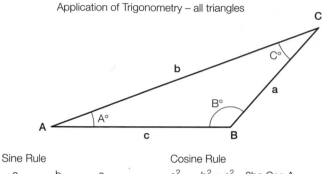

Sine Rule

$$\frac{a}{\text{Sin A}} = \frac{b}{\text{Sin B}} = \frac{c}{\text{Sin C}}$$

Cosine Rule

$$a^2 = b^2 + c^2 - 2bc \text{ Cos A}$$
$$b^2 = a^2 + c^2 - 2ac \text{ Cos B}$$
$$c^2 = a^2 + b^2 - 2ab \text{ Cos C}$$

Example 1

Using the sine rule in the following example solve the problem for the length of the sides **a** and **b**.

Example 1 – solution

A = 28°
C = 32°
c = 23 cm

Calculate the lengths of the sides **a** and **b**

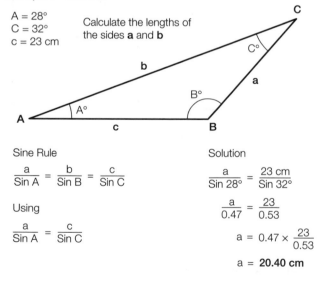

Sine Rule

$$\frac{a}{\text{Sin A}} = \frac{b}{\text{Sin B}} = \frac{c}{\text{Sin C}}$$

Using

$$\frac{a}{\text{Sin A}} = \frac{c}{\text{Sin C}}$$

Solution

$$\frac{a}{\text{Sin 28°}} = \frac{23 \text{ cm}}{\text{Sin 32°}}$$

$$\frac{a}{0.47} = \frac{23}{0.53}$$

$$a = 0.47 \times \frac{23}{0.53}$$

$$a = \textbf{20.40 cm}$$

Solution – distance b

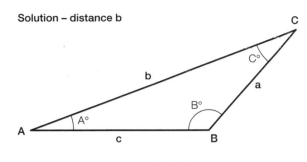

Solution using the Sine Rule

Angle B = 180° – (28 + 32)

Angle B = 120°

$$\frac{b}{\text{Sin }120°} = \frac{23\text{ cm}}{\text{Sin }32°}$$

$$\frac{b}{0.866} = \frac{23}{0.53}$$

$$b = 0.886 \times \frac{23}{0.53}$$

$$b = \textbf{37.58 cm}$$

INDEX

Bold type is used to indicate the main entry where there are several.